T0146235

Subverting Aristotle

Subverting Aristotle

*Religion, History, and Philosophy
in Early Modern Science*

CRAIG MARTIN

Johns Hopkins University Press
Baltimore

This book has been brought to publication with the generous assistance
of the Lila Acheson Wallace Publication Subsidy at Villa I Tatti.

2 4 6 8 9 7 5 3 1

Johns Hopkins University Press
2715 North Charles Street
Baltimore, Maryland 21218-4363
www.press.jhu.edu

Library of Congress Cataloging-in-Publication Data
Martin, Craig, 1972–
Subverting Aristotle : religion, history, and philosophy in early modern
science / Craig Martin.
pages cm
Includes bibliographical references and index.
ISBN-13: 978-1-4214-1316-7 (hardcover : alk. paper)
ISBN-13: 978-1-4214-1317-4 (electronic)
ISBN-10: 1-4214-1316-7 (hardcover : alk. paper)
ISBN-10: 1-4214-1317-5 (electronic)
1. Aristotle. 2. Religion and science—History. 3. Science—History.
I. Title.
B485.M267 2014
149'.91—dc23   2013031642

A catalog record for this book is available from the British Library.

Special discounts are available for bulk purchases of this book.
For more information, please contact Special Sales at 410-516-6936
or specialsales@press.jhu.edu.

Contents

Subverting Aristotle

Introduction

Even if Aristotle was not an atheist in the sense that he directly
and openly attacked the divine . . . one could say that he was one
in a broader sense, because his ideas on divinity indirectly tend to
undermine it and destroy it.

Denis Diderot and Jean Le Rond d'Alembert, Encyclopédie

The belief that Aristotle's philosophy is incompatible with Christianity is hardly
controversial today. The conviction that his views about religion and society
might be best understood placed in the context of Greek pagan culture is not
likely to evoke strong reactions. What is true today, however, was not always the
case. For centuries, Christian culture embraced Aristotelian thought as its own,
reconciling his philosophy with theology and ecclesiastical doctrine. The image
of Aristotle as source of religious truth withered in the seventeenth century, the
same century in which he ceased being an authority for natural philosophy. The
coincidence of Aristotle's loss of authority for theology and for natural philosophy
was not accidental. Aristotle's transformation from ancient sage into an impious
pagan, who espoused dogmatically dubious doctrines, was part of the general
rejection of Aristotelianism that accompanied the scientific revolution.

The transformations of natural philosophy during the seventeenth century
were tied to understandings of the past. In the words of William Ashworth, "the
Scientific Revolution was, after all, itself a historical revolution."[1] Early modern
rejections of Scholasticism were also historical evaluations of ancient thought and
thinkers, most important among them Aristotle. Nearly by definition, the novel
philosophies of the seventeenth century, those traditionally identified with the
scientific revolution, shared a critical view toward Aristotelianism and its past.
Some promoters of novel philosophies were vitriolic in their hostility, even if they
borrowed terminology and concepts from the philosophy of the schools.[2] Never-

theless, most seventeenth-century natural philosophers believed that breaking with Aristotle's philosophy was crucial to the improvement or renewal of human knowledge. As a result historians have, for the last century, if not longer, identified the emergence of modern science with the loosening grip of Aristotle's thought.[3]

The mechanisms and dynamics that caused the downfall of Aristotelian thought were complex. They are hidden behind well-known caricatures of university professors, such as Galileo Galilei's slavish and foolish character Simplicio or Francis Bacon's worshippers of idols. The traditional story tells that Aristotle's philosophy, despite its entrenched positions in universities and churches, fell because its tenets did not conform to the discoveries of the seventeenth century. Aristotle's philosophy failed conceptually, methodologically, and experientially. Evidence taken from astronomical observations undermined geocentric cosmology. Promoters of corpuscular or atomistic philosophies criticized the metaphysical nature of Aristotelianism, contending that key concepts such as substantial forms and potency and act were unintelligible or chimerical. Increasingly, the cultures of experimentation were seen as inconsistent with the bookish textually based musings found in commentaries and *quaestiones* on Aristotle's works, notwithstanding the centuries-old tradition of alchemical testing that was tied to Aristotelian matter theory.[4]

Common to René Descartes' speculations, Galileo's astronomical pronouncements, Nicholas Copernicus's mathematical theorizing, Pierre Gassendi's atomism, Robert Boyle's experimentation with air pumps, and Bacon's promotion of new methods was their open hostility toward Aristotelianism. In this light, historians have pointed to general causes to the eventual decline of Aristotle's influence. One is conceptual. The ideas put forth in new cosmological theories, mechanical models of nature, and corpuscular or atomistic philosophies rendered the metaphysical niceties of Aristotle's thought irrelevant and inaccurate. The second general cause put forth by historians of science is institutional or social. New scientific societies and academies became more capable of making collective inquiries into nature, limiting discord, and publicizing discoveries than the conservative universities had been. As a result, group experimentation replaced the methods of commentary and textual exegesis that were practiced among Aristotelians, and the authority of the ancients gradually vanished. All these parts of the traditional story are true to a certain extent. Nevertheless, they are only parts, not the entire story, in which religion played a significant role.

Positivist historians of the nineteenth century, and some afterward, conceived modern science as having arisen in step with secularism.[5] For them, Galileo's clash with the Catholic Church, Giordano Bruno's trial and execution, bans of

Cartesian teachings at Paris, Leiden, and Utrecht pointed to the incompatibility of free inquiry into nature and the rigid authoritarian nature of churches and churchmen. Despite these controversies, religion motivated those who proffered alternatives to Aristotelianism. Descartes produced what he considered a proof of the existence of God. Robert Boyle wrote Christian-themed treatises and attacked atheism. Neoplatonists and Hermeticists, among others, aligned themselves with a Mosaic philosophy that sought to combine truths of ancient scripture with meditations on nature. Perhaps somewhat enthusiastically, one recent commentator has concluded that all new natural philosophies of the seventeenth century were "necessarily Mosaic," even if this term does little to help categorize Galileo and Descartes.[6] The desire to assimilate natural philosophy with ancient theology was intense for many during the early modern period, although, purveyors of Hermetic thought did not have a monopoly on this outlook. A handful of Aristotelians, for example, the seventeenth-century Paduan professor Fortunio Liceti, also tried to connect Aristotle to Judaism and ancient scripture.[7]

The remarkable changes in natural philosophy that took place during the sixteenth and seventeenth centuries were not part of the secularization of thought but rather were closely tied to religion. Recent scholars have described these ties in numerous ways, showing that what once positivists perceived as secularism was in fact religious. Natural philosophy did not become secularized, instead laymen began to speculate on theology.[8] Revivers of skepticism in the early modern period intended to bolster religious belief rather than undermine it with doubt.[9] Although there are dissenters, some scholars have pointed to the politics of courts and courtiers rather than hostility between free thought and religion as a major cause of Galileo's problems with Rome.[10] The atomist philosopher Pierre Gassendi was neither a libertine nor hedonist but rather adapted Epicureanism to the demands of Catholic theology.[11] Bacon's proposed models for natural philosophy and its application were devised to improve humankind's material conditions in a Christian society still recovering from the Adamic Fall.[12] New hermeneutical techniques ushered in by the Reformation altered attitudes to reading the book of nature.[13] Even if it is possible to find numerous examples of the importance of religion to the scientific revolution, there is still not a clear answer to the question of what was the role of religion in the downfall of Aristotelianism.

THE IMPIETY OF ARISTOTLE

The narratives of the scientific revolution that emphasize the conceptual and methodological breaks enhance the perception that modern science was much different than the natural philosophy of medieval and Renaissance universities, where, for many but not all, natural philosophy was ancillary to theology. The battles between the Catholic Church and proponents of new natural philosophies appear to be clashes between ecclesiastical Aristotelians and lay neoterics. Theologians attacked Cartesian matter theory, and perhaps Galileo's atomism as well, because it implicitly undermined Thomistic explanations of the Eucharist that depended on the concepts of matter and form, that is, on Aristotelian hylomorphism.[14] The unwillingness of churchmen to accept heliocentric models depended, at least partially, on a commitment to Aristotelian physics. After the Council of Trent, many Catholic theologians endorsed a version of Aristotelianism based on Thomas Aquinas's thought; and subsequently Scotists and Thomists dominated French theology during the seventeenth century. And despite Martin Luther's negative views toward the Scholastic tradition, both Lutherans and Calvinists introduced Aristotelian physics and metaphysics into their schools' curricula in order to strengthen theological arguments used in sectarian clashes.[15] But if the religious motivations of the *novatores* are considered a major spur for developing alternatives to Aristotelianism, then it is necessary to consider the theological problems that Aristotle's thought posed for sixteenth- and seventeenth-century natural philosophers.

The image of Aristotle drastically changed over the course of the early modern period. The medieval integration of Aristotle's thought with Catholic theology led David Lindberg to describe the success of thirteenth- and fourteenth-century Scholastic philosophers and theologians as being so great "that by the sixteenth century Aristotle had taken on the appearance of a Christian saint."[16] Although it should be conceded that Lindberg overlooked medieval opposition to Aristotle, nevertheless a number of seventeenth-century depictions of Aristotle were very distant from Lindberg's characterization of their predecessors. In the seventeenth century, Aristotle was no longer an emblem for piety. To the contrary, a number of thinkers described him and his thought as impious and even atheistic. For many, Aristotle's philosophy became a symbol of incorrect thinking about God and religious doctrine in general. Perceptions of Aristotle's orthodoxy or unorthodoxy shifted during the late Renaissance to such a degree that Pierre Bayle's entry on Aristotle in his *Dictionnaire critique et historique* focuses on religious beliefs, rather than his philosophical ones, as he noted that Aristotle "clashed with Christianity . . . in points of the greatest consequence."[17]

This remarkable change in the presentation of Aristotle reflects not just hostility toward his natural philosophy but also the divided sectarian nature of seventeenth-century Europe. Concerns with impiety were central to political and intellectual life. In particular, fears of atheism arose, if not for the first time, then, in a way markedly distinct from earlier eras.[18] In the seventeenth century, atheism could merely refer to those of another religious persuasion or specifically to someone who denied the existence of the divine. Aristotelians potentially fit into both groups, because, on the one hand, their adoption by both Protestant and Catholic theologians placed Aristotle in the middle of sectarian strife, or, on the other hand, because prominent Aristotelians promoted a philosophy that had no recourse to supernatural explanations. The specific versions of Aristotelianism that circulated during the late Renaissance aided the potential for describing Aristotle as impious, a heathen, or even an atheist.

KINDS OF ARISTOTELIANISM

Even though many of these polemical writings focused specifically on Aristotle, such attacks targeted not just the ancient philosopher but also those who used his writings as a basis for natural philosophical discussion. The plasticity of the corpus lent itself to a divergence of positions and schools of thought. Charles B. Schmitt argued that there was not one Aristotelianism but rather a multiplicity of them.[19] If *Aristotelian* refers to a family or species of thinkers, it is a family that shows wide variation and an ample ability for adaptation.[20] The "hard core principles," including matter and form, potency and act, and the four elements, as described by J. M. M. Hans Thijssen, might possess continuity throughout the Aristotelian tradition, although other more peripheral principles were ephemeral, transforming over time and from one thinker or school to another.[21]

The range of transformation or adaption among Aristotelians was broad. Niccolò Cabeo, a Jesuit who wrote a commentary on the *Meteorology* in the middle of the seventeenth century, believed the explication of Aristotle's texts should be the starting point into investigations into nature, yet also believed there were only three elements; the celestial realms experienced change; and substantial forms were physical, composed of small, spirituous particles.[22] Similarly, the capacity of Aristotelians for adapting the natural philosophical principles of Aristotle was remarkable. Over the course of the Middle Ages and Renaissance his ideas supported both Catholic and Protestant theology. Physicians and medical authors cited his writings as the foundation for physiology.[23] Alchemical theorists blended Aristotelian matter theory with the results of laboratory tests.[24] Astrology depended largely on Aristotelian cosmology, which was reconciled with the Ptolemaic system.[25] Although the flexibility and adaptability of Aristotelianism might be seen as a positive, it was also potentially a negative, as the numerous schools and diversity of opinions at times led to fracturing rather than consensus.[26]

Methods of reading Aristotle varied as well. The picture of Aristotelians believing everything written in the corpus to be true without further inspection is inaccurate. Numerous medieval and Renaissance scholars pointed out errors in his thought or in matters of fact.[27] These failures included the occasional lack of correspondence to experiential evidence as well as to religious doctrine. The purpose of expounding on his writings was twofold: to explain Aristotle's thought and to search for answers about the natural or spiritual world.[28] Increasingly, however, over the course of the sixteenth century, scholars presented themselves as searching for accurate interpretations of Aristotle, regardless of their veracity. Institutions adopted such approaches to Aristotle. The 1607 statutes of the University of Padua demanded that professors explain the assigned texts, which were often Aristotle's, without broader amplification.[29] Oxford had already done so twenty years earlier.[30] The Paduan philosopher Cesare Cremonini stated this meant that he was obliged to explain Aristotle's intent regardless of its conformity to truth or faith.[31]

Seventeenth-century religious attacks on Aristotle and implicitly his followers targeted specific kinds or versions of Aristotelianism, specifically two versions. In an attempt to thwart the spread of the Protestant Reformation, Catholic authorities at the Council of Trent reaffirmed the Thomistic synthesis of Aristotelian metaphysics and Catholic doctrine. The decisions taken there led to a renewal of Thomas's philosophy. Jesuit Colleges developed curricula based on Thomas's teachings, Dominican inquisitors defined orthodoxy in accordance with Thomism, and theologians wrote numerous commentaries on Aristotle that emphasized metaphysics and the possible concord with Christian faith. In this version, Aristotle was presented if not as pious himself, than as a handmaiden to true religion. A number of English anti-Aristotelian polemicists, such as Thomas Hobbes, attacked the Thomistic version, seeing it as popish or as religious dogma masquerading as philosophy.[32]

The second version of Aristotelianism that was attacked in the seventeenth century saw Aristotle's philosophy not as an aid to theology but rather as distinct from it. This version, which was taught primarily in Italian universities by famed professors such as Pietro Pomponazzi, Simone Porzio, Giacomo Zabarella, and Cremonini, contended that discussions of the natural world should be separate from considerations of faith. These professors maintained that they followed Aristotelian principles, or what were sometimes called arguments by natural reasoning (rationibus naturalibus), in order to uncover explanations of the natural world, regardless of their correspondence to religion or theology. By the end of the sixteenth century, a number of both Protestant and Catholic authors distrusted these Italian authors, considering them impious because of their discussions of naturalistic explanations of miracles or their casting of doubt over the immortality of the human soul. Moreover, the sincerity of their claims that they were only interpreting Aristotle or following the rational conclusions of his principles were met with doubt, suspicion, but only in a few cases prosecution.

Despite being the subject of controversy and at times the object of scorn, the version of Aristotelian thought that promoted following Aristotle without recourse to religious doctrine also assisted anti-Aristotelians in presenting a more historically accurate understanding of Aristotle. In their efforts to determine what Aristotle really thought or what was true according to Aristotle's natural principles, Italian philosophers, such as Zabarella and Pomponazzi, concluded that some of Aristotle's doctrines did not conform to religious doctrine. Pointing out that there are differences between Aristotle and Christianity was not new. Almost all medieval Aristotelians, even Thomas Aquinas, who attempted to reconcile Aristotle with Christian doctrine, recognized that Aristotle believed the universe to be eternal. Yet sixteenth-century attempts to find Aristotle's true opinion demonstrated how difficult it was to pin down Aristotle's view on many issues, among them the extent of God's providence and the nature of the human soul.

Seemingly contradictory passages made, and continue to make, determining Aristotle's true position on a number of issues difficult if not impossible. Nevertheless, Renaissance Italians were increasingly sensitive to textual and historical issues that promoted assent to the idea that Aristotle could not be reconciled with faith. A number of them thought that Thomas Aquinas and others forced natural philosophy to agree with theology by introducing concepts and arguments that were fundamentally alien to natural philosophy or Aristotle's concepts. Paradoxically, the determinations of Aristotelian university professors that Aristotle's true thought was at odds with Christianity became a tool for anti-Aristotelians. Both Renaissance Platonists and later promoters of novel natural philosophies, who wished to portray traditional natural philosophy as inherently irreligious, accepted their interpretations of Aristotle as accurate and thus evidence for Aristotle's impiety.

HUMANISM AND HISTORY

The connections between the scientific revolution and humanism are numerous. The availability of new scientific sources, the development of a sophisticated historical chronology, and the cultivation of observation all link innovations in natural philosophy to Renaissance humanism.[33] Humanists' ideals played a significant role in the changes and eventual downfall of Renaissance Aristotelianism. Humanism conditioned sixteenth-century Aristotelians who aspired to determine what Aristotle thought regardless of its truth. Although some medieval commentators on Aristotle, such as Walter Burley, professed a historical understanding of Aristotle as one of their goals, the extent of such an outlook was far greater in the sixteenth century.[34] In the 1490s, Francesco Cavalli brought the study of the Greek texts of Aristotle into the lecture halls of Padua, most likely for the first time in that university's history.[35] Throughout the sixteenth century an increasing number of scholars at Italian universities read Greek. Philological methods of establishing texts and the growth of knowledge of the ancient Greek language altered interpretations of Aristotle. New translations proliferated that strove to render more accurately Aristotle's thought into Latin. As a result, linguistic sophistication grew. The suggested textual emendations of Francesco Vimercati, an Italian-trained professor who taught at the Collège Royal in Paris, and of Zabarella remain cited in present-day critical editions of Aristotle.[36]

The attention to linguistic fidelity was only one of several innovations of the Renaissance that changed interpretations of ancient philosophy. Evaluations of what was authentically written by Aristotle changed the Aristotelian corpus itself.[37] The discovery, translation, and interpretation of the ancient Greek commentaries on Aristotle, such as those of Alexander of Aphrodisias, Simplicius, Themistius, and Philoponus, provided a surfeit of authoritative interpretations that greatly influenced Renaissance Aristotelians.[38] All these Greek commentators were pagan, with the exception of Philoponus, who was Christian. Philoponus, however, was not a model for medieval and Renaissance theologians because his Christological views were condemned posthumously at the Third Council of Constantinople in 680–81.[39] Consequently, the views of the Greek commentators only infrequently coincided with medieval Christian theologians. The lack of concord between Christian thought and those of the Greek commentators soon became obvious during the Renaissance. Still, that these writers wrote in Greek and lived in an era closer to Aristotle's gave them significant authority for Renaissance authors. Accordingly, their non-Christian interpretations influenced early modern considerations of Aristotle's views.

The ancient Greek commentaries were just one set of texts read in the Renaissance that changed perceptions of Aristotle. Medieval authors still exerted influence, among Latin authors Thomas Aquinas, Albertus Magnus, and John Duns Scotus counted among the most influential. The Arabic-writing authors Ibn Sīnā (circa 980–1037) and Ibn Rushd (1126–1198)—whose names were Latinized as Avicenna and Averroes, respectively—enjoyed wide readership. Averroes' works, in particular, experienced a resurgence that was related to humanism. Because Averroes had modeled himself on the Greek commentators and professed to try to understand Aristotle's true thought, numerous Aristotelians, influenced by humanists' ideals and methods, prized Averroes' commentaries, reading them in tandem with Aristotle's own corpus. Beside these medieval authors, a number of ancient works, including Diogenes Laertius's biographies of the philosophers, Cicero's philosophical dialogues, and early patristic writings, altered perceptions not just of Aristotle but also of the history of philosophy as a whole.

Renaissance models for the history of philosophy followed the patterns of Renaissance historiography more generally.[40] Lorenzo Valla's recognition of periodization, according to which he separated his time period from the Middle Ages and the Middle Ages from antiquity, altered ecclesiastical, political, and intellectual history. Humanists, including Valla, questioned Aristotle's dominance in the universities of his day by examining Aristotle's position in the ancient world.[41] By Valla's time, university professors lectured on Aristotle because of tradition and the demands of university statutes, making their acceptance of the importance of Aristotle seem reflexive to some. Investigations into ancient history uncovered a multiplicity of schools, showing the waning strength of Aristotelianism during the last years of the Roman republic and placing Aristotle in the context of Athens of the fourth-century BC. By historicizing Aristotle, those influenced by humanism were more apt to note that he was pagan and less likely to expect him to agree with Church doctrine.

Historicizing Aristotle accompanied evaluations of the history of Aristotelianism. By the seventeenth century, a number of scholars wrote treatises that examined the position of Aristotle's philosophy in medieval universities. Often these historical analyses were part of religious polemics. For example, the Frenchman Jean de Launoy uncovered various condemnations of Aristotelian works and principles in thirteenth-century Paris as part of his effort to show that Jesuits incorrectly interpreted Aristotle and Thomas as authoritative throughout the Catholic Church's tradition.[42] Polemicists of the seventeenth century also reinterpreted sixteenth-century Italian Aristotelians such as Pomponazzi, oftentimes emphasizing, or even exaggerating, potentially heretical elements in their works.[43] Historical analysis does not put forth unequivocal results. The development of objective tools for analyzing the past did not eliminate the use of history for partisan advantage. Opponents of Aristotle were not always sensitive to the changing definitions of orthodoxy in their condemnations of his philosophy. Champions of Aristotle also interpreted the past to combat their opponents, offering historical interpretations of why patristic authors were mistaken in their preference for Plato over Aristotle or, as Liceti, among others, did, arguing that the Old Testament influenced Aristotle.[44]

Humanism fostered the development of philological and antiquarian methods for making judgments about Aristotle and his followers, yet its influence extended even further in debates about Aristotle's piety. Humanism developed in the late Middle Ages as an oppositional force to the universities and Scholasticism. Francesco Petrarca, Coluccio Salutati, and Lorenzo Valla wrote invectives attacking Aristotle and Aristotelians. Many of their complaints pointed to linguistic issues. They attacked the Latin style of Scholastics, their deviations from classical vocabulary, and their use of translations from the Arabic. They also pointed to the alleged impiety of Aristotle and his followers, accusing them of replacing Christianity with the false idol of pagan philosophy.

Averroes played a key role in humanists' invectives. Averroes, a twelfth-century philosopher, jurist, and physician, lived most of his life in al-Andalus. Among his most influential writings were commentaries on large portions of the Aristotelian corpus. Beginning in the thirteenth century, university professors relied on Latin translations of these commentaries. Because Averroes wrote in Arabic and was ignorant of Greek, his writings became an emblem for linguistic impurity. That he was Muslim, coupled with the fact that he promoted a reading of Aristotle that on several accounts was irreconcilable with Christianity, made him a conspicuous target for humanists. Valla among others used Averroes' professed devotion to Aristotle as evidence for the lack of liberty of thought among Aristotelians.

Averroes frequently became a point of reference in attacks on Aristotelianism in general during the seventeenth century, despite his no longer being representative of the dominant forms of Aristotelianism. These attacks arose out of a variety of motivations, which reflected the new intellectual and religious outlooks of the 1600s. Thus to understand the rejection of Aristotle and how he was separated from Christian theology in the seventeenth century, it is necessary to begin with the polemics that surrounded his writings during the late Middle Ages. Early modern scholars appropriated the content of medieval writings while engaging in polemics that resonated with earlier centuries but served their own interests.

The religious motivations of promoters of new natural philosophies who attacked Aristotelianism were sincere. Most natural philosophers of the seventeenth century sought to discover explanations of nature that corresponded to their visions of God and the divine. As a result they were, in a sense, obliged to question the relation between organized religion and Aristotle. In order to demonstrate that their innovations were not impious, they pointed out the spiritual flaws——the lack of coherence with faith——of traditional natural philosophy. The religious battles might seem secondary or tertiary to disputes over concepts or methods, but, because religion was intertwined with nearly all those disputes, subverting Aristotle's authority was necessary for and concomitant to the ascent of modern science.

CHAPTER ONE

Scholasticism, Appropriation, and Censure

The introduction and integration of Aristotelian thought into the Latin West during the twelfth and thirteenth centuries is among the most remarkable movements of European intellectual history. The success that Aristotelianism enjoyed for more than four centuries obscures the originality and determination of those first thinkers who interpreted this newly translated corpus and adapted the writings of a long-dead pagan Greek to the needs of Christian medieval society. These needs were not just religious but also medical, and, perhaps most important, dialectical, as Aristotle's logic became a tool used in all fields that required learned argument.

Some of Aristotle's writings, such as *De interpretatione*, had long been available in Latin. Masters at Cathedral Schools taught logic, demonstrating its applicability for organizing argument in general. Many of his works, however, were translated into Latin for the first time at the end of the twelfth century or during the thirteenth century, oftentimes first from Arabic translations and later from Greek.[1] The existence of these translation movements depended on high esteem for the contents. For many medieval scholars, the worth of Aristotle's works stemmed from their comprehensive conceptual framework for understanding nature and the cosmos. Extensive discussions of metaphysics and epistemology clarified and expanded this framework. Despite the comprehensiveness of his corpus, understanding Aristotle posed a number of difficulties. His extant writings derive from lecture notes, full of awkward wording and unclear phrasing. The content is conceptually challenging because he gave complex solutions for fundamental paradoxes about nature, being, and knowledge. Incorporating Aristotle's works into what would develop into the Scholastic culture of the medieval universities required numerous aids.

The translators of Aristotle rendered not just his works into Latin but also made available the most sophisticated recent philosophical writings in order to clarify his philosophy. Two authors were especially significant: Avicenna and Averroes. Both wrote in Arabic and were Muslim. Avicenna lived in the eastern part of the Islamic world, and Averroes spent most of his life in al-Andalus. Avicenna's and Averroes' works were valued for multiple reasons. Both were outstanding philosophers, deeply engaged with Aristotelian thought. Both were also authors of medical writings, and the desire for more effective medical knowledge was one of the impetuses that brought Aristotle's writings on nature to the Latin world.[2] Avicenna's *Canon* became the primary introductory medical textbook for universities in Italy throughout the late Middle Ages and Renaissance.[3] Numerous professors of medicine read and cited Averroes' *Colliget*, a work on medicine, although its diffusion was smaller than the *Canon*'s. Averroes' worth stemmed, in large part, from his extensive commentaries on Aristotle. These commentaries had the primary goal of explaining and clarifying the words and doctrines of Aristotle's texts.

Avicenna's and Averroes' appropriations of Aristotle were part of the Islamic intellectual movement *falsafa*, which was in debt to the late-antique Greek commentators on Aristotle. Practitioners of *falsafa* and the philosophy of the late Academy shared techniques, methods of exegesis, styles of presentation, and an interest in Plato and Neoplatonic authors. The Academy's curriculum, as established by Iamblichus (circa AD 240–325), started with Aristotelian logic and natural philosophy leading to Platonic metaphysics, taught by lecturing on the dialogue *Parmenides*. Iamblichus gave a historical justification for his combining of Aristotle and Plato, contending that Aristotle had taken the material that comprised the *Categories* from the Pythagorean philosopher Archytas of Tarentum. Thus both Aristotle and Plato had Pythagorean roots.[4] Despite Iamblichus's reliance on a spurious text for this contention, later Greek philosophers, such as Simplicius, Proclus, and Ammonius, followed him and interpreted Aristotle in light of their Neoplatonism, seeking to reconcile or harmonize these two strands of thought, while simultaneously making a systematic and consistent interpretation of Aristotle.[5] Avicenna, Averroes, and other proponents of *falsafa* appropriated these ideals of reconciliation and system building while adapting Greek philosophy to their own culture.[6] Avicenna played an extremely important role in transforming Greek philosophy by incorporating discussions from *kalām* into his metaphysics and theories of the soul. *Kalām* was a doctrinal theology that arose independently from the Greek philosophical tradition during the first centuries of Islam. After Avicenna's death, *kalām* and *falsafa* overlapped, whereas before they had been largely distinct.[7]

Averroes, however, was hostile to the development of a philosophical *kalām*.[8] He attacked the philosophies of Avicenna and al-Ghazālī (1058–1111), a Persian philosopher and theologian, for refraining from imitating Aristotle and for mixing Aristotelian thought with the theology of *kalām* and the metaphysics of Plato. Averroes endorsed what he considered a purer interpretation of Aristotle. His view of Aristotle was unequivocal. Justifying his attempts to discover Aristotle's mind in order to uncover philosophical truth, he contended Aristotle was "a rule and exemplar which nature devised to show the final perfection of man. . . . The teaching of Aristotle is the supreme truth, because his mind was the final expression of the human mind. Wherefore it has been well said that he was created and given to us by divine providence so that we might know all that is to be known."[9] In the prologue to his commentary on the *Physics*, Averroes dismissed post-Aristotelian philosophy, maintaining that none of Aristotle's followers, "from then until now, that is in 1500 years, added anything or found any mistake in his words," and elevating Aristotle's stature so that "it is more worthy of the divine than human."[10] In his middle commentary on the *Meteorology*, Averroes justified his commentaries by the argument that "through a divine virtue," Aristotle was able to become the "inventor of science," which he "completed or perfected."[11] As a result of this extremely high assessment of Aristotle, Averroes' commentaries attempted to systematize, paraphrase, and explain Aristotle's writings in order to make his thought more easily understood and present philosophical truth in a clearer manner.[12]

Although Averroes' assessment of Aristotle deviated from Avicenna and other immediate predecessors, like much of the thought of *falsafa*, it has its origins in the texts produced by late-antique philosophers. Averroes' literalism may have reflected the ideologies of the twelfth-century Almohad rulers who established dominion over Northern Africa and Spain.[13] Even if this is the case, Averroes' high estimation of Aristotle coincided with views found in the works of the ancient Greek commentators. In a brief biography of Aristotle, which has been attributed to the circle of Ammonius, an Alexandrian Neoplatonist philosopher of the fifth century AD, Aristotle is described as having "surpassed human genius in his studies of philosophy," so that "he touched upon no part of philosophy, which he did not treat perfectly."[14] Accordingly, Averroes might have found authority for the near infallibility of Aristotle in the Greek commentary tradition that defined much of the agenda for *falsafa*.

Although many medieval thinkers, both Muslim and Christian, likely thought that Aristotle agreed or would have agreed with their philosophical views, Averroes' view of Aristotle was more than a mere employment of ancient authority.

As part of an Andalusian movement that countered the philosophical and theological syntheses of the Eastern parts of the Islamic world, Averroes promoted an
unmixed version of Peripatetic philosophy.[15] He wrote that Avicenna's philosophy
occupied "almost a midpoint between the Peripatetics and the *mutakallimūn*,"
the theologically minded dialecticians who practiced *kalām*.[16] Throughout his
works, perhaps most notably in the *Long Commentary on the Metaphysics* and
in the *Destruction of the Destruction*, Averroes disputed the positions of the *mutakallimūn* and of philosophers who sought to combine *kalām* with Greek philosophy, such as Avicenna, and on occasion even Christian theologians.[17] In his
comments on the *Metaphysics* he noted that creation ex nihilo is both an Islamic
and Christian doctrine, even if it is not Aristotelian.[18] In a dual attack on theologians and Avicenna's doctrine of *dator formarum*, a doctrine that posited that
God endows created substances with forms, Averroes wrote that just as imagining
forms being created led some men to say forms exist as does a "giver of forms," the
imagination had led the theologians "of the three laws which are today," that is,
Judaism, Islam, and Christianity, to posit creation ex nihilo.[19] In the same work,
he also contended that God is one in number and being, noting that others had
thought there was a trinity in God's substance.[20] Despite these dismissive comments toward Christianity and perhaps all revealed religion, Averroes' real target
was not Christianity but al-Ghazālī, Avicenna, and others who rejected his vision
of a pure Aristotelianism.

For Averroes, deviations from Aristotle represented failures in philosophical
acumen. In his commentary on *De anima*, he chastised his contemporaries who
erred in putting "down the books of Aristotle [and in] believing that this book is
impossible to understand."[21] He continued, "Avicenna does not imitate Aristotle,"
thereby revealing his own belief that philosophy is partly the imitation or the restoration of Aristotle's positions and arguments.[22]

According to Averroes, Avicenna at times endorsed Plato's view "because it
is similar to what the *mutakallimūn* think about our laws."[23] Averroes' condemnations of the fusion of Platonic and Aristotelian stemmed from the mixing of
theology with philosophy and from Plato's limited consideration of sense experience. He believed Plato's love of geometry and devotion to Socrates, whom Plato
portrayed as dismissive of earlier investigations into nature, prevented him from
using sense experience in his philosophy. Similarly, Avicenna's errors were the
result of his "lack of investigations in natural things and his confidence in his
own genius."[24] That is not to say that Plato's works were of no use to Averroes,
just of limited use. In the *Destruction of the Destruction*, he linked Plato to Aristotle, writing, "Their philosophy is the highest point human philosophy can

reach."[25] Furthermore, he apparently justified writing a commentary on Plato's *Republic* by explaining that Aristotle's *Politics* were unavailable to the Arabic-reading world.[26] Moreover, because of the strong links between late-antique Neo-platonism and *falsafa*, certain doctrines that Averroes put forward, such as his theory of the intellect, can be understood to have Platonic elements, even if he was unaware of their Platonism.

Avicenna's and Averroes' writings and interpretations greatly colored the reception and appropriation Aristotle in the Latin West. Their value stemmed largely from their philosophical acumen and dependability. Averroes' commentaries were especially useful in light of their comprehensive nature. That these philosophers were Muslim, initially, was not considered threatening.[27] Controversies, however, arose in the thirteenth century over the reconciliation of philosophy and theology in general among a number of Christian theologians. Such concerns predated the translation and even existence of Arabic writings, as a number of early patristic authors were suspicious of Aristotle and other Greek philosophers. In the thirteenth century, Averroes' and Avicenna's thought played a large role in controversies, because of the overwhelming influence they had on the reception of Aristotelianism. These polemics against Averroes demonstrate the difficulty, or impossibility, of reconciling Aristotelian thought with Christian theology as well as the broad diffusion and influence of his writings.

RECONCILIATION, CENSURE, AND POLEMIC

The arrival of a new philosophical tradition provoked a number of reactions in western Christendom. The value of these newly translated works was significant for many, as universities incorporated these writings into their curricula. Despite the rapid proliferation of these treatises and commentaries, there was not broad agreement on the role Aristotelian thought should have in Christian society. Although some welcomed his writings as a source of knowledge and even wisdom, others pointed to the dangers inherent in a number of the ideas of a pagan philosopher and of his Muslim followers. The disagreements, polemics, and censures point to the diversity of positions among medieval philosophers and theologians as well as to the difficulty of defining orthodoxy.

Coupled with problem of defining orthodoxy was the problem of enforcing specific ecclesiastical doctrines throughout the entire Catholic world. The various condemnations and censures of philosophy handed down during the thirteenth and fourteenth century were often local and frequently held force only for short periods of times. As a result, a broad range of positions about Aristotle's thought emerged. Although some scholars and ecclesiastical authorities were virulently opposed to Aristotle or Averroes, their views did not define the official doctrine of the Catholic Church. During the thirteenth century, no definitive orthodoxy directed the attitudes toward the relation of philosophy and theology. Consequently, interpretations of the acceptability of Aristotle were heterogeneous, as some medieval scholars were positively, others negatively, disposed toward Aristotelian natural philosophy. The views of Bonaventure (circa 1217–1274), Thomas Aquinas (circa 1225–1274), and Albertus Magnus (circa 1193–1280), all active in the second half of the thirteenth century, all clerics who were canonized, illustrate three distinct positions that had significant influence for centuries.

Bonaventure, influenced by the Augustinian tradition saw Aristotle's rejection of Platonism as a source of his impiety. He thought that it follows from Aristotle's rejection of Plato's theory of forms that God only knows himself and not individual particulars. Therefore implicitly, Aristotle rejected the possibility of God's having providence over individual substances. Bonaventure believed this line of thought induced unnamed "Arabs," presumably Avicenna and Averroes, to posit a deterministic model for the world, whereby the substances that move the celestial orbs are the cause of everything in the universe, leaving no place for the acts of demons, angels, or other supernatural influences. The rejection of Plato's theory of forms and the concomitant denial of providence was by no means Aristotle's only error, according to Bonaventure. Unlike Plato, Aristotle believed the world to be uncreated and eternal. Bonaventure cited the early Greek patristic writers and Arabic-writing commentators as witnesses that Aristotle thought this way. One mistake led to another. Bonaventure maintained that an eternal world demands heterodoxy with regard to the human soul. If the world were eternal, then human souls must be infinite in number, which was considered conceptually impossible; corruptible; or, as Averroes argued, one in number. A consequence of both of the last two positions is that there would be no immortal individual soul and therefore no rewards or punishments after death, a conclusion that contradicted a core Christian belief.[28] Accordingly for Bonaventure, Aristotle's thought cannot be reconciled with Christianity.

Like Bonaventure, Albertus Magnus saw the conclusions of natural philosophy as separate from discussions of the supernatural or theology. Unlike Bonaventure, the separation did not entail his rejecting Aristotle. Instead he proposed that nature should be investigated independently from religious concerns.[29] In his commentary on Aristotle's *De generatione et corruptione*, Albertus addressed the question of what is the cause of continual natural generation such that "generation according to nature [secundum naturam] has never stopped and never will."[30] The suppositions of the problem, however, are not in accord with Christian conceptions of creation. He recognized the discord, separating his discussion of nature from considerations of divine will. He wrote, "If someone should say, that at some time generation will be stopped by God's will, just as at some time there was not [generation] and after that time it began, I would say God's miracles do not pertain to me, because I am speaking about naturals" (de naturalibus).[31] So Albertus saw his discussion of generation, which according to nature is eternal, to be suppositional and distinct from the truths of scripture.

Albertus took a similar approach to his discussion of the biblical universal flood. Others, he wrote, "attribute all these things [related to flood] to divine disposition and say that we should not look into any cause of these matters except divine will."[32] Although admitting that the natural world is caused by God, Albertus distinguished his view from theological understandings of the world, "we do not seek to find the causes of his will, but natural causes, which are like his tools by which his will produces an effect in these cases."[33] In natural philosophy, therefore, God's will is beside the point. The focus should be on the powers of nature, which are the instruments of God's will. As a result, Albertus attempted to show that the natural causes of the universal flood were celestial, resulting from a conjunction of planets in the house of Pisces.[34]

Siding with those he defined as Peripatetic, Albertus contrasted his chosen method with his contemporaries, or *moderni* as he called them. These unnamed *moderni* were evidently Latin writers. Albertus minimized differences among Islamic philosophers and emphasized their adherence to Aristotle. Thus he identified Avicenna, Averroes, al-Farabi, and al-Ghazālī as the Peripatetics. The advantage the Arabic-writing Peripatetics held over the *moderni*, according to Albertus, was that they spoke about only those things that can be considered according to physics (considerare physicum) and omitted discussions where modern theologians should be consulted.[35] In his discussion of the possibility of prophecy through dreaming, he said "it was in no way possible to understand this [prophecy] from physical reasons" (ex physicis rationibus), to the contrary, he intended to "follow the sentence of the Peripatetics" and speak only about natural substances, even though the position that derives from theology should be preferred over the one taken from physics.[36]

This approach of the Peripatetics, which Albertus admired and adopted, however, was not confined to physical bodies but also pertained to discussions of insensible and immaterial substances. In the *Metaphysics*, Albertus wrote that his goal was to "declare the opinion of the Peripatetics on these substances [i.e., the insensible substances that are the subject of the book], withholding judgment on whether what they said is true or false."[37] Regardless of the veracity of his conclusions, the method chosen "to explain according to Peripatetic way" was in contrast with theology. He wrote, "Theological matters do not agree with philosophical principles, because they are based on revelation and inspiration and not on reason, and therefore we cannot dispute these matters in philosophy."[38] Albertus concluded that philosophy and theology are distinct and rely on different sources of knowledge. He also implied that Aristotle and those who most closely followed his thought, the Peripatetics, were not in agreement with Christian theology. Albertus's goal in natural philosophy was not to reconcile Aristotle's view with Christianity, but, following its Aristotelian principles, he put forth probable answers while withholding judgment on the ultimate truth.

That philosophy should be distinct from theology did not please all Albertus's colleagues and contemporaries. Thomas Aquinas put forth an alternative view of philosophy's relation to theology. Thomas was responsible for the longest lasting and most thorough integration of Aristotelian thought with Catholic theology. As opposed to Albertus, he believed that reconciliation was not only possible but also necessary, because "demonstrative truth agrees with the faith of the Christian religion."[39] Thus much of Thomas's writings developed philosophical arguments for religious tenets, as he offered proof for the existence of God, developed the concept of transubstantiation while explaining the Eucharist using Aristotelian hylomorphism, and argued that the rational soul is immortal and the form of the human body.[40] Just as many of his contemporaries did, Thomas read large portions of Averroes' commentaries, citing him as the "Commentator" and as an authority. This authority had limits, as Thomas disagreed with specific doctrines, including Averroes' view that the passive intellect is one for all humans.

In the short treatise *On the Unity of the Intellect against the Averroistas*, Thomas set out to show that Averroes' view that the passive intellect is one numerically is both wrong philosophically (contra philosophiae principia) and as an interpretation of Aristotle. Thomas wrote that he does not need to show that the position is contrary to Christian faith, since it should be manifestly clear.[41] It is difficult to know precisely whether Thomas had living opponents in mind or who those were. He wrote that "many" of his contemporaries held this view.[42] His allegation might be true, pointing to Siger of Brabant, a philosopher in the Faculty of Arts at Paris, or others, or it might be hyperbole. A few years earlier, Roger Bacon (circa 1214–1294) suggested that there were few if any who endorsed Averroes' understanding of the soul. He wrote that in his times "there was no mention of these errors because it was clear they were heretical."[43] Even if he did not name his contemporary opponents, Thomas specified their position. Their view is based not on Latin authors. Rather, "they say the words of the Peripatetics are followed, whose books on this subject they have never seen."[44] What they take to be the Peripatetic view comes from Averroes, because they trust the citations of Greek authors contained in Averroes' comments. Those Greek authors were Themistius, a late-antique commentator on Aristotle, and Theophrastus, who was Aristotle's successor at the Lyceum. Thomas, here, revealed the importance of textual interpretation to this exercise. For one, his jab points to the fact that his opponents have never actually seen the material they refer to as Peripatetic while at this point in time Thomas possessed Themistius's paraphrase of the *De anima*.[45]

Knowledge of the texts written by Greek philosophers that were cited as authorities for Averroes' psychology gave Thomas an advantage over his "Averroist" opponents because it was not sufficient to show that Averroes' view was incorrect according to "philosophical principles." Thomas wanted to demonstrate that the unicity of the passive intellect was "completely opposed to the words and doctrine" of Aristotle as well as being philosophical incorrect, so his exercise combined philosophical argument with textual interpretation.[46] In the example of the intellect, Thomas aimed to show that Aristotle agreed with Christianity.

Thomas prized correctly interpreting Aristotle and used textual arguments to dismiss Averroes for several issues. For example, in the first book of his commentary on the *Sentences*, he discussed whether God's providence extends to everything. Thomas divided those who denied providence into two camps, those, such as Democritus, who deny all providence, and those, namely Averroes, who admit providence with regard to species but deny that particulars are governed as such. In comment 37 of book 12 of the *Long Commentary on the Metaphysics*, after explaining that the celestial bodies, having been moved by the unmoved mover, cause generation and corruption in the sublunary region, Averroes contended that "[t]he view of those who think that God's providence extends to every person is right in a sense and wrong in another. It is right insofar as nobody is in a condition peculiar to him, but (this condition) belongs to the class of this species. If this is so, it is correct to say that God takes care of individuals in this way; but providence for an individual, in which nobody else shares, is something that the divine bounty does not necessitate."[47] Here Averroes contended that God has some concern with individuals, because all individuals obtain their disposition from their species, which are connected to God, yet God does not directly communicate with sublunary individuals.

This comment became the source for the widely held view that Averroes did not believe there was providence in the sublunary world. Thomas was careful to distinguish between Averroes' position and Aristotle's, maintaining that despite Averroes' attribution of this view to Aristotle, Aristotle did not in fact hold this position. Moreover, according to Thomas, Averroes' conclusion was not in accordance to Christian belief, as it "removes God's judgment from the works of humankind."[48] Thus Thomas believed Averroes erred twice. He misinterpreted Aristotle and was incorrect about the nature of providence.

Despite these polemics, in other cases Thomas frequently used Averroes' commentaries as authoritative. Vehemently condemning Averroes' views on the intellect made Thomas' use of Averroes in other contexts less threatening since it was well known that he recognized the doctrinally erroneous nature of portions of Averroes' philosophy. The attack on Averroes and "Averroists" was a means for Thomas to disassociate Aristotle and his own philosophy from Averroes' dangerous propositions. There were limits to Thomas' ability to reconcile philosophy and theology. According to Thomas, "holy scripture, which is infallible, testifies" that the world was created in time.[49] Still, it could still be argued that it is possible for it to be eternal, even if it is not the case. Thomas contended that, God being both omnipotent and eternal, "it cannot be demonstrated" that God could not have created an eternal world even if an eternal world entails an actualized infinite, a possibility that Aristotle rejected.[50] Consequently, the creation of the world in time can only be known through faith, not through philosophical argument, even if the eternity of the world is not philosophically necessary, as Thomas also argued.[51]

Later medieval thinkers increased the number of doctrines that they believed could not be understood through reason. John Duns Scotus (circa 1265–1308) included the immortality of the soul in this category. Furthermore, he wrote that reason and arguments could not demonstrate the resurrection, "which must be held only through faith."[52] William of Ockham (1287–1347) even cast doubt on his own proof of the existence of God. Reason cannot demonstrate the existence of God and God's attributes. He wrote, "[I]t is not possible to be known clearly [evidenter] that God is."[53] Thus in Scholastic philosophy, human reason's ability to attain theological truths was subject to dispute.

Similarly, theology's ability to attain natural truths was questioned in decades around 1300. Siger of Brabant, most likely one of Thomas's philosophical opponents, described the goal of his treatise on the intellective soul, as seeking "only the meaning [intentio] of philosophers and especially Aristotle, even if the Philosopher [i.e., Aristotle] believed other than the truth and that some of the beliefs made through revelation of the soul could not be discovered through discussions of nature" (rationes naturales). Echoing Albertus, Siger wrote that, "nothing of God's miracles pertains to this discussion, because we are speaking naturally about natural things" (de naturalibus naturaliter).[54] Siger equated Aristotle's and other philosophers' actual view with discussions of nature that had no recourse to revelation. Thus in his work on the eternity of the world, he described his approach as following the "path of Aristotle."[55] In this treatise he considered only the views of Aristotle, Themistius, and Averroes but not scripture or theological writings. On the one hand, Siger's approach can be seen as dividing philosophy from revelation or theology; on the other hand, his approach emphasized a hermeneutics that privileged attempts to discover precisely what Aristotle actually thought. Siger's approach could be applied to a wide range of questions about naturals as is shown by some of his contemporaries and successors.

One of the most influential medieval medical authors, Pietro d'Abano, studied in Paris, possibly with those who had been influenced by Siger, before returning to Padua. Philosophically sophisticated, Pietro wrote a commentary on the Aristotelian *Problemata*, an astrological work called the *Lucidator*, and the extremely influential *Conciliator*, which attempted to reconcile the findings of medicine with those of philosophy. Pietro had an ambivalent view of Averroes. At the beginning of the *Conciliator*, he condemned promoters of the unicity of the passive intellect as depraved but meanwhile used Averroes' views on a large number of issues throughout the rest of the work and supported the separation of natural philosophy from theology.[56]

In the *Lucidator*, Pietro disagreed with theologians who posit additional spheres beyond the spheres of the stars. According to Pietro, these additional spheres, such as the crystalline and the empyrean, are "known by revelation rather than by reason [ratio] or experience [experientia]," the two most important tools of natural philosophy.[57] Justifying his approach he cited Albertus as a model. In the *Conciliator*, he contended that "when nature is being discussed" (de naturalibus disseratur) there should be no consideration of miracles or God's will.[58] Pietro made a further epistemological distinction, as he contended that religious "laws have only a force of persuasion, therefore the truth of the matter should be investigated by the words of astrologers, philosophers, and physicians."[59] Consequently, he considered the universal flood in terms of astrology, just as Albertus had done, and discussed the possibility of cycles of human history, distinct from biblical accounts of the creation of mankind.[60] Going beyond Albertus, Pietro also considered the resurrection of Lazarus in natural terms, although he decided that the unusual circumstances of the case demanded that the resurrection "should be attributed more to a miracle than nature."[61]

Pietro's naturalistic approach provoked conflict. In the *Conciliator* he described his successful evasions of Dominican theologians in Paris who had been investigating him for teaching that the intellective soul could arise from the potency of matter. He recounted that he escaped the prosecutions "thanks to God and an intervening Church official."[62] The intervention of a Church official illustrates the divisions of views about nature and about the correct interpretation of Aristotle. Dominicans were likely to favor a Thomistic approach. In Santa Maria Novella in Florence and Santa Caterina in Pisa, Dominicans decorated their chapels with depictions of Thomas triumphing over Averroes.[63] Others saw such investigations into nature that avoid theology as harmless. Such differences were also apparent in a trial, in which, according to one account, Pietro d'Abano was posthumously convicted of heresy for deriding "miracles of Christ" in his account of Lazarus's resurrection, even though in the written account he affirmed the miraculous character of the event.[64] An inscription placed in 1420 at the Palazzo della Ragione in Padua, however, reported that Pietro was later absolved.[65] If these accounts are to be trusted, it points to rifts and disagreements about definitions of heresy and the acceptability of distinguishing philosophy from religious truths.

CENSURE

The proceedings against Pietro d'Abano were only one of several official processes or censures made against Aristotelians during the Middle Ages. Beginning shortly after the introduction of Aristotelian natural philosophy into the medieval universities, Aristotle's thought was the object of censure from ecclesiastical authorities, even while simultaneously being integrated with Christian theology. The earliest documented condemnations came from Paris. The precise impact and meaning of these censures and condemnations are subject to dispute, yet they show at minimum some ambivalence toward Aristotelian natural philosophy or at least toward specific interpretations of it.

In 1210, the Bishop of Paris, Petrus de Corbolio condemned specific philosophers and their works. The decree ordered that Amaury of Chartres' body be removed from its original resting place and be buried in unconsecrated ground and that the works of David of Dinant be burned. In the same year, the Council of Sens also condemned David. Although Amaury and David were philosophers, their heresies stemmed from deviations from Aristotelianism. David rejected key concepts of Aristotelianism, such as substantial change and the four elements, and posited indestructible particles and metaphysical theories inclined toward pantheism.[66] Despite David's denial of basic Aristotelian tenets, the bishop further decreed, "Neither the books of Aristotle on natural philosophy, nor the commentaries, should be read in Paris publically or privately, under the punishment of excommunication."[67] Here, the word "read" (legere) refers to actual lecturing on the books, meaning the decree effectively forbade the teaching of Aristotle's natural philosophy in Paris.

Five years later, a papal legate, Robert de Courçon, repeated and expanded the earlier proscription. After authorizing lecturing on Aristotle's dialectic in the Faculty of Arts at Paris, Robert ordered that they not read "the books of Aristotle on metaphysics and on natural philosophy, or epitomes of them" and then repeated the condemnation of the works of David of Dinant, Amaury of Chartres, and Mauricius of Spain.[68] Ernest Renan believed "Mauricius" somehow resulted from a corruption of Averroes' name, although his contention remains unproven and unlikely.[69]

Despite these local condemnations, Aristotelian thought grew in importance in western Europe and possibly gained influence at Paris as well, because it was possible for professors in the Faculty of Theology to make use of Aristotle's metaphysics and natural philosophy. Consequently, for some the threat of pagan philosophy to religion remained. In 1228, Pope Gregory IX warned the regents of the Faculty of Theology at Paris that those "inclined toward natural philosophy" were a threat to the true understanding of religion.[70] Three years later Gregory IX decreed that "[t]hose books of natural philosophy," which had earlier been condemned in the provincial council in 1210, "should not be used at Paris, until they have been examined and purged of all suspicion of error."[71] It seems probable that these condemnations affected teachings at Paris. Roger Bacon, who interpreted these condemnations as being against "Aristotle as set forth by Avicenna and Averroes," maintained that those in Paris who used their books "were excommunicated for quite long periods."[72] In 1245, however, Albertus Magnus obtained direct permission from the papacy to examine these books in order to determine their acceptability.[73] The result was that according to the university statutes of 1254, Aristotle's books on natural philosophy and metaphysics largely composed the curriculum at Paris.[74] Aristotle was restored.

His restoration, however, did not please all. In the late 1260s or early 1270s, Giles of Rome listed what he considered to be doctrinal errors of pagan and Islamic philosophers. He included fourteen propositions attributed to Aristotle and twelve to Averroes.[75] In 1277, Etienne Tempier, the Bishop of Paris, condemned 219 propositions, mostly surrounding natural philosophy, as erroneous. Interpretations of the impact of the condemnations of 1277 have held an important place in the historiography of medieval natural philosophy. A century ago, Pierre Duhem argued that the condemnations sparked modern science by encouraging natural philosophers to depart from Aristotle's authority.[76] Duhem's thesis is no longer a contender; these condemnations were limited to the Faculty of Arts at

Paris and therefore did not affect the Faculty of Theology or universities else-where. Perhaps, most important, Aristotelianism, in one form or another, did not diminish in strength after the condemnations but remained dominant in univer-sities for centuries.

Because Tempier's condemnations did not specifically name those who put forward the condemned propositions, it is difficult to identify precisely his targets. Roland Hissette has traced a number of the propositions to Siger and his contem-porary Boethius of Dacia, although he notes that many of the condemned prop-ositions can be traced back only to Aristotle or the ancient commentators.[77] That is, it is possible or even likely that no contemporary of Tempier taught or held a number of the propositions, even if investigations are to include the Faculty of Theology, as Hans Thijssen has proposed doing.[78] Tempier's concerns were numerous and included pronouncements that concerned limitations on God's power or knowledge, causation, the eternity of the world, the human intellect, and human happiness. The condemned propositions were not only philosophi-cal; one also governed the interpretation of Aristotle, forbidding teaching, "that it is not found in Aristotle, that the intellective [soul] remains after separation [from the body]."[79] Tempier intended to restrict the interpretation of ancient authorities in addition to limiting what conclusions could be drawn from them.

Although the 1277 condemnation's precise object is unidentified, Tempier's targeting of Siger is perhaps evident in the condemned proposition that forbade teaching: "That contraries cannot at the same time be true in a different sub-ject."[80] From this condemned proposition derives the notorious concept of *dou-ble truth* (duplex veritas) that has been attributed to Averroes and other Aristo-telians. Rather than accurately representing the views of Averroes, Siger, and Boethius, here Tempier appears to have been ridiculing their investigations *in naturalibus*. Neither Averroes nor Siger actually endorsed the concept of *double truth*. Thirteenth-century Christian philosophers contended that faith provided the truth whenever conclusions based on philosophical principles appeared to contradict religion.[81]

Notwithstanding the condemnations of 1277, the threat of pagan philosophy persisted, at least in the mind of Ramon Llull. Llull was a mystic as well as a missionary, who attempted to convert Muslims using philosophical arguments. Around the year 1311, he wrote a series of treatises that attacked either Averroes or Christian followers of Averroes. These polemical treatises reflect many of the concerns of Tempier, including the alleged positing of a double truth theory. In *On Contradictory Syllogisms*, he contended that the *Averroista* held forty-four propositions that were either "against God or against the holy Catholic faith." This hypothetical Averroist believes that these propositions "are intelligible philosophically," yet theologically false, "even though there is but one truth, philosophically and theologically." Llull then set out syllogisms or arguments that he believed the Averroists "could not resist." The propositions Llull found objectionable cover a number of subjects. Many relate to God's power or knowledge, such as, "God does not have infinite power;" or, "God does not and cannot act directly on lower bodies [inferiora]," or, that God does not have infinite knowledge or knowledge of particulars." Other propositions relate to the creation or eternity of the world, motion, and matter. A few concern important matters of faith: denials of hell, heaven, demons, the creation of angels, and the possibility of virgin birth. Emphasis on questions surrounding the intellect is relatively slight. Only three propositions mention it, with one being the Averroistic contention that, "The intellect is one in number in all humans."[82]

Many of Llull's arguments were repeated in his other works, such as the *Sermons against the Errors of Averroes*, which he sent to Philip the Fair and the university at Paris, "so that the errors of Averroes might be cleared away from the city of Paris." The crusading character of Llull shines out in the prologue. Not satisfied with syllogisms, he resorted to ad hominem attacks, falsely writing, "The Saracens, however, even though they are infidels, stoned [lapidaverunt] Averroes, who was also a Saracen, because of the errors that he introduced against the faith."[83] In Llull's eyes, Averroes' ideas were so pernicious that even infidels found them a danger to religion. His crusading was directed at Averroism, yet it made a distinction between Averroes and Aristotle. That Llull valued aspects of Aristotelianism is evident by the mere fact that he used syllogism to fight against what he saw as heresy.

At the same time Llull was writing diatribes against Averroists, a church council addressed the philosophical interpretation of the human soul, an important step in defining Catholic dogma. The Council of Vienne (1311–1312) approved a decree that made "asserting, defending, or holding pertinaciously, that the rational or intellective soul is not the form of the human body per se and essentially" punishable by the charge of heresy.[84] This decree was likely directed at the teachings of Peter John Olivi (1248–1298), who, inspired by Augustinianism and Platonism, had become an important figure in debates within the Franciscan order over the practice of poverty. Conventual Franciscans advocated a more relaxed approach, while Spirituals, who promoted stricter observance of poverty, fought for Peter Olivi's canonization and held high regard for his philosophical positions. Clement V's decree on the rational soul was part of his attempt to unite the order by reigning in those inspired by Peter Olivi.[85]

Although this decree made a philosophical position a matter of orthodoxy, that is, the officially accepted doctrine of the Catholic Church, its effects were not especially broad. Llull had petitioned the Council of Vienne to decree, "No philosophy should be read against theology but natural philosophy should be read that agrees with theology." He hoped to stop those from following "ancient philosophers who said many things against the faith."[86] Llull's petition was directed toward the suppositional approaches to natural philosophy that he had attacked in his polemics against Averroists, approaches similar to those of Siger or Albertus, who investigated the natural world following the path of Aristotle (via Aristotelis) without recourse to supernatural concepts (de naturalibus). Regardless, the council did not adopt Llull's suggestions, thereby implicitly permitting such considerations if they did not run counter to the proposition that the rational soul is the form of the human body.

The Council of Vienne defined the orthodox theory of the soul as one that was in agreement with Thomas's thought, but it did not specifically rule out teaching the mortality of the soul, perhaps because no one maintained that position. Scotus's contention that the immortality of the soul could not be proven by philosophy also lay beyond the scope of the decree. An English Carmelite named John Baconthorpe tried to reconcile Averroes' view that the passive intellect is one numerically with the doctrine that the intellect is the substantial form of the human body. He suggested that Averroes held this theory only hypothetically or, alternatively, that the passive intellect is one in number only from certain perspectives.[87] In sum, the Council could demand that a philosophical proposition be orthodox, but it could not force philosophers to agree that philosophical demonstrations proved the truth of that doctrine.

CONCLUSION

The Council of Vienne signaled the ascendancy of Aristotle's thought in theology. Using the term "form" as part of the definition of the soul, it linked Aristotelian hylomorphism with Christian doctrine. Evidence for the acceptance of Aristotelianism in theology is found elsewhere. Thomas Aquinas was canonized in 1323. Possibly related to his canonization was the retraction of the portions of the 1277 condemnations that appeared related to Thomas's doctrines.[88] Concerns over heresy and false teachings at Paris continued into the fourteenth century, but these worries shifted toward the teachings of William of Ockham, Nicholas of Autrecourt, and John of Mirecourt.[89] As doubts over the orthodoxy of Ockhamists increased, Thomas's interpretations of Aristotle gained surer footing. The statutes of 1339 at Paris point not just to the acceptability of Aristotle but also to the usefulness and even acceptability of Averroes' comments. Bachelors of arts were required to swear that they would not follow Ockham or his followers but rather would sustain "the knowledge (scientia) of Aristotle and his commentator Averroes and other ancient commentators and expositors of the words of Aristotle, except where they are contrary to faith."[90] Thus the statutes recognized that Aristotle as well as his Islamic and ancient commentators, even if at times they contradicted Christian dogma, as a whole bolstered theology.

Although it was conceded that certain Aristotelian doctrines did not conform to Christianity, it remained possible to merely excise these propositions and accept the rest of the corpus. Aristotle's view of the eternity of the world, for example, was not taken to be a foundation on which the rest of his natural philosophy supported itself, suggestive of an inherently non-Christian understanding of reality. Rather, it was merely one proposition that could be discarded because of its incompatibility to faith. When interpreted correctly, Aristotle was a handmaiden to theology. Deviations from this correct interpretation could present potential dangers to the proper reconciliation of Greek philosophy with Christianity, yet, for Scholastic thinkers, the possibility of using Aristotle and his interpreters for creating Christian thought was not only accepted but actually formed the starting point of their philosophical project.

CHAPTER TWO

Humanists' Invectives and Aristotle's Impiety

Giles of Rome, Thomas Aquinas, Ramon Llull, and other ecclesiastical author-ities critiqued Aristotelianism from within, believing that, if his texts were in-terpreted in the correct way, they were necessary for considerations of nature and theology. As a result, church decrees institutionalized Aristotelianism. The Council of Vienne endorsed a doctrine of the human soul based on the Aris-totelian distinction of *forma per se*; and the Fourth Lateran Council endorsed the concept of transubstantiation, providing a theological understanding of the Eucharist that was in accord with Aristotelian concepts.[1] Beginning in the four-teenth century, assaults on Aristotle's thought came from humanists and men of letters outside of universities. Unlike earlier Scholastic polemicists, these human-ists largely saw the study of nature as meaningless—futile for gaining wisdom or accessing religious truth. For many of them Aristotle was not an authority to be followed but rather the fount of misguided approaches to learning that privileged syllogistic logic over rhetoric and the study of nature over ethics.

Although the specific complaints about Aristotle's followers varied among hu-manists, in general, they frequently found their contemporaries in universities abhorrent because of their allegedly barbarous or nonclassical writing style and vocabulary, their dependence on Muslim Arabic-writing authors, and their sup-posed impiety. Furthermore, humanists considered the centrality of Aristotle and his commentators to university instruction to have shackled Scholastic thought, thereby rendering it narrow in its scope and beyond humanistic visions of eru-dition, Christianity, and human nature. Some Scholastics thought that careful interpretations of Aristotle could be reconciled with Christianity, but many humanists were more skeptical of that possibility, believing dependence on Aristotle and his commentators rendered university education pernicious and pointless.

PETRARCA

One of the earliest and most influential humanists, Francesco Petrarca (1304–1374) pioneered an approach to attacking Aristotle that conflated Scholasticism, natural philosophy, and Aristotelian thought with one person: Averroes. Two of Petrarca's works, *Invectives against a Physician* and *On His Own and Others' Ignorance* addressed the use of Averroes among university-trained physicians. In immoderate fashion, characteristic of humanist invective, Petrarca offered complaints that addressed primarily two issues: Averroes' impiety and his chosen genre, the commentary. In *Invectives against a Physician*, Petrarca accused his unnamed and most likely fictional addressee of secretly wishing "to challenge Christ, over whom you privately prefer Averroes." The anonymous physician, against whom Petrarca rants, is like the fool of Psalms who says in his heart, "There is no God." In his attack on this physician, Petrarca noted that the conflicting theories of ancient Greek philosophers, such as Epicurus, Democritus, Pythagoras, have put forward remarkable opinions on the number of worlds or the nature of the soul, but the physician's "master," that is Averroes, asserted even more amazingly "the unity of the intellect," one of the same doctrines to which Thomas, Etienne Tempier, and Llull objected.[2]

Thomas and Llull used argument and textual analysis in their attempts to counter Averroes, Petrarca looked beyond specific philosophical convictions in his attacks. According to Petrarca, Averroes' danger comes not just in his philosophy but also in his being an example of someone who is openly impious. In a letter to Ludovico Marsili, Petrarca urged him to write against "the rabid dog Averroes . . . who barks against the Lord Christ, against the Catholic faith."[3] In a similar light, Petrarca has his hypothetical physician say, "How can someone named Christ threaten me? Averroes defamed him with impunity, as no poet or indeed any mortal has done." He concluded: "You physicians worship Averroes; you love him; and you follow him simply because you oppose and hate Christ, who is the living truth."[4] Averroes has replaced Christ as an object of worship, according to Petrarca.

Somewhat improbably, Petrarca believed reading or following Averroes' positions was a disguised way of undermining the Christian faith. He wrote, "Since you do not dare to blaspheme publicly the one [Christ] whom the world worships, you practically worship his sacrilegious and blasphemous enemy [Averroes]. This is the way of spite and cowardly malevolence: when you are afraid to disparage someone, you applaud the detractors."[5] In *On His Own and Others' Ignorance*, Petrarca linked Averroes' impiety to his chosen method of writing the commentary. By commenting on Aristotle's works, he "in a way made them his own," but Averroes is suspect, because his praise of Aristotle is a clever way of hiding the pride and vanity that that praise truly shows.[6] If Tempier offered an inaccurate, yet damning, portrayal of natural philosophers in his condemnation of the doctrine of double truth, Petrarca's rants seem even further removed from the stated motives and methods of Aristotelians. Using overblown rhetoric, Petrarca portrayed readers of Averroes in the worst possible manner; however, it seems likely that Petrarca's familiarity with Scholastic discussions was limited.[7] Despite the hyperbole and willful inaccuracy, the picture of an Averroist secretly plotting against Christianity lasted for centuries.

DANTE AND HIS INTERPRETERS

Notwithstanding Petrarca's invectives, consensus about Averroes was as difficult to find among literary figures of fourteenth-century Italy as it was for Scholastics. In *Inferno*, Dante Alghieri (1265–1321) placed both Aristotle and Averroes in limbo, grouped with those who had not sinned but could not enter paradise because they were not baptized.[8] Dante did not directly name Aristotle but simply called him "the master of those who know" (maestro di color che sanno) and portrayed him as honored by the rest of the family of philosophers" (filosofica famiglia). The members of the family are all ancients except two: Avicenna and Averroes. Dante described the latter as having "made the great commentary" (Averoìs che 'l gran comento feo), thereby identifying Averroes by his commentaries on Aristotle. In the *Convivio*, Dante echoed Averroes' high assessment of Aristotle's mind in his conclusion that "nature had placed in Aristotle a nearly divine genius."[9] Aristotle's Christian followers fared even better than Aristotle did in the *Divine Comedy*. Albertus Magnus, Thomas Aquinas, and even Siger of Brabant reached paradise, suggesting that Dante had familiarity with the philosophical milieu of thirteenth-century Paris.[10] Scholars have long noted the influence of Aristotle and Averroes on Dante. Aristotelian cosmology influenced his conception of inferno's levels formed of concentric circles that mirror the celestial spheres. Some have even seen his description of the heavens in *Paradiso* as corresponding to Averroes' denial that the substance of the heavens has both matter and form.[11] In *Monarchia*, Averroes is an authority for Dante's view on the purpose of civilization, although he conceded that he did not agree with the Averroist view of the potential intellect.[12] Much like Scholastic philosophers, Dante found value

in Averroes' commentaries, although he distanced himself from the most controversial views about the soul.

Early interpreters of Dante puzzled over the seemingly positive assessment of
Averroes in the *Divine Comedy*. Much as Averroes had commented upon Aristotle, a number of literary figures of the fourteenth century wrote commentaries
on Dante. They explained the meaning of each line and provided information
on the historical persons Dante portrayed. The *Ottimo commento* (circa 1333–34),
one of the earliest commentaries, presented little information on Averroes. Despite maintaining incorrectly that he was from Morocco, the brief portrayal of
Averroes is favorable. Perhaps indicating a lack of awareness of any controversy
surrounding Averroes, it describes him as a great genius who "expounded on
many of Aristotle's books . . . and wrote many very useful books on medicine."[13]
Three decades later, Giovanni Boccaccio, most famous for the *Decameron*, described the confused status of Averroes in his *Expositions on Dante's Comedy*.
Boccaccio, who knew Petrarca and was evidently better informed than the author
of the *Ottimo commento*, confessed that there was disagreement about whether
Averroes was Spanish or an "Arab" who lived in Spain. Regardless, according to
Boccaccio, Averroes possessed "such a lofty intellect that . . . [his commentaries]
succeeded so well" that Aristotle's works, "which seemed never to have been
understood before him and were therefore left untouched, assumed wondrous importance." In his eyes, therefore, the integration of Aristotle in Scholastic thought
depended on Averroes' project. Averroes' significant role in intellectual history,
however, should not ensure him an agreeable afterlife. Boccaccio disagreed with
Dante's placement of Averroes in limbo, arguing that anyone who lived after the
Gospels had been preached and was not Christian "deserves a great penalty."[14]

Benvenuto da Imola (1338–1388), who knew both Boccaccio and Petrarca, interpreted Dante's description of limbo in the context of fourteenth-century universities.[15] In this light, he explicated the line where Dante wrote that when visiting the pagan philosophers they were "speaking of matters about which it is good
to be quiet."[16] Benvenuto considered it acceptable for Dante and Virgil to speak
about pagan philosophers while in limbo, giving an example based on the practices of contemporary philosophers. According to Benvenuto, these masters of
arts "visit some excellent and famous doctor of theology, and privately in his room
or study, using naturalistic theories and demonstrations discuss natural matters
[de rebus naturalibus cum rationibus et demonstrationibus naturalibus], such as
the origin of the soul, the production of the world, human happiness, the eternity
of motion, and other such things."[17] Afterward, the theologian climbs up to the
pulpit and preaches publicly to a broader audience and addresses topics differ-

ently from what he discussed in his study with the philosopher. Benvenuto concluded that Dante meant that it was licit to discuss such topics with the pagans in limbo, but it would be dishonest and useless to speak about these subjects among Christians. Analogously, Benvenuto suggested that it is legitimate to speak about controversial subjects using naturalistic arguments but only privately and not to the general population.[18]

Aristotle fared well in Benvenuto's commentary. He explicated the phrase "maestro di color che sanno" by citing Aristotle's supposed expertise in medicine, theology, law, morals, poetry, and oratory and concluding that he is "the best master of all masters."[19] As might be expected from an acquaintance of Petrarca, Benvenuto gave Averroes less sympathetic treatment than he did Aristotle. Despite Dante's apparent esteem for him and notwithstanding his provenance from Cordova, which produced the classical authors Seneca and Lucan, Benvenuto emphasized Averroes' impiety. In opposition to Avicenna, Benvenuto wrote that Averroes "cursed every religion."[20] Benvenuto then wondered how Dante "placed him without punishment" in limbo given that Averroes "impudently and impiously blasphemed Christ," by saying "that there were three impostors [baratores] in the world, that is, Christ, Moses, and Muhammad, one of whom, namely Christ, because he was young and ignorant, was crucified."[21] Averroes wrote no such thing, although he did refer to the "three laws" that hold that the world was created.[22] Instead of Averroes' writings, the source for the accusation that the prophets of the three major religions were impostors emerged most likely from Pope Gregory IX's polemic against Holy Roman Emperor Frederick II, although others accused the twelfth-century Parisian master Simon of Tournai of introducing the idea of the three impostors to the Christian world.[23] Even though translators of Averroes and even possibly Averroes' sons were present at Frederick II's court, the papal charges, most likely imaginary, only vaguely point to anything that Averroes wrote, namely that Frederick II sustained that it is necessary to believe something only after it is proven by natural reason.[24] Regardless of their accuracy, these accusations of sustaining that religion was imposture were transferred to Averroes, adding to his image of impiety and unbelief.

SALUTATI AND BRUNI

For Italian humanists and men of letters, impiety became a hallmark of Averroes' reputation and, thus, at least implicitly, also of some Aristotelians. The Florentine humanist Coluccio Salutati (1331–1406), in a letter probably from 1397, wrote that he "marvels not without laughing at the most irreligious Averroes, who thought very badly about God and the eternity of the soul."[25] For Salutati, the poor thinking that led Averroes to his views about the soul deserves ridicule and is symptomatic of the weakness of his thought in general. Salutati pointed to a passage from Averroes' medical writings that he thought was proof of Averroes' lack of intellectual discrimination. In this passage, Averroes referred to a maiden who testified that she conceived by being in a bath in which there remained "semen that had been emitted *contra naturam*."[26] In Salutati's eyes, Averroes' credulity in believing this unwed mother's tale matched his incredulity of the dogma of all religion. The use of this example in support of Salutati's accusation that Averroes is irreligious is strange, given its lack of relevance to religion. Citing this story of unplanned pregnancy for evidence of Averroes' impiety perhaps belies a shallow familiarity with Averroes' works; he could have cited his belief in the unicity of the passive intellect or the eternity of the world. Nevertheless, contrary to the "three impostors" attribution given by Benvenuto, Salutati's tale of conception through inseminated bathwater is found in Averroes' medical work the *Collectanea*. Furthermore, he cited Averroes' commentary on the *Physics* in his 1396 *On fate and fortune*.[27] Thus, unlike some humanists, Salutati had firsthand knowledge of Averroes' work and Scholastic culture in general, which he saw as opposed to true wisdom.[28]

Similar to Petrarca, Salutati distinguished unnamed philosophers' obedience to Aristotle or Averroes from following true religion. In a letter from 1406, he wrote, "The throng of philosophers would follow Aristotle or Plato, they would follow the poisonous Averroes . . . even if they had someone better. For me only Jesus Christ would be pleasing."[29] Instead of philosophy, Salutati argued one should study grammar, which he claimed to be the gateway to "all doctrine whether divine or human."[30] His letters championed the study of grammar more than they pinpointed problems with Averroes' thought, as is also true for his *On the Labors of Hercules*, where he attacked schools of Aristotelians. There he distinguished Aristotle from contemporary Aristotelians, whom he associated with English scholars, most likely from the terminist school, using Virgil's line from the *Eclogues*, "Britain completely divided from the whole globe," to describe the state of Aristotelianism.[31] He accused these Aristotelians of being overly interested in logic; using corrupt vocabulary; not reading Aristotle's texts; and, juggling propositions, corollaries, and conclusions.[32] Since Averroes' influence was limited in these English schools, the actual importance of Averroes in universities is irrelevant to Salutati's critique. Averroes' reputation for impiety made him useful for undermining Scholastic culture in general.

The humanist Leonardo Bruni (1369–1444), who, like Salutati, served as chancellor of Florence, made a similar distinction between Aristotle and Aristotelians, maintaining they slavishly clung to "harsh, awkward, dissonant words," which they identified as Aristotle's. For Bruni's version of Aristotelians, Aristotle had become an authority, comparable to a divine source, "as fixed as the Pythian Apollo," so that "it is impious to contradict him."[33] Bruni implied that by revering Aristotle they do not worship true religion, since their versions of natural philosophy had replaced religion for these unnamed thinkers. For that matter, according to Bruni, they do not even revere the true Aristotle, because "those books which they say are Aristotle's have suffered such a great transformation that . . . he would not recognize them as his own."[34] Bruni left unclear precisely what transformations had rendered Aristotle's work unrecognizable, but part of the problem for Bruni was the ahistorical nature of Scholastic interpretation.

The Scholastic devotion to Aristotle is at odds with an accurate understanding of the history of philosophy, Bruni contended. Unlike in the universities of the fifteenth century, in ancient Roman and Greek cultures numerous schools of philosophy flourished. Bruni pointed out that during Cicero's time, when philosophy was "watered by that golden stream of eloquence," there were not just Peripatetics but also Stoics, Academics, and Epicureans; and, in fact, "not very many people knew Aristotle."[35] Cicero was a model for rhetoric and prose style for humanists such as Bruni. Consequently, a revival of Cicero's thought meant a diminished, or at least altered, role for Aristotle, especially since his philosophical writings were little influenced by Aristotelianism.

Yet Bruni did not intend to eliminate Aristotle's thought. He wished only to censure contemporary Aristotelians, while retaining the sections of Aristotle's philosophy that he found useful. Bruni used historical inquiry to remold Aristotle so that he fit more closely with his vision of philosophy that promoted rhetoric and ethics over the study of nature. Understanding the historical Aristotle was a potential remedy against what he considered to be distorted versions of ancient thought taught in universities.

Bruni was correct that medieval Scholastic philosophers knew relatively little about the history of ancient philosophers and their schools. Few historical analyses of Aristotle circulated among Scholastics, despite the extensive knowledge of Aristotle's philosophical writings during the thirteenth century. Scholastic philosophy, for better or worse, operated in an ahistorical mode. The views and arguments of authorities formed the basis of discussions with little concern for their chronology or cultural circumstances. The opinions of ancient, Islamic, and contemporary authors were discussed in the present tense and as though thinkers separated by centuries were conversing with each other. Reflecting this ahistorical stance, only a few biographies of Aristotle circulated in western Europe during the late Middle Ages. One was based on the work attributed to the circle of late-antique Neoplatonic commentators and is called the *Vita latina*. Many manuscripts of this biography are extant.[36] The *Vita marciana*, which is extant in only one Greek manuscript, is very similar to the *Vita latina*.[37]

Only a few medieval Christians wrote biographies of Aristotle. The thirteenth-century Franciscan John of Wales wrote a brief biography of Aristotle using the *Vita latina* and a number of ancient Roman sources, such as Cicero and Seneca, as well as medieval authorities, including Avicenna and Maimonides. This biography of Aristotle is part of a larger collection of considerations of the lives of ancient philosophers. In John's account, Aristotle held moderate positions in philosophy, was of singular genius, and eminent in all areas of learning.[38] He was also vainglorious, eager to win fame through his subtle thought and smooth style of speaking. In John's account, Aristotle's desire to understand the causes of nature caused his death. Unable to understand the cause of the tides, Aristotle threw himself into the Euripus and drowned.[39] Despite the evident moral lessons of John's life of Aristotle, there is little if any concern with Aristotle's religious practices or belief.

Another biography, which dates from the first decades of the 1300s and circulated in numerous manuscripts during the fifteenth century, was attributed, most likely incorrectly, to the fourteenth-century philosopher Walter Burley.[40] Like John of Wales's work, the Pseudo-Burley biography is part of a collection of short biographies of philosophers. The depiction of Aristotle is extremely similar to what is found in the *Vita latina* and is a largely favorable portrait, even while describing him as a pagan. Both biographies portray Aristotle in glowing terms as a philosopher, much as Averroes had done—"in philosophy he transcended human measure"—and recognize him as the inventor of logic and of parts of natural philosophy.[41] Aristotle's contributions to theology were less significant, according to this biographical tradition. The *Vita latina* maintains that "he [Aristotle] added nothing to its study," betraying perhaps the author's own preference for Platonic metaphysics. Yet the author of the *Vita latina* conceded that in *Physics* 8, he proved "the divine is neither body, nor undergoes change," suggesting that Aristotle had at least some concerns for understandings of God.[42]

Portions of this biography were potentially unflattering to Aristotle, especially seen by those coming from a Christian culture. According to the *Vita latina*, Aristotle took up philosophy at the age of 17, upon being warned by the Pythic oracle. Thus his studies stemmed from his pagan roots, even if the role of the Pythia in his education could be, and no doubt was, ignored. The Pseudo-Burley biography made Aristotle closer to Christian ideals. The anecdote about the oracle is missing, but of greater importance is that Aristotle's interest in the divine becomes more prominent. The author of this biography pointed to a book of secrets of Aristotle that contained esoteric knowledge revealed to him through holy proph-ets.[43] In this version of his biography, Aristotle "inquired into the scriptures of all wisdom" and urged his followers to understand the immortal and divine.[44]

Both the *Vita latina* and Pseudo-Burley's *Vita* are extremely different from Diogenes Laertius's biography of Aristotle, which formed part of his *Lives and Opinions of Eminent Philosophers.* Diogenes lived in the third century AD and was by no means a partisan of Aristotelian thought, or even reverent toward the philosopher. He included in his biography stories of Aristotle's alleged marriage to the former concubine of Hermias, the tyrant of Atarneus who was alleged to be a eunuch; of indictments for impiety; and of his death by suicide (which Diogenes later denied).[45] Despite these blemishes of comportment, Diogenes also determined that "in the investigation of the causes of natural beings, he [Aristo-tle] went further than everyone else."[46] Thus Diogenes allowed for the possibility that a man whose life was marked by scandal could also possess great knowledge.

The Florentine humanist Ambrogio Traversari translated Diogenes' *Lives of the Philosophers* into Latin by the 1430s, making it more broadly available, espe-cially to an Italian audience.[47] Bruni's *Life of Aristotle* was intended to restore Aris-totle's reputation from the tarnish of Diogenes' and the Latin biographies in order to justify the Latin translation of Aristotle's *Ethics,* which dealt with topics central to the concerns of humanists such as Bruni but was not part of most universities' curricula.[48] In Bruni's biography, gone is any reference to the oracle. He men-tioned Aristotle's sojourn at Hermias's court but, instead of a marriage to a former concubine, it is with first the tyrant's daughter and then with his niece. Jealousy among his rivals explained the charge of impiety, Bruni concluded. Rumors of his death by suicide were false and nonsensical. In Bruni's description, Aristotle was dedicated to family, public religion, his teacher Plato, and his disciples. Unlike in the earlier biographies, Aristotle did not start his own school until after Plato's death. There was no need for a splintering because, according to Bruni, their philosophies were largely in agreement, especially about ethics, "the nature of the universe, and the immortality of the soul."[49] Aristotle disagreed with Plato in places where the latter depended more on "revelation than demonstration," such as in Plato's very non-Christian belief in metempsychosis, and where Plato's views were "utterly abhorrent to our customs" such as the belief in that "wives should be held in common," as expressed in *The Republic.*[50]

Most remarkable, perhaps, was Bruni's transformation of Aristotle into an orator. In order to distinguish Aristotle from the Aristotelians of the universities, who in the eyes of humanists wrote poorly and used barbaric vocabulary, Bruni emphasized the eloquence of Aristotle, which he contended was unknown among Renaissance and medieval scholars because of "adulterated translations," resulting from "simply the nonsense of translators." Whereas, the earlier biographies wrote of Aristotle's perfection and nearly superhuman ability in philosophy, Bruni, playfully transformed this trope: "But in Aristotle, all is perfection. Indeed, no one was his equal for taking care in composition."[51] Linking virtue with oratory, Bruni saw Aristotle's perfection in his skill as a writer. His idealization of Aristotle is not that of Averroes' exemplar for the study of nature but rather as the ideal rhetorician.[52] Thus Bruni rehabilitated Aristotle from the slurs of previous biographers, turning him into orator, and separated him from the Scholastics that saw him as primarily a natural philosopher.

Bruni's historical treatment attempted to render Aristotle acceptable to Christians and, perhaps of greater significance, to humanists. Making Aristotle an ideal rhetorician, in turn, promoted Bruni's other Aristotelian endeavors, namely translating the *Politics*, the *Nicomachean Ethics*, and the *Economics*. Although known to Scholastics, these Aristotelian texts were closer in spirit to the teachings of the *studia humanitatis* that promoted the study of ethics and politics as part of the preparation for civic engagement. Bruni's biography attempted to make at least some of Aristotle's writings part of the standards of humanism, shifting his importance away from the study of nature.

LORENZO VALLA

Bruni used biography to mold Aristotle into a rhetorician palatable to Christians, but other humanists used the study of ancient history to attack Aristotelians. Lorenzo Valla (1407–1457), perhaps most famous for his philological unmasking of the "Donation of Constantine" as a forgery, used his knowledge of classical literature and history to subvert the status of Aristotle among philosophers and especially theologians of the universities. Valla believed that Aristotle was a source of medieval intellectual corruption, although not the only one. In his view, Boethius introduced numerous improper translations of the Greek and created improper philosophical definitions.[53] Valla's invectives against Aristotle evoked themes similar to those of many of his humanist predecessors. In the *Dialectical Disputations*, he argued: "Recent theologians flushed with ancient Aristotelian precepts insult and tease; they are armed with Aristotelian doctrine, like the healthy are armed with the weak, armed like the rich with the poor, since they know metaphysics, logic, and the modes of meaning." The tools of Aristotelian thought, that is, metaphysics and logic, according to Valla, prevent those who use them from understanding theology. In his view, these unnamed Aristotelian theologians should be objects of derision rather than authorities because, "they hold the master Aristotle to be like God." Moreover, their failings are not just a misplaced devotion to Aristotle. Their lack of linguistic skills limits their ability to obtain wisdom. Valla maintained they do not know Aristotle well. They are "inexperienced in Greek letters" and they cannot explain the doctrine well because they are "little experienced" in their own language, that is, Latin.[54]

The predominance of Aristotelian thought among theologians was a novelty that did not reflect the attitudes of the ancient world or early Christians, according to Valla. In particular, ancients and early Christians endorsed the "liberty," which once philosophers possessed to discover and speak the truth. To the contrary, Valla believed "[r]ecent Peripatetics . . . forbade the liberty to dissent from Aristotle."[55] The demand of philosophical agreement runs counter to the views of ancient philosophers, who since the time of Pythagoras have used the word "philosopher" not for those who followed other men but rather for those who followed "truth and virtue." As a result, Valla believed that "there had always been a liberty of philosophers for bravely saying what they judged, not only against the leaders of other sects, but against their own leader, so that they were not bound to any sect."[56] Consequently, unlike recent Peripatetics, the ancients freely dissented from Aristotle. Among those who disagreed with Aristotle were the Platonists at the Academy and Aristotle's own successor at the Lyceum, Theophrastus. Ancient Greeks admired Homer, Plato, and Demosthenes more than they did Aristotle, and Romans, such as Cicero, Seneca, and Apuleius, preferred the Stoics or Plato.

Valla distinguished contemporary Aristotelian theologians from the philosophical allegiances of early Christian thinkers. Jerome thought Stoicism was the school of philosophy closest to Christianity. Ambrosius was a Stoic and follower of Cicero, while Augustine and Hilarius were partisans of Platonic philosophy. For Valla, Boethius, despite his faults, represented the end of ancient philosophy. He too was "more Platonic than Aristotelian." After Boethius, ignorance and barbaric Latin reigned, as is evident by the fact that "Avicenna and Averroes were clearly barbarians, completely ignorant of our language [i.e., Latin] and barely touched by Greek."[57] Valla wondered how it was possible that their writings held any authority.

Despite his reference to Averroes and Avicenna, Valla's attack on Scholastic philosophy did not concentrate on the strands of Aristotelianism most influenced by these Islamic thinkers. Despite some scholars having seen parallels between Valla and terminist philosophers such as Ockham, it is more likely that he had only shallow knowledge of such works, being more concerned with rhetorical traditions than the philosophy of the schools.[58] Yet the concerns with terminology allowed him to point out where Aristotle and his followers posited what he believed were impious theories. In his discussion of the soul he wrote: "Part of the soul cannot be called a 'vessel' of the other part of the soul," for that suggests that the soul "is either generated from within or not generated, both of which are false."[59] Use of such a metaphor, according to Valla, made it unclear whether the rational soul was a "part of God" and whether it "lives when the body is dead." As a result, "Aristotle most stupidly pronounced on the soul." Similarly, Aristotle "disputes most insipidly" on God because he used the name "prime mover." Such a name is lacking in dignity for God, because, according to Valla, he is more than the mover of the universe but created it ex nihilo.[60] Citing *Metaphysics* Lambda, Valla accused Aristotle of confusing nature with the divine and conversely denigrating God by called him animated, thereby suggesting God is corporeal.[61] With these examples, Valla endeavored to demonstrate the importance of correct terminology for producing an accurate and pious understanding of humanity and the divine, as well as pointing out Aristotle's and Aristotelians' failure to do so.

THE PLATO-ARISTOTLE CONTROVERSY

Early humanists from Petrarca to Valla adopted an overwhelmingly negative attitude toward Aristotelians, while their Scholastic counterparts defended Aristotle as a learned source for the study of nature and metaphysics that could be reconciled, for the most part, with Christian thought. During the middle of the fifteenth century, a group of emigrants to the Italian peninsula from the Greek-speaking world confronted these themes in decades-long debates over the superiority of either Plato or Aristotle over the other, the possibility of reconciling Plato and Aristotle, and their relation to Christianity.[62] Because these emigrants participated in humanist culture while being involved with ecclesiastical affairs, their specialized knowledge of ancient philosophy, based on their reading of Greek texts, many of which were previously unavailable to the Latin West, greatly influenced assessments of Aristotle as well as the methods for making these evaluations. In a sense the Plato-Aristotle controversy rehashed the debates of late antiquity, when Neoplatonists attempted to reconcile Aristotle, as well as the struggles of the thirteenth century, when Thomas Aquinas and others combined Aristotelian thought with Catholic dogma. The Greeks in Italy, however, had access to a larger corpus of ancient writings and thus addressed these questions with a level of philosophical and historical sophistication that surpassed many earlier humanists and Scholastics.

In 1438 Georgios Gemistos Pletho (1360–1452) accompanied John VIII Palae-
ologus to the Ecumenical Council at Ferrara and then Florence, where the possible unification of Eastern and Western churches was discussed. At the Council,
Pletho argued that Plato's philosophy should form the basis of Christian theology. He composed a treatise that contained these arguments called *On the
Differences*. Pletho's treatment of Aristotle and Aristotelians is reminiscent of humanist invective, despite his providing philosophical grievances with a greater
degree of specificity than found in the works of many earlier humanists. *On the
Differences* begins by chastising Pletho's contemporaries who "admire Aristotle
more than Plato," ignorant of the fact that "[o]ur ancestors, both Hellenes and
Romans, esteemed Plato much more highly than Aristotle." These Aristotelians
have been misled by Averroes' belief that "Aristotle alone has achieved a complete
account of natural philosophy." According to Pletho, Averroes cannot be taken
seriously on any matter because he was "so misguided as to suppose that the soul
is mortal . . . when even Aristotle himself is evidently not guilty of this particular
error."[63] Although Averroes did not actually argue that the soul is mortal, Pletho
used an *ad hominem* approach that associated Aristotelians with the seemingly
perverse views of Averroes, an approach common to many humanists.

Despite Pletho's lack of precision in his treatment of Averroes, he had specific philosophical disagreements with Aristotle, which he believed pointed to
a greater acceptability for Platonic thought. In particular, Pletho considered Aristotle's treatment of God insufficient. Looking largely at *Metaphysics* Lambda,
Pletho complained that, while Plato's God is the demiurge of all beings, Aristotle does not consider God a creator "but only the motive force of the universe,"
thereby making God the final cause only of movement rather than for individual particulars.[64] Additional evidence that Aristotle's God is not exalted highly
enough is found in his assignation of "a sphere and the movement of it to God
himself, thus placing him on a level with the minds dependent on him" that also
have their own spheres.[65]

Aristotle's mistakes about mankind accompany his errors about the divine, according to Pletho. Whereas Pletho believed that individual humans are created
for "the sake of human nature as a whole, and human nature itself for the benefit
of rational nature as a whole," Aristotle's conviction that individual particulars
are primary substances, making universals secondary and inferior, leads to the
incorrect view that God orders human nature in general for the sake of individuals.[66] Pletho interpreted Aristotle as believing the rational soul survives death,
basing himself on his reading of *De anima* and *Metaphysics*. Still, he chastised
Aristotle for not explicitly linking the immortality of the intellect to "moral theory

and to discussions of virtue" in the *Nicomachean Ethics*; Pletho found Aristotle's
ethical theory in general to fail because it does not posit rewards in the afterlife.[67]
For Pletho, Aristotle deviated from Christian conceptions of the divine and the
human.

Another Greek scholar, George Trapezuntius, defended Aristotle from Pletho's
attacks, attempting to show that "Plato was extremely distant from the truth, Aristotle greatly conformed to it, and the Church's dogmata were much helped by
his writings and were openly in opposition to Platonic ones."[68] Trapezuntius went
beyond Thomas's synthesis contending not just that Aristotle's theories of the soul
and of divine providence conformed to Christian belief but also that Aristotle's
philosophy endorsed a Trinitarian conception of God as well as God's creation
of the universe out of nothing. Thomas and other late medieval Scholastics had
been unable to uncover the latter two doctrines in Aristotle's writings.

Invectives against Plato accompanied the positive assessment of Aristotle.
Among other charges, Trapezuntius maintained that Plato advocated wicked
acts, including the "most revolting loves," wondering how it was possible that
others "admire, revere, carry to the stars with praise, and call the prince of philosophers this enemy of nature, destroyer of good morals, who clings to the buttocks
of boys."[69] Perhaps equally incendiary were Trapezuntius's judgment that Plato
was more similar to Epicurus and Muhammad than he was to Aristotle and his
contention that Muhammad exceeded Plato with respect to intelligence.[70] With
these proclamations, Trapezuntius indicated that reconciliations of Plato with
Aristotle or with Christianity were in vain.

Trapezuntius's invectives demanded a defense of Plato from those who sought
philosophical and theological reconciliation. Cardinal Basilios Bessarion, also a
Greek emigrant to Italy, wrote the most significant Platonic apology. His response
to Trapezuntius was measured in tone, especially in comparison to Trapezuntius's writings and to other refutations, such as Nicola Perotti's vitriolic *Refutation
of the Insanities of the Trapezuntius*. Bessarion was more interested in defending
Plato and his accord his Christianity than in attacking Aristotle.[71] Despite his
desire to find agreement among Plato, Aristotle, and Christian dogma where
possible, Bessarion did not believe Aristotelian philosophy could demonstrate all
fundamental religious truths, casting doubts on the synthesis endorsed by Trapezuntius. In *On the Procession of the Holy Spirit*, Bessarion wrote that the words
of scripture could be resolved of difficulties of interpretation without philosophy,
"since it seems to me that syllogisms, probabilities, and demonstrations do not
persuade, but rather do the plain words of scripture themselves."[72] He echoed this
contention in his *On the Calumnies against Plato*, maintaining that the human

mind cannot attain religious truth. He wrote, "The proof of divine matters should not be demanded, but what we believe, must be accepted by only faith and divine institution."[73] The persuasiveness of scripture and ecclesiastical tradition is independent of the rigors of logical argument.

Aristotle's deviations from Christian thought further impeded the possibility of Aristotelian theology. Bessarion believed Plato was more easily assimilated to Christianity. The relative conformity of Platonic philosophy to Christianity emerged in spite of his paganism. In his discussion of the trinity, Bessarion stated, "No one should think that Plato or Aristotle should be praised because they wrote about and revered the trinity. Both were deprived of true belief and ignorant of our religion." As a result, Aristotle never wrote at all about the trinity, and although Plato spoke about the trinity, he did so in a manner removed from Christianity.[74] The trinity was just one of several issues where Bessarion thought the reconciliation between Aristotle and true religion was impossible.

The third book of On the Calumnies against Plato, which was based on the work of the Dominican Giovanni Gatti, contains a discussion of divine providence.[75] Implicitly disagreeing with Thomas, the defense of Plato adopted the position that Averroes correctly interpreted Aristotle's view that God's providence extends only to heavenly substances. Bessarion believed that Aristotle's view, however, was incorrect and its impiety shows the difficulty of reconciling Aristotle with Christianity and the superiority of Plato over Aristotle.[76] What emerged from these debates was a dual question over the question of providence. The first was philosophical or theological—whether God's providence extends to particulars. The second was historical or interpretative. Did Averroes correctly interpret Aristotle? Did Aristotle hold that God's providence was limited to the celestial realms? Although there are exceptions, the schema that emerged was perhaps unlikely. The Platonists, Pletho and Bessarion, sided with Averroes over Thomas, believing that Aristotle held a limited conception of providence, while the Aristotelian Trapezuntius dismissed Averroes' view.

Similarly, when Bessarion discussed the immortality of the soul, he noted Plato's and Aristotle's assent to this proposition. The question of the soul's immortality, however, is tied to eternity of the world. According to Bessarion, those who hold that the world is uncreated and incorruptible and that no infinite can be realized must accept metempsychosis, if they also accept the human soul to be immortal. Otherwise, an infinite number of human souls would exist. Aristotle's solution to this conundrum differed from Plato's, according Bessarion. He wrote that in order to avoid this contradiction, Aristotle "embraced that impiety," which Alexander of Aphrodisias and Averroes attributed to him, that "there is one common intellect for all humans."[77] By adopting the view that the universe is created and will have an end, as Christianity teaches, Bessarion concluded, the transmigration of souls and the unicity of the intellect can be rejected. Just as in his discussion of providence, Bessarion agreed with Averroes' interpretation of Aristotle on a hermeneutic level but rejected the truth of that interpretation according to philosophy and theology. Bessarion's association of Alexander with Averroes' theory of the soul is unexpected, considering that Averroes rejected Alexander, attributing to him the position that the soul is material and mortal.[78] In any case, Bessarion contended Averroes correctly interpreted Aristotle's psychology and therefore Aristotle could not be reconciled with Christian dogma.

MARSILIO FICINO

Though the Plato-Aristotle controversy largely died out after Bessarion's death in 1472, the polemics altered the practice of philosophy in Italy.[79] Among those affected by humanism, interest grew in Plato's texts because they provided an alternative to the Aristotelianism taught in universities. Marsilio Ficino spearheaded this movement, translating Plato into Latin and writing the lengthy work *Platonic Theology*, which tried to establish Plato's thought as a foundation for Christian theology.[80] Like earlier humanists, Ficino cited opposition to adherents of Averroes as his motivation, contending that their philosophy deviated from Christianity.

Precise identification of these *Averroists* is perhaps impossible, but, just as in the previous two centuries, a number of university professors relied on Averroes' commentaries. Professors at Bologna and Padua, such as Paul of Venice and Gaetano of Thiene, favorably interpreted parts of Averroes' texts.[81] Lecturers, closer to Ficino at Florence and the Studio of Pisa, also had the reputation for adherence to Averroes. John Argyropoulos, who taught at Florence from 1457 to 1471 and from 1477 to 1481, followed Averroes' teachings about the unicity of the intellect.[82] In 1473, Alamanno Rinuccini, an official at Pisa, wrote to Dominicus of Flanders, a Dominican friar who wrote commentaries on Thomas, that he had proceeded to lecture on the *De anima* "not according to the common opinion of philosophers [i.e., Averroes] but from the sentences of Thomas and Giles of Rome," thereby suggesting that it was standard to follow Averroes.[83] During the academic year 1473–74, Niccolò Tignosi da Foligno, also a professor at Pisa, distinguished between arguments based purely on naturalistic arguments and those that relied on faith. He wrote that Averroes and Aristotle agreed the soul was one in number even if their view contradicted what is "most true" according to the Catholic faith, noting that the argument that the intellect separates from the body after death so that the good are rewarded and the bad punished was unavailable to Aristotle and Averroes.[84] *Circoli scholastici* from the same year show that students and professors debated Thomistic and Averroistic conceptions of the intellect.[85]

The documentary evidence points to a university divided between Thomists and followers of Averroes, but Ficino interpreted Scholastic culture differently. In a letter to the humanist Ermolao Barbaro, Ficino wrote he was working on the books of the Neoplatonist Plotinus so that "Peripatetics would be warned not to treat religion like it was old wives' tales."[86] He warned, "The entire orb of the earth occupied by Peripatetics is divided into two sects: Alexandrists and Averroists."[87] He identified these sects by their views on the intellect: the Alexandrists considering it to be mortal, the Averroists believing it to be one in number, "both equally strip away completely all religion."[88] For Ficino, Platonic texts represented a cure for the impiety of the Peripatetics who occupied universities.

Yet his desire to find convincing arguments that would undermine Averroes' views arose also from personal considerations. In a letter to Antonio Vinciguerra, Ficino expressed hope they might convert their "Averroicus" friend, Bernardo Bembo, a Venetian diplomat and man of letters as well as the father of Pietro Bembo, from the "heresy of Averroes," which he evidently adopted while studying at Padua.[89] With these motivations, Ficino wrote book 15 of the *Platonic Theology* in order to dismiss Averroes and his psychology. In an earlier book of the *Platonic Theology*, Ficino offered the explanation that Averroes suffered from melancholy and therefore despaired over the immortality of the soul.[90] In this book, he echoed humanist commonplaces about language as the cause of Averroes' failings and made the interpretation of texts central to his attacks. Arguing that Averroes' faults were due to textual and linguistic issues—reading books that had been "perverted rather than converted . . . into a barbarous tongue"—Ficino wrote that the "words of Aristotle in Greek contradict Averroes."[91]

According to Ficino, Averroes' ignorance of Greek undermined his ability to interpret ancient texts and thus his authority. Consequently, "Averroes himself did not arrive at his opinion through the free judgment of his own mind, but because he did not know how to interpret the books of Aristotle in any other way, [given] their corrupt translation into Arabic."[92] Ficino also attacked Averroes' views about providence as well as his interpretation of Aristotle. Associating him with Epicurus, Ficino called Averroes impious because of his lack of awareness that Aristotle "claims that individual parts are led back to the good of the order which is in the whole as to their end."[93] Although it might be unclear how this view undermines Averroes' view of providence, it shows that Ficino believed Averroes misinterpreted Aristotle.

Stressing mistranslation and misinterpretation as the cause of Averroes' failing belies Ficino's desire to salvage Aristotle, at least minimally. Ficino pointed to supposed historical reasons that Aristotle's thought might conform to Christianity. He believed that Clearchus of Soli, a follower of Aristotle who lived in the fourth and third centuries BC, wrote that Aristotle was Jew. Ficino, however, based himself on Trapezuntius's incorrect translation of Eusebius Caesarensis, the patristic author of the third century. In fact when correctly punctuated the fragment asserts only that Clearchus knew a Jew who had been on familiar terms with Aristotle.[94] Philosophically speaking, Thomas provided much for Ficino's treatment of Averroes, which defended the position that the substantial form of the human body is the intellect, a doctrine established by the Council of Vienne.[95] At this time, however, many Catholic doctrines had not yet been established as orthodoxy, and Ficino thought that explorations of Platonic philosophy provided the most promising path to reforming theology.[96] His proposed solutions point not to a development of a new Platonic orthodoxy but to growing dissent about the existing philosophical underpinnings to theology.

SYNCRETISM AND SKEPTICISM

Platonism remained a viable choice for humanists after Ficino, although those who sought the true theology of the ancients, that is, the *prisca theologia*, adopted a variety of paths. Ideals of philosophical reconciliation, such as those found in the ancient Neoplatonists and Bessarion, attracted numerous thinkers. Perhaps the most ambitious of whom, Giovanni Pico della Mirandola, sought to reconcile not just Plato and Aristotle but all thought, including Jewish Cabalism, Pythagoreanism, Islamic philosophy, and Hermetic Philosophy in his 900 *Theses* (1486). Since Pico's goal was reconciliation, the antagonism of earlier humanists is absent from his work. Not only do Aristotle and Plato receive generous treatments, but Avicenna and Averroes do as well.[97] For Pico the humanists' initiative to uncover old texts extended not just to antiquity but also to Arabic writers, as he commissioned Elia del Medigo to translate some of Averroes' writings that had previously been unavailable in Latin.[98] Pico's conciliatory ambitions, however, failed to convince. One year after its publication, a commission of theologians and a separate group of inquisitors investigated the work, resulting in its condemnation.[99]

Pico attempted a grand reconciliation of all philosophy and theology, yet others doubted the value of philosophy for religious knowledge. Possibly inspired by the rediscovery of Sextus Empiricus and ancient skepticism, the Dominican friar and mystic, Girolamo Savonarola, questioned Aristotle's and Plato's piety.[100] Pietro Crinito wrote that, in a disputation between Pico and Savonarola, the latter said, "Plato prepares the soul for pride, Aristotle truly for impiety."[101] Gian-francesco Pico della Mirandola, Giovanni's nephew and a follower of Savonarola, repeating a trope of earlier humanists, listed "greater adherence to Aristotle than to Christ" as one of the vanities that Savonarola preached against.[102]

Savonarola's sermons and writings show a nuanced view of the relation be-tween ancient philosophy and Christian theology. The relative merits of Plato and Aristotle form part of the subject for Savonarola's sermons on Exodus given in March of 1488. He preached that even though "Aristotle has a few positions contrary to the faith, the positions of the Platonists are more contrary to the faith." Savonarola explained Augustine's adherence to Plato as resulting from the lack of knowledge of Aristotle during that time. Yet, in his eyes, it is a mistake to force Plato's or Aristotle's philosophy into Christianity because they were not Chris-tians. Emphatically, Savonarola wrote that it is best if "philosophers are philoso-phers, and Christians are Christians."[103] In *Triumph of the Cross*, 1497, Savonarola maintained that the religious views of philosophers contain numerous defects. Accordingly, Averroes' theory of the soul is "irrational and very erroneous" but stems also from Aristotle's extremely obscure treatment of the separated intellect. Expressing skepticism about philosophy in general, Savonarola wrote that Aris-totle, despite his genius, "seeing that the natural light could not arrive at perfect knowledge of this material, spoke cautiously so he would not be mistaken."[104] The vagueness of his pronouncements on the soul resulted in discord and confusion among Aristotle's philosophers, which would still be the status quo "if the faith of Christ had not illuminated the world."[105] For Savonarola, revelation, not philoso-phy, is the source for truth about the human intellect.

Gianfrancesco Pico developed the skeptical or fideistic tendencies found in Savonarola. Gianfrancesco Pico based himself on Sextus Empiricus's writings and a detailed analysis of Aristotle, his ancient commentators, and early patris-tic writers.[106] He used historical analysis to discredit Aristotelianism, which he believed irreconcilably differs from Christianity. In his account, Aristotelianism became heavily influenced by Averroes, after Parisians adopted his interpreta-tions in the thirteenth century.[107] Gianfrancesco Pico dismissed Averroes for his untenable view that no one had found an error in Aristotle's works in the previous fifteen hundred years—noting that many Greek followers such as Theophrastus, Andronicus of Rhodes, Philoponus, and Simplicius deviated from Aristotle. He also pointed to Averroes' adherence to Islam and the "barbarous" translations that he used as causes of his alleged failure to understand accurately Aristotle.[108] But Averroes was not alone in his use of uncertain texts.

As a promoter of skepticism, Gianfranceso Pico exploited uncertainty. He noted that it is uncertain whether the books ascribed to Aristotle are genuine, a tactic that Francesco Patrizi, Petrus Ramus, and Pierre Gassendi would later use. Aristotle's doctrine is uncertain because of his obscure style of writing. Early ecclesiastical authors, including Augustine, Dionysius Aereopagite, Lactantius, Justin Martyr, and Origen, judged Aristotle's philosophy not just uncertain but also false in numerous cases. Despite the opinion of the Greek Neoplatonists, such as Ammonius, Philoponus, and Simplicius (whom Gianfrancesco Pico identified as Peripatetics), that Aristotle believed the human soul to be immortal, Gianfrancesco Pico argued that it is just as easy to interpret Aristotle as advocating the mortality of the soul, as Alexander of Aphrodisias did. In his view, the lack of clarity of Aristotle's writings led medieval Christian theologians, who wrote *quaestiones* in the style of Parisian universities, to foment so many disagreements, which provide additional evidence for the uncertainty of Aristotelian doctrine. What seemed certain to Gianfrancesco Pico was that Aristotle's conception of the divine deviated from Christianity, since Aristotle believed God was an animal and did not possess infinite power.[109]

Gianfrancesco Pico's critique of Aristotle was novel. Using humanist ideals of textual investigation and historical research, he put forth arguments that questioned the validity of Aristotelian philosophy as a whole in addition to its conformity with Christianity as conceived by the Church Fathers. Even though there are traces of the early humanist invectives in his attack on Aristotle, he relied more on philology than rhetoric in his promotion of faith and scripture rather than philosophy as the source for religious truth.

CONCLUSION

Masters of philosophy in universities, for the most part, ignored Gianfrancesco Pico's complete dismissal of Aristotle. Platonism failed to replace the Aristotelianism of universities' curricula. Although Ficino and Giovanni Pico influenced a number of Italian professors, natural philosophy, theology, and medicine retained their Aristotelian underpinnings.[110] The continued dominance of Aristotelians meant that their philosophy maintained its place in humanist invectives during the years around 1500.

These attacks spread out from Italy. For example, Heinrich Cornelius Agrippa von Nettesheim (1486–1535), most famous for his writings on natural magic, criticized Aristotle for holding that beatitude is possible through contemplating God in his *On the Uncertainty and Vanity of the Sciences*. This notorious work was widely distributed in France, the Netherlands, Germany, and England, in addition to being condemned at Paris and Louvain.[111] Agrippa believed that Aristotle held the human soul to be immortal, but, echoing Savonarola and Gianfrancesco Pico, he complained that Aristotle's lack of clarity on the issue rendered it possible for interpreters, including Alexander of Aphrodisias and Gregory of Nazianus, to read Aristotle as having promoted a mortalist position, a complaint that a number of seventeenth-century authors would repeat.[112] Turning to ad hominem arguments, Agrippa added that Aristotle was wicked, ungrateful to Plato, and made sacrifices to demons, who in turn "taught him to know."[113] He concluded by questioning whether wisdom could be drawn from a number of mainly ancient philosophers, including from "the impious Aristotle and the perfidious Averroes."[114]

As Agrippa's work illustrates, one way to criticize the intellectual culture of the universities was to target Aristotle's "Commentator." Textual and historical interpretation remained prominent issues for humanists' attacks on Averroes. Juan Luis Vives (1492–1540), a Spaniard who studied at Paris and spent much of his life in the Low Countries, wrote sustained polemics against Aristotelianism and medieval logic. Juan Luis Vives's invectives against Averroes accompany his dismissal of Aristotle's philosophy and his seemingly unassailable position in Scholastic philosophy that excludes the consideration of other ancient schools of natural philosophy, the works of Plato, Plotinus, Pliny, Theophrastus, Cicero, and Seneca. Vives's critique of Aristotle is multifaceted, accusing both him and his followers of ignoring the light of nature and using artificial constructions of knowledge such as syllogisms.

Vives targeted Averroes' linguistic faults, echoing Petrarca's critiques. Not unlike earlier humanists' critiques, he faulted those who relied on his commentaries and thought that Averroes was "nearly equal to Aristotle's authority."[115] Deficiencies in Averroes' linguistic and historical knowledge especially piqued Vives. He complained that Averroes' knowledge of the history of philosophy largely derived from Aristotle's citations. Consequently, Averroes' understanding of the elements in the *Timaeus* was limited, as he referred to the equal angles but not the triangles that Plato hypothesized were the ultimate building blocks of the material world. Averroes' inability to read Latin and Greek, and the resulting inaccessibility of Greek language sources, lowered Vives's esteem for him. For example, Averroes confused "Protagoras" and "Pythagoras," and did not know what Vives considered to be the proper titles for Plato's books. Averroes' knowledge of ancient commentators, such as Alexander of Aphrodisias, Themistius, and Nicolaus of Damascus, was based on writings "most perversely and corruptly translated into Arabic."[116] Vives emphasized failures in Averroes' historical knowledge. Averroes wrongly identified the earliest school of Italian philosophers, making no distinction between the Ionian Presocratics such as Anaxagoras and Democritus and Empedocles and Pythagoras who hailed from Southern Italy.

Vives also accentuated, and perhaps exaggerated, the linguistic failings of Averroes' commentaries. For Vives, the problem centered on the difficulties in using the writings of an author, Averroes, who wrote and read in Arabic to interpret Aristotle who used Greek. Averroes, in his view, used corrupt and inaccurate texts. Vives erroneously believed that Averroes based his commentaries on poor quality Arabic translations of poor quality Latin translations of Aristotle's Greek. As a result he made numerous errors—perhaps 600,000 suggested Vives.[117]

If that number seems impossibly large, an examination of Ambrogio Leone's *Castigations against Averroes* suggests otherwise. Leone, a native of Nola in southern Italy, corresponded with Erasmus and wrote an anticommentary on Averroes' commentaries on Aristotle's logical works and *Physics*. This work prints each comment from Averroes' *Physics* commentaries and then rebuts them with a *castigatio*. Most of Leone's *castigationes* do not arise from grammatical or linguistic disagreements but still reveal Leone's humanist leanings and his interests in hermeneutics and ethics. For example, when Averroes declared that his intention in his commentary on the *Physics* was to "gloss" the book, Leone contended that Averroes was naive in not recognizing that to "gloss" a text entails "interpretation" and "exposition" that give "sense" to the text.[118] Leone also rejected Averroes' attempt to separate natural philosophy from metaphysics, or what Leone called *philosophia divina*. Leone saw this desire to consider nature in a pure sense not just a perversion of the goals of philosophy, which as Socrates saw should be directed toward understand the divine and concomitant moral philosophy. Explicit in these *castigationes* is the suggestion that Averroes' anti-Platonism, which was directed at Avicenna, whom Averroes alleged to have not just mixed theology and philosophy but also combined Plato and Aristotle, was a historical anomaly. "All who followed Aristotle, with the exception of Aphrodisias," wrote Leone, "followed Plato and Socrates."[119] The curriculum of the Academy of late antiquity, which began with Aristotle's logic and natural philosophy and ended with the metaphysics in Plato's *Parmenides*, shows that Averroes' supposed goal to interpret Aristotle without Plato alienates his work from the true purposes of ancient philosophical tradition. Leone suggested that Averroes' conception of physics is flawed and that seeking to separate the philosophy of nature from mathematics and the divine is not just bad philosophy but also "impious three times abominable" (kaì trìs katáraton), as Erasmus had described in 1518 letter that he wrote to Leone after receiving news of this book.[120]

For early sixteenth-century humanists, Averroes and Averroists were synonymous with heresy, barbarism, and dimwitted Scholasticism. For example, Erasmus abused Scholastics by equating them with Averroists.[121] Lamenting that some "sit at the feet of Aristotle [and] Averroes" instead of listening to the eternal truth that emanates from Christ, he wrote that the simplicity of earlier theologians was now clouded with "inane arguments and Averroistic dogma."[122] In a letter to Erasmus, Agostino Steuco proposed to strip away the interpretation of Averroists from contemporary readings of Aristotle.[123] Pietro Bembo, employing an early example of the word "Averroista" in Italian, described Marino Giorgio, one of the *riformatori* of the Studio of Padua, as someone who "has a doctrine completely barbarous and confused, and is simply an Averroist."[124] For Bembo, whose father was an "Averroist" according to Ficino, following Averroes was problematic as much for religious reasons as for its style of writing and thinking. Muddled syllogisms, linguistic barbarism, impiety, and poor Latin writing style were among the prime characteristics of "Averroists" and Aristotle for these humanists.

CHAPTER THREE

Renaissance Aristotle, Renaissance Averroes

A number of fifteenth-century humanists reviled Aristotle; others, such as Bruni, prized at least portions of Aristotelian thought, valuing it as part of antiquity's intellectual heritage. Just as Ficino and his circle sought out new texts to interpret Plato's thought, Aristotelians influenced by humanism attempted to find more accurate and complete interpretations of Aristotle's philosophy. Although the texts of the Aristotelian corpus remained for the most part stable, the goal of finding a purer Aristotle changed the ways that scholars interpreted his writings, affecting their perceived relation to Christianity.

Lauro Quirini, a Venetian noble who studied at Padua and associated with Bruni and Bessarion, expressed his desire to understand a pure version of Aristotle. In a letter from 1440, addressed to the captain of Padua, Andrea Morosini, Quirini wrote that it was valuable to understand the differences between Peripatetics and Platonists, differences that he contended were "most unknown in our age."[1] Writing that he had distanced himself from the views of contemporary Platonists, he distinguished between "the orthodox Christian faith" and the views of ancient philosophers.[2] Although he argued that faith is superior to philosophy, he also maintained that "We should not dispute against philosophers using extraneous arguments," that is arguments based on subjects beyond philosophy, such as theology.[3] Instead of reconciling theology and philosophy, Quirini's goal was to understand the true views of ancient philosophers. In this light, Quirini concluded, "We should be content to go back to Aristotle himself, rather than attributing faith to his authority."[4]

Quirini continued the letter by describing a dream in which he encountered the ghost of Aristotle. In this narrative, Aristotle's ghost told Quirini that he has taken special pleasure in three of his followers and that Quirini will be the fourth. Quirini's three predecessors were Theophrastus, Alexander of Aphrodisias, and Averroes. Aristotle's ghost contrasted these three with a long list of late-antique philosophers—Iamblichus, Syrianus, Proclus, Plotinus, Porphyry, Ammonius, Aspasius, Themistius, Damascius, and Simplicius. Even though these philosophers, some of whom wrote commentaries on Aristotle, possessed subtlety, they were Platonists.[5] Quirini, thereby, suggested that the best path to understanding Aristotle is to avoid the Neoplatonists and follow Theophrastus, Alexander, and Averroes.

Quirini's account of a nocturnal visit by Aristotle's ghost illustrates the influence of humanism among Renaissance Aristotelians. Numerous university professors, who, despite not being humanists in a strict sense, embraced some of the ideals of humanism. Among these ideals were an interest in an historical understanding of ancient Greek texts and the desire to understand the literal meaning of Aristotle's writings, even if this true meaning did not correspond to philosophical or religious truth. They wished to interpret Aristotle as Theophrastus, Alexander, and Averroes had done, in a manner they believed to be free of Platonism and devoid of considerations of faith.

LITERAL ARISTOTELIANISM AND AVERROES

Growing concern for literal interpretations of Aristotle recommended Averroes' works. Averroes believed his goal was to interpret Aristotle, who had exceeded all others in his understanding of nature. Averroes' objective corresponded to the desire of many scholars, influenced by humanism: to establish Aristotle's thought rather than use it as a tool for supporting religion. The desire to make new interpretations of Aristotle consistent with the Greek commentaries on Aristotle, some of which were translated into Latin for the first time in the Renaissance, gave Italian commentators reason to consult the works of Averroes.[6] Just as Quirini linked Averroes to Alexander, a number of scholars duly noted the concurrence of Averroes and his Greek predecessors. Their agreement testified to Averroes' reliability and in turn encouraged the continued use of his works, because his deliberate and extensive consideration of the Greek commentators on Aristotle appealed to those who privileged ancient sources.[7] The connections between the reception of Averroes' works and the ideals of humanism reveal overlapping interests of seemingly opposed schools of Renaissance thought.[8]

Averroes' attempt to recover Aristotle did not strictly follow Quirini's assessment of late-antique Greek philosophy. Not just Alexander, but also Themistius, Simplicius, and Olympiodorus, whom Quirini assessed as Neoplatonists, influenced Averroes. Yet for Averroes, they represented a purer form of Aristotelianism free from Platonism and *kalām*.[9] They were sources of interpretations and provided models for philosophical writing. Averroes structured his works so they paralleled those of the Greek commentators. The organization of his short commentaries is reminiscent of Themistius's works. Agostino Nifo, a famed professor who wrote in the years around 1500, believed, not implausibly, that Averroes' adoption of short, middle, and long commentaries came from the Greeks.[10] Regardless of the accuracy of Nifo's remark, Averroes likely saw his own works as a continuation of this earlier tradition. In the prooemium to his long commentary on the *Physics*, Averroes explained the rationale for writing this commentary. He noted that, because Alexander of Aphrodisias's commentary on the *Physics* stopped short, there was no complete account of this work.[11] Averroes aimed to finish Alexander's job.

In the decades around 1500, authors of Aristotelian commentaries embraced humanist emphases on discovering ancient texts. Growing knowledge of ancient Greek and the desire to find ancient sources promoted an interest in the works of the Greek commentators. Many of their works were unknown in the Latin West until the Renaissance, although a few had been translated in the thirteenth century.[12] These works appealed to Renaissance scholars. Their authors were native speakers of Aristotle's mother tongue and, as a result, were seen as the best guides to reading Aristotle in Greek, which was done in universities for the first time at the end of the fifteenth century.[13] Like Averroes, Renaissance scholars concerned with Aristotelian philosophy saw the Greek commentators as models for their own writings. Unlike Quirini, many of these scholars were either unaware of or undisturbed by the ancient commentators' frequent recourse to Platonic concepts. For example, Jacques Lefèvre wrote paraphrases of Aristotle, having been inspired by Ermolao Barbaro's fifteenth-century translations of Themistius, whom Quirini reasonably identified as a Platonist.[14]

For some Renaissance Aristotelians, it seemed that Averroes' knowledge of Greek commentators was one of his prominent traits. Nifo in the first pages of his commentary on *De substantia orbis* wrote, "When we Latins did not have the Greeks, we relied on this man [Averroes], because of the fragments of the Greeks, which he compiled."[15] Averroes' greatness thus depended on his reliance on Greek fragments. Expanding on his own goals as interpreter of Aristotle, Nifo aligned himself with the Greek commentators, Alexander, Simplicius, and Themistius, all of whom he believed attempted to give literal expositions (pro expositione litterae) of Aristotle. Nifo likened Averroes favourably to Themistius, "whom Averroes followed *in toto*."[16] Themistius unfolded Aristotle's words paraphrastically; but Averroes did so "by commenting and expanding."[17] Nifo thereby reasoned that commenting on a book written by a Muslim was warranted because of his literal expositions of Aristotle's words. His commentary on Averroes' long commentary on *Metaphysics* Lambda repeated the idea that Averroes, although *barbarus*, was an admirable collector of relevant ancient passages, having scoured the works of Alexander, Themistius, and others. Averroes, according to Nifo, had "sufficiently brought [these passages] if not to the words, at least to the ears of Aristotle."[18] As a result of Averroes' talent, Nifo claimed that he "was so famous, that no one seemed to be Peripatetic unless he was an Averroist."[19] Averroes and Aristotelianism were one.

The view that Averroes faithfully followed Alexander and other ancient commentators colored considerations of the Latin translation of Alexander's commentary on *De anima*. Girolamo Donato, the Venetian patrician who first translated Alexander's *Narration of* De anima into Latin in 1495, gave a brief history of philosophy in the preface. In the centuries after his death, Donato recounted, Aristotle lost his following because he was considered more difficult than Plato.[20] After Greece and Rome had fallen, the "Mauritians," according to Donato, lived in Spain and reinvigorated the study of medicine and philosophy. Avicenna, who in fact never lived in Spain, won merit for his expertise in medicine; and Averroes for philosophy, because he relied on the "best authors," namely Alexander, Themistius, and Simplicius. Donato contrasted Averroes with more recent philosophers from England, France, and Italy, who philosophized "more from religion than from the doctrine of Aristotle."[21] Accordingly, Donato's translation aimed to help restore a more accurate interpretation of Aristotle than was found in theologically minded university *quaestiones*. Alexander's interpretation, for Donato, paralleled Averroes'; Donato distinguished Averroes from the culture of universities, which he saw, like many of his humanist contemporaries, to be of little value for understanding ancient thought.[22]

PURE ARISTOTLE

One of the effects of the desire to uncover literal or historical interpretations of Aristotle was the growth of doubts over the possibility of using Aristotle's philosophy as a foundation for theology, as Thomas and others had done in the thirteenth century. That Averroes and Alexander both established views that contradict Christian dogma and were considered, by some, to be the most accurate interpreters of Aristotle fueled the growth of these doubts. During the Plato-Aristotle controversy, Renaissance Neoplatonists, hostile toward some versions of Aristotelianism, questioned the possibility of reconciling Aristotle and Christianity, providing additional evidence for the incompatibility of Aristotle and Christianity. Bessarion dented the Thomistic synthesis by impugning the accuracy of Thomas's interpretations of Aristotle and casting doubt over reason's ability to bolster faith. Later Aristotelians arrived at opinions similar to Bessarion's in their attempts to find the true opinions of Aristotle. By the years around 1500, some Aristotelian philosophers openly promoted the unlikelihood that Aristotle supported a number of Catholic positions, most notably with respect to the human intellect, the universe's creation, and the causes of miracles. These admissions arose not out of anticlerical attitudes but from attempts to uncover Aristotle's true opinion and to find satisfying solutions to outstanding problems of natural philosophy. Nevertheless, some of these philosophical discussions worried ecclesiastical officials, provoked an official declaration about the relation between philosophy and Catholic dogma at the Fifth Lateran Council, and instigated interventions into the teachings of philosophers.

In the 1480s at Padua, the professor of philosophy Nicoletto Vernia explicitly defended Averroes' unicity position. Vernia contended that Averroes' doctrine was the same as Aristotle's, namely that the intellective soul is one in number and eternal but that it is not the substantial form of the human body. Rather, the rational soul acts without being dependent on a human body, but being connected to it by the phantasmata of sensation. The rational soul is an "assisting form;" its relation to the body is "like a sailor in a ship" (tanquam nauta navi).[23] This issue was only one among many for which Vernia believed Aristotle differed from Christianity. For example, he argued Averroes had correctly interpreted Aristotle by rejecting creation in a Christian sense.[24] He rejected the Thomistic and Scotist positions that knowledge of God is impossible for the human intellect when joined to the body.[25] Rather, following Averroes, he maintained that Aristotle believed that human happiness was available to the living through the knowledge of divine substances. To argue that Aristotle was Christian and followed the doctrine of original sin would be an absurdity, Vernia maintained. He conceded that Averroes and Aristotle were wrong on this subject but failed because they used only arguments based on nature for an issue that required faith.[26] Vernia believed Averroes and Aristotle were in essential agreement, that the medieval Latins had misinterpreted Aristotle by reconciling his views with Christianity, and that Aristotelian philosophy should consider nature without concern for religious doctrine.

In 1489, the bishop of Padua, Pietro Barozzi, reacted to Vernia's teachings. In a decree that had only local jurisdiction, wielding the threat of excommunication, he forbade "public disputations in which there were questions on the oneness of the intellect."[27] Barozzi justified this decree on the grounds that these

disputations threatened morality. Those who lecture on the unicity of the intellect remove rewards for the virtuous and punishment for the wicked and "think
they can commit freely the most depraved acts."[28] Barozzi believed that Averroes
himself was wicked. Moreover, he attacked the entire method of university disputation. Citing a verse from Colossians that warns not to be deceived "through
philosophy and inane fallacy according to the tradition of men and the elements
of the world and not according to Christ," he accused professors of "accepting
what they know is false as the truth" and ignoring Christian philosophy.[29]

The decree affected Vernia and perhaps a few others. Antonio Trombetta, a
theologian at Padua, wrote a treatise that attacked the unicity thesis, despite his
seeing some worth in Averroes.[30] Vernia appears to have altered his position in
a treatise written in 1492 but not printed until 1505. In the preface to this work,
which was dedicated to Bishop Domenico Grimani, he asserted that he had never
been convinced by Averroes' "insane position" on the soul. Now after reading
Greek and Latin commentators on Aristotle, he was able to write his new work,
in which he "refuted the evil opinion of the worthless and perfidious Averroes
with very certain arguments."[31] In the body of the argument, he considered the
views of Simplicius, Themistius, Alexander, as well as those of the "three lights
of Christian Religion," namely Thomas, Albertus, and Scotus.[32] Vernia nearly
recapitulated the text of the Council of Vienne, writing, "I say according to the
holy Roman Church and according to the truth that the intellective soul is the
substantial form of the human body."[33] He then clarified his position: the intellective soul is "formally and intrinsically created by God and infused in the human
body, multiplied according to the number of bodies and individuated in these."[34]
His argument, he believed, surpassed Scotus's view in its conformity to faith, in
that he maintained that, "I do not believe all these matters just from faith, but I
say that they can be proven by physics as well."[35] He did not explain whether this
view was truly Aristotle's.

Vernia's apparent about face has been attributed alternatively to his being
convinced by newly available ancient authorities, to pressure and threats from
Barozzi, and to intellectual dishonesty.[36] However, Dag Nikolaus Hasse inter-
prets Vernia as introducing Averroes "through the back door," by using Albertus's
theory of the intellect that posits that universals held by the passive intellect are
not affected by their location and thus are not individual, a view that approaches
Averroes' unicity position.[37] Nevertheless, Vernia's revised position, with its firm
determination of the existence of a multiplicity of intellects, even if acts of intel-
lection remained one and universal, appears not to have worried Church officials.
Barozzi, in a letter addressed to Vernia in 1499, applauded the new treatise and

Vernia's attempt to prove immortality through natural principles. Pride emanates as Barozzi recounted how earlier Vernia explained the intellect erroneously, "just as nearly all of Italy" had.[38] Now, according to Barozzi, Vernia not only adopts the position opposite to what he previously wrote "but he also proves it."[39] Three theologians—Trombetta, Vincenzo Merlino, and Maurice O'Fihely—approved Vernia's discourse before its printing. It is possible that Vernia fooled four theologians, but it is perhaps more likely that his new arguments were theologically acceptable and found to conform to the Council of Vienne.

Although Barozzi seems to have influenced Vernia, in no way did his decree rid Italy of Averroes' influence, even with regard to the question of the intellective soul. Early in his career as a professor at Padua, Agostino Nifo put forward a number of theses on the subject that coincided with Averroes' views. Writing three years after Barozzi's decree in 1492 and clearly aware of humanists' discourse, he admitted that Averroes' writings contained "innumerable errors" caused by translation. But he still justified his comments on Averroes' *On the Beatitude of the Soul* on the grounds that there were no other extant sources of Aristotle's view on this topic.[40] Nifo defended Averroes' limited agreement with Christian orthodoxy, pointing to his defense of the survival of the human intellect, which he thought satisfied "Catholic custom" (mos Catholicus).[41] Nifo, however, not surprisingly, saw differences between Averroes' views and Christianity. He noted, for example, that Averroes' "active intellect" could be called a "holy spirit" but should not be confused with "the holy spirit that is the third person in the divinity that Catholics describe."[42]

Throughout the 1490s, Nifo continued to assert that Averroes was in essential agreement with Aristotle, even if their beliefs were contrary to faith. In his 1497 commentary on Averroes' *Destruction of the Destruction*, a work that attacked the theologizing polemics of al-Ghazālī, Nifo chastised medieval Latin philosophers for differing from Aristotle. He praised Averroes in terms similar to Averroes' praise for Aristotle. "Because Averroes' foundations correspond to Aristotle's principles," Nifo wrote, "it is therefore my custom to tell my students that Averroes is Aristotle transposed. Because, if someone should consider Averroes' foundations and assemble these perfectly with the words of Aristotle, no discrepancy would be found."[43] Nifo openly stated that a number of Averroes' and Aristotle's position were contrary to faith. For example, he noted that Averroes and Aristotle differed from Christian belief about the intellective soul.[44] Yet Nifo excused Averroes for not recognizing the doctrine of the trinity, as it "was not famous" in his setting, and maintained that Averroes' understanding of divinity agreed with the principles of Islam.[45]

The overarching cause of Averroes' and Aristotle's lack of agreement with Christianity was their method, not their religion. According to Nifo, Averroes formed a number of arguments that are "against our law (i.e., Christianity)" because "philosophers accept all doctrine from the senses; and therefore because their conclusions are based on the senses, they can be said to be according to natural reason."[46] Averroes' conclusions based on sensation, however, are not true. In fact, in Nifo's eyes, they are "simply false," and moreover, "all of Averroes' arguments are fatuous," since the principles are contradicted by religious truth.[47] Religious law exceeds the senses and, as it is based on "the testimony of prophets, it is much more certain than what is grasped through the senses."[48] Philosophers therefore must accept religious law, even when it contradicts the senses, because it possesses a higher degree of rationality. Nifo noted that the "senses often contradict reason," providing the slightly risqué yet moralizing example, "the senses [sensus] desire to fornicate, while reason desires to be continent."[49] Although, in the next twenty years, after having learned Greek, Nifo altered his view on the accuracy of Averroes' interpretation of Aristotle. These changes stemmed from his newfound ability to read Greek texts and had little if anything to do with Barozzi's decree.[50]

Outside of Padua, philosophers also wrote about Averroes in positive terms. Alessandro Achillini taught primarily at Bologna from 1484 until his death in 1512. Achillini agreed with Averroes' interpretations of Aristotle on a number of issues, including the unicity thesis. In his *On the Elements*, he maintained that Averroes understood Aristotle correctly that "the intellect is one according to natural reason."[51] Conceding that Averroes' and Alexander's, as well as Aristotle's, position is contrary to the faith, he maintained that it is necessary to "give up Aristotle" [Philosophus] and use the criterion of probability to choose the most likely of "these two false opinions."[52] Achillini sided with Averroes, finding his theory "more probable" than Alexander's. Here, Achillini relied on the epistemological standard that holds natural philosophy cannot provide certain answers, as theology or religion can, but only probable ones. Therefore the best solution is probable, and Aristotle should be rejected when his tenets contradict religion.

Achillini gained the reputation of an adherent of Averroes among sixteenth-century scholars. He was perhaps the predominant example of the Renaissance. Paolo Giovio, Francesco Vimercati, Francesco Vernier, Rinaldo Odoni, Jacopo Mazzoni, and Giacomo Zabarella labelled him as such because of his views on the intellect.[53] Indeed, in many places, Achillini found Averroes' reading of Aristotle "more probable." He argued that the specific forms of the elements remained in a remissive state in mixtures, believing that both Averroes and Aristotle held this view.[54] He also rejected the use of eccentric circles to explain planetary motion, writing that this conclusion "was confirmed by the most solid foundations of Averroes."[55] Averroes along with al-Bitrūjī had judged that Ptolemaic inventions that replaced the homocentric orbs of Aristotle's cosmology with eccentric circles and epicycles violated the principles of natural philosophy.[56] Averroes' and Achillini's purer versions of Aristotle demanded a rejection of portions of Ptolemaic astronomy.

Achillini elaborated on Aristotle's differences from Catholic theology. For example, he believed that Aristotle opposed Christian theology by denying that an "intensive infinity" was one of God's properties.[57] In a related disputation on the extent of God's knowledge, Achillini determined that Aristotle held that divine knowledge extends only to those things that God is concerned with. According to Achillini, both Aristotle and Averroes believed God's providence extended only to individuals as they related to their species and did not extend to their properties that were accidental or irregular, such as, in Achillini's example, a hunched back. What occurs by chance is outside of God's providence and therefore beyond God's knowledge. Achillini noted that Aristotle was in error and "stands very far from the truth of faith" but conceded that "[i]n natural philosophy, he does not appear to be wrong."[58] Faith and philosophy were at times irreconcilable.

Barozzi's decree appears even weaker in its effects if interpreted broadly, that it attempted to prevent disputations on the unicity of the intellect, even if the judgments were negative. Despite or because of the unflagging attention that Averroes' psychology gained, a number of theologians sought to define more clearly Church doctrine regarding the teaching of philosophy. On December 19, 1513, in its eighth session, the Fifth Lateran Council issued the bull "Apostolici regiminis" that specifically addressed teaching on the rational soul as well as philosophy's relation to faith more generally.

THE FIFTH LATERAN COUNCIL

The bull "Apostolici regiminis" was in a sense an extension of humanist polemics against Aristotle and Averroes into papal policy. Giles of Viterbo likely had a hand in the crafting of this bull. Having studied in Padua with Nifo as a young man, Giles moved up the hierarchy of the Augustinian order, becoming prior general in 1507. Influenced by apocalyptic views of history, he believed that the Church was beset by corruption and wickedness and that numerous Christians had adopted impious beliefs and immoral practices.[59] His objections to Aristotelianism as taught in universities were shared by others in the Augustinian order.[60] Conditioned by the rhetoric of humanist invective, he targeted the teachings of universities as a source for heresy and depravity. He followed Ficino and others who promoted the *prisca theologia,* by seeking knowledge and wisdom from Plato and Orphic writings, trying to harmonize Aristotle and Plato, and simultaneously making a list of Aristotle's errors.[61] The Plato-Aristotle controversies of the fifteenth century informed Giles and his protégé Nicolaus Scutellius, who made a loose translation of Pletho's *On the Differences,* which added material intended to bolster refutations of Aristotelians.[62] Giles was not the only author of an outline of Aristotle's errors to have been present at the eighth session of the council. Gianfrancesco Pico della Mirandola also attended.[63] Consequently, the text of the bull corresponds to some degree with earlier attempts to discredit Scholastic Aristotelian philosophy.

The bull refers to a "mischief-maker" (seminator zizaniae) who has spread "the most pernicious errors" of late, especially about the rational soul, arguing either that it is "mortal or one in all humans" or at minimum "assert[ed] that this is true at least according to philosophy."[64] The bull then recapitulates and expands on the Council of Vienne. Whereas in the fourteenth-century council, it was merely asserted that it was heretical to defend that "the rational or intellective soul is not the form of the human body per se and essentially," the 1513 bull also condemned asserting or putting in doubt that the rational soul is immortal and multiplied in each of the bodies in which it is infused, thereby explicitly condemning the theories of Averroes, Alexander, and Epicurus.[65]

The "Apostolici regiminis" most likely was intended to assert control over teaching in general rather than target someone specific, such as Pietro Pomponazzi, whose controversial book on the immortality of the soul was published three years later. The bull's remedies potentially also address the teaching of poetry, yet they were directed to a greater degree at philosophers.[66] The bull ordered that those who lecture publicly in universities or elsewhere and use the "principles and conclusions of philosophy" when explaining matters they know deviate from the faith, such as the "mortality or oneness of the soul and the eternity of the world," must clarify and teach the "truth of the Christian religion with every effort, persuading as much as possible."[67] The bull also asserts that philosophy "without the light of revealed truth, leads more toward error than toward the illumination of truth" and ordered that all studies beyond grammar schools be supervised by canons and rectors in order to enforce the bull.[68]

There was not complete agreement about the advisability of the bull among those present at the council. Niccolò Lippomano had reservations. A Venetian patrician, he had studied philosophy at Padua. Among his associates were Donato, the translator of Alexander's *Narration of the* De anima, and Barbaro, who translated Themistius. In 1513, Lippomano was the bishop of Bergamo. He objected to theologians forcing philosophers to dispute the truth about the intellect rather than the uncovering Aristotle's "mind" or Averroes' position.[69] Another dissenting view among those present arose from Tommaso de Vio, who took the name Caietano, the general superior of the Dominican order.[70] Tommaso de Vio objected to the requirement that those teaching philosophy should be required to teach the "truth of faith."[71] He had maintained in his 1509 treatise on the soul that Aristotle held that the soul was corruptible, even if this view contradicted both Christian doctrine and Thomas's reading of Aristotle. In this treatise, he begged the reader not to consider him a follower (sectator) of some school of Aristotelianism, because in this work he is explaining the manner "according to the mind of Aristotle, regardless of its relation to the truth."[72] Although sympathetic to Aristotelianism, Tommaso de Vio believed, as did his Augustinian counterparts, that Aristotle had erred on significant issues. In spite of this, he did not see the need for the Church to control philosophers' lectures.

Even though it was far more specific than the decree of the Council of Vienne, the language of "Apostolici regiminis" was potentially ambiguous. For example, the judgment against those who argued that the mortality or unicity of the intellect was "true at least according to philosophy," fails to take in account that the most famed followers of Averroes did not argue that his view was true according to philosophy. Achillini characterized the unicity thesis as "more probable," and Nifo called it "plainly false." Pomponazzi thought that Averroes agreed with Aristotle, but that both were wrong. For them, divine light held greater authority than the evidence of the senses, philosophy, or Aristotelian authority. Furthermore, the bull allowed philosophers to present conclusions that were contrary to faith so long as they explained the truth and presented persuasive arguments that supported it. Standard philosophical practice at the time already demanded the presentation of contrary arguments as well as the recognition of positions that were contrary to faith. So, if the "Apostolici regiminis" was an attempt to control university teaching, it was a poor one.

PIETRO POMPONAZZI

In the decade after the promulgation of "Apostolici regiminis," Pomponazzi's writings coupled with the diverse reactions to them illustrate the absence of a uniform interpretation of the bull. Pomponazzi was one of the most highly paid and famous philosophers during his lifetime. He taught at Padua and Bologna, as well as being employed by Prince Alberto Pio of Carpi and later by the Gonzaga family of his native Mantua.[73] In 1515, Florence offered him the princely sum of five hundred gold florins a year to teach at Pisa. He eventually refused the offer, choosing to remain at Bologna.[74] The wit, learnedness, and careful argumentation evident in the numerous student *reportationes* of his lectures explain his success as a teacher. Pomponazzi's lectures contain irony and often end in professed uncertainty, thereby making it difficult to pinpoint his actual position.[75] Added to the difficulty of interpreting his works, is that his views on a number of issues changed over the course of his career and that he, like many of his contemporaries, believed that natural philosophy often provided only probable solutions.[76]

Some scholars have assumed that Pomponazzi was the target of the "Apostolici regiminis." There is no direct evidence for this position and it would seem strange if Florence, led by Lorenzo di Piero de' Medici, had attempted to bring him to Pisa shortly after Lorenzo di Piero's uncle, Pope Leo X, had presided over a council that intended to condemn him. Giulio de' Medici, a cardinal and papal legate to Bologna, was aware of the 1515 contract offer.[77] Nevertheless, the bull's phrase "placing in doubt" (in dubium vertentes) seems potentially applicable to what Pomponazzi was doing.

Before the eighth session of the Fifth Lateran Council, Pomponazzi had lectured on the soul on several occasions. In Padua, during the academic year 1503–4 he presented several *quaestiones*, which went unprinted until the twentieth century. The *quaestiones* show Pomponazzi to have been unconvinced by Averroes' arguments about the soul. He believed the Thomistic and Alexandrian alternatives were flawed as well. In the *quaestio*, "Whether the soul is immortal?," Pomponazzi limited his disputation to arguments based on "natural principles," supposedly due to considerations of space.[78] Elsewhere in the *quaestio* he referred to this approach as the "Peripatetic method," a method that he contended Scotus used when he argued that the immortality of the soul was a neutral question (problema neutrum) and could only be known according to faith.[79] Pomponazzi, however, did not think that Scotus correctly understood Aristotle. Rather he thought that Aristotle thought the rational soul is immortal, and that "Averroes' opinion is without fail the opinion of Aristotle."[80] The coincidence of Averroes' view with Aristotle's does not, however, ensure coincidence with the truth.

Pomponazzi described his solution as being syncretic, ironically likening himself to Muhammad, "who aggregated and accepted the sayings of these people and the sayings of those people and established laws."[81] In a similar fashion, Pomponazzi wrote, "I accept some of the sayings of Christians and some of the sayings of Alexander, and I find the answer."[82] Accordingly, his solution is dual. The intellect is material because of its relation with its subjects, and, because the soul is form, "[i]t elevates above matter . . . and therefore is immaterial and abstracted."[83]

The entirety of the *quaestio* is inconclusive. After going through numerous objections and responses to them, he closed by writing that all responses to the passages in Aristotle that interpret him to be a materialist are "false and impossible." All arguments Aristotle put forward also present insoluble puzzles, "therefore they must be rejected and it should be accepted on the grounds of consistency that the rational soul is immortal."[84] The exercise presented skepticism and doubt as much as a determination of an answer.

In the next *quaestio*, Pomponazzi addressed in what way the rational soul is the form of the human body. One of the potential answers, attributed to Averroes, was that the rational soul is an "assisting form" (forma assistens) to the body. Under this theory, which Vernia had endorsed, the relation between rational soul and body is like the "sailor to a ship," the soul guides the rational body but remains distinct and therefore separable. Repeatedly in this *quaestio*, Pomponazzi maintained that Averroes had correctly interpreted Aristotle. Applying the trope "Aristotle was a man and therefore could err," Pomponazzi contended Aristotle's and Averroes' opinions, which he thought coincided with Theophrastus's and Themistius's, were false, unimaginable, chimerical, fatuous, bestial, erroneous, and not true.[85] The incompatibility of Aristotelian philosophy with Christian philosophy emerged from several other points. Pomponazzi pointed out that Aristotle did not believe in bodily resurrection, which can only be known through faith.[86] Furthermore, he conceded that considering only nature (stando in puris naturalibus), it is possible to sustain Alexander's view that the rational soul is material, "although with regard to the truth of the matter, this opinion is false" or even "very false."[87]

The Christian religion provides the truth on this matter, but its theologians have erred in attempting to reconcile Aristotle with Christianity, according to Pomponazzi. In a discussion about how the intellective soul changes its location after death from the body to paradise, hell, or purgatory, he sided with Scotus's position that it is transported by locomotion. Yet he noted that this position runs counter to the Aristotelian principle that indivisible substances cannot move. Pomponazzi expressed incredulity that Thomas "did not see why this opinion is not Aristotle's" and posited that Thomas was misled "perhaps by the zeal of faith."[88] In this manner Pomponazzi pointed to Thomas's exegetical failings, namely his attribution of Christian views to Aristotle in order to soften differences between faith and pagan philosophy.

Pomponazzi's doubting of Thomas is similar to Bessarion's. The truth of Christianity is maintained, even when Aristotelian philosophy presents solutions that deviate from faith. In a *quaestio* on the unicity of the intellect disputed in

1506, Pomponazzi summed up this position and wrote, "Thomas, having been compelled, spoke the truth, which was not the opinion of Aristotle."[89] According to Pomponazzi, "Thomas moved by zeal for the faith tried to say that this agreed with Aristotle's mind."[90] Thomas's failure in correctly interpreting Aristotle should not bother us because, according to Pomponazzi, "Good Christians will not be moved by the authority of Aristotle."[91] Over the centuries, many have interpreted Pomponazzi as disingenuously asserting the veracity of Christian dogma by undermining the ability to reconcile it with philosophy or Aristotle's view. Yet his view bordered Bessarion's and Tommaso de Vio's, whose faith has never been questioned.

After the promulgation of "Apostolici regiminis," Pomponazzi continued to endorse similar positions on the soul and on philosophy's relation to Christian theology, despite inquisitors' interest in his lectures. According to the seventeenth-century historian, Giacomo Filippo Tomasini, Pomponazzi was investigated for heresy during the summer of 1514. Although little is known about this encounter, the charges evidently were not pursued and Pomponazzi continued his lectures in the autumn.[92] Two years later, in 1516, his *On the Immortality of the Soul* was published. This work evoked numerous responses, which required Pomponazzi to defend his position and seek the protection of patrons. By this time, Pomponazzi had altered some of his positions. First, he was more negatively disposed to Averroes' unicity thesis, which he considered not just "very false" as he had a decade earlier but now also "monstrous and completely foreign from Aristotle."[93] Second, he favored Alexander's position to a greater degree, believing that using philosophical principles and according to the mind of Aristotle, it is probable that the human soul is corruptible and therefore mortal.

His argument is largely philosophical, based on the necessity of sensation for the development of the intellect. Nevertheless, Pomponazzi pointed to ecclesiastical authorities who agreed with his position, namely, the Greek patristic writers Gregory of Nazianus and Gregory of Nyssa maintained that Aristotle believed that the "soul is the actualization of the body" and therefore corruptible and mortal.[94] Pomponazzi was seemingly aware of a need to defend his conception of philosophy's and religion's relations to the truth. Citing the Aristotelian precept that each field must use only its own principles, also known as the prohibition of *metabasis*, he pointed to the instability of philosophical argument, which is always in flux.[95] The only stable road to truth is faith. Because the immortality of the soul is an "article of faith," Pomponazzi contended that "it must be proved by the principles of faith," which come from revelation and scripture; "other arguments," such as philosophical ones, "are extraneous."[96] He concluded by stating his submission to the Church in regard to this issue and others, a phrase that he repeated a number of times in his lectures and writings.[97]

Within a few years, several scholars had responded negatively to Pomponazzi's treatment of the soul. In a 1518 point-by-point refutation of Pomponazzi dedicated to Leo X, Nifo rejected the idea that the soul's immortality was a *problema neutrum* and could not be proven by philosophy. Nifo and Pomponazzi had been on negative terms when both were professors at Padua in the 1490s. The ill will generated by their encounters contributed to Pomponazzi's departure to the court of Alberto Pio.[98] Their mutual hostility endured the decades. In the 1518 treatise, Nifo maintained that Pomponazzi had ignored passages in Aristotle that show he believed in the immortality of the soul and therefore Pomponazzi had not faithfully followed the *via peripatetica*.[99] Nifo's argument at times bordered on sophistry. Begging the question, he wrote: "We should consider articles of faith in a non-neutral way, since we believe them firmly. Therefore if the immortality of the soul is a neutral problem we would not believe it firmly, thus it would not be an article of faith."[100] Aristotle was not the only authority to whom Nifo appealed. He elaborated Plato's arguments for the immortality of the soul, noting that Bessarion, in his exchanges with Trapezuntius, had "asserted that Plato firmly believed and demonstrated the immortality of the soul."[101] Nifo's response pleased Leo X and the Medici family. In 1519, he took a position at Pisa. One year later Leo X named him Count Palatine.

Pomponazzi had his protectors as well. In 1518, he published an *Apology* for his earlier work on the immortality of the soul. In his effort to muster support, Pomponazzi dedicated the book to Sigismondo Gonzaga, who was at this time a cardinal; the Gonzaga family hailed from Mantua, Pomponazzi's birthplace. The coda of the *Apology* thanked Pietro Bembo for helping him escape ecclesiastical prosecution.[102] Bembo, who earlier had complained of the Averroists for their forms of argumentation, was a Venetian patrician, humanist man of letters, and former student of Pomponazzi. At this time he was also a cardinal as well as secretary to Leo X and intervened within the papal curia to stop the prosecution of Pomponazzi that had Leo X ordered. The pope had issued a statement that maintained that Pomponazzi's teaching that, "The rational soul is mortal according to the principles of philosophy and mind of Aristotle is contrary to the decision of the Lateran Council," even though the Lateran Council did not stipulate such a thing.[103] The pope requested that Pomponazzi retract his statement. Evidently Pomponazzi failed to follow the papal order, but suffered no direct harm presumably likely due to Bembo's influence in the papal curia. Bembo's intervention reveals that within the higher echelons of the Church there were divisions over his book.[104]

Meanwhile, in December of 1518, the Studio of Bologna signed a contract with Pomponazzi for eight years at the rate of six hundred ducats per year.[105] The new terms offered by Bologna granted Pomponazzi great freedom in his teaching, allowing him "to read and interpret all the books of Aristotle or only the more useful and necessary parts of those books," regardless of whether his choice followed the customary pattern of lecturing.[106] This provision might explain his shift to lecturing on *Meteorology*, *De animalibus*, and *De generatione et corruptione*, works not typically read by ordinary professors.[107] Struggles over the acceptability of his positions continued to be waged both on philosophical and legal levels. Even before the new offer was agreed upon, Pomponazzi, the Florentines, and the university of Bologna had been entangled in a legal process in Rome. One of the concerns was that Pomponazzi had "published certain works that are contrary to faith," without the appropriate license.[108] The process had not been resolved by December 20, 1519, and it remains unclear how the process ended, except that Pomponazzi was not convicted of heresy.[109] The *riformatori* of the Studio of Bologna took interest in this process, defending Pomponazzi, presumably because they feared losing a famed and valued professor but perhaps also reflecting a tradition of separating its university from religious concerns. The 1507 statutes of the college of arts at Bologna prohibited priests of any order from lecturing on topics other than metaphysics, theology, or moral philosophy, giving the rationale that laymen would not want to study there if they saw the university filled with clergymen.[110]

In 1519, Bartolomeo Spina, a Dominican, wrote a series of treatises that attacked not just Pomponazzi but also Tommaso de Vio for their interpretations of Aristotle.[111] In his response to Pomponazzi's *Apology*, Spina maintained that to argue "the immortality of the soul is repugnant to natural principles is heretical," rejecting Pomponazzi's contention that natural philosophy should not be mixed with theology.[112] Pomponazzi responded to Spina's charges of heresy, writing that the university regulations demanded that he "faithfully uncover the mind of Aristotle."[113] Following that mandate meant that "[i]t is not up to our will to say that Aristotle held or did not hold this or that" but rather that "our judgment arises from arguments and his words."[114] Since it was his duty to interpret Aristotle, maintaining the opposite of his interpretation would be dishonest and illegal. Therefore he begged those who wanted to condemn him of heresy to "put an end to their barking," because they "put forth a far greater heresy, since they accuse an innocent man."[115] By linking his arguments for the mortality of the soul to "the mind of Aristotle," Pomponazzi emphasized the separation of natural philosophy from theology in addition to the differences between an accurate interpretation of Aristotle and Christian dogma. Pomponazzi's desire to uncover Aristotle's mind was not merely a ruse that allowed him to suggest the reasonableness of irreligious sentiments. This desire emerged in other areas besides those that touch theology. For example, he attacked the deviations from Aristotle of Richard Swineshead, who followed the calculatory tradition, a method of Aristotelian natural philosophy that arose among fourteenth-century English philosophers that relied on mathematical analysis not found in Aristotle's writings.[116]

Although these controversies over the immortality of the soul threatened Pomponazzi with prosecution, he continued to pursue similar arguments until his death in 1525. In his *On Nutrition and Growth*, first printed in 1520, he maintained that Aristotle believed the rational soul was material, even though this view is "in itself false and contrary to our truest religion."[117] In 1525, his *On the Immortality of the Soul*, *Apology*, and his *Defense* were reprinted, accompanied by Crisostomo Javelli's *Solutions of Reason*, which were first printed in 1519. A member of the Dominican order, Javelli was convinced of Pomponazzi's sincerity toward religion, although unconvinced of his philosophical arguments. Diplomatically, describing himself as a "faithful follower of Thomas," Javelli confessed that he did not find Pomponazzi's deductions "clear," yet he wished to avoid discussing interpretations of Aristotle.[118] Instead he set forth arguments for the rational soul's immortality not necessarily based on Aristotle but rather based on Catholic theology and true philosophy. In this manner, he found true philosophy to be linked to metaphysics and subservient to theology.[119] Thus gentle corrections, from a sympathetic Thomist, accompanied the second printing of Pomponazzi's writings of the soul in an approved edition.

In Pomponazzi's unpublished works written during the years surrounding the publication of his *On the Immortality of the Soul*, he continued to argue that the probable conclusions of natural philosophy based on Aristotle's principles deviated, at times, from the dogma of faith. He distanced himself from religious thinkers whom he accused of confounding theology with philosophy, or "mixing different soups."[120] Still, Pomponazzi, during these years, despite his proffering arguments based on natural principles, wrote that when it came to matter of faith he "always submits to ecclesiastical correction."[121]

Pomponazzi's two major writings of the 1520s, *On Fate* and *On Incantations*, went unprinted in his life. In *On Fate*, he used Stoic concepts to argue for determinism. For a number of issues, such as whether God's providence extends to monsters, Pomponazzi indicated that the conclusions *ex puris naturalibus* were inferior and that the question "seems to be better determined by our theologians than these Peripatetics."[122] He deployed similar tactics in *On Incantations*. Some have postulated that *On Incantations* was circulated only among clandestine circles before its 1556 printing.[123] Analysis of the extant manuscripts, however, suggests a broad and relatively open circulation. Both students and professional scribes copied the work, especially in northern Italy.[124] The starting point for *On Incantations* was that Aristotle did not admit the existence of demons that scripture posits and ecclesiastical officials use in their explanations that "save the appearances" of a number of extraordinary phenomena.[125] Pomponazzi therefore reasoned it might be possible to apply Aristotelian principles, as well as those from Platonic and Stoic schools, as the basis for potential natural explanations for phenomena that some had considered miraculous or caused by demons and other supernatural forces. Pomponazzi argued that his explanations were not necessarily true but that many purportedly miraculous events "can on the surface be reduced to natural causes, although there are many which only minimally can be reduced to such a cause . . . and for infinite others it is not possible to reduce them to a natural cause."[126] Thus Pomponazzi limited the phenomena that could be explained even "on the surface" by natural explanations. Defining miracles as what "are beyond created nature's order and can only be done by God," he wrote that miracles "show the insufficiency of Aristotle's doctrine . . . and declare the truth and firmness of the Christian religion."[127] Providing an example, he wrote there are no available natural explanations for the long-lasting eclipse that took place when Jesus was crucified because it defied Aristotle's belief that celestial bodies cannot change their path.[128]

There is little reason to believe Pomponazzi denied the existence of miracles completely. In 1520, at roughly the same time he was composing *On Incantations*, Pomponazzi was called as an expert to examine the corpse of Elena Duglioli, a Bolognese woman, whom some believed to be holy. At an autopsy designed to determine whether there were miraculous signs in her body that confirmed her sanctity, Pomponazzi testified that the presence of milk in the chaste woman's breasts and her breasts' resistance to rotting were indicative of supernatural influences.[129] The sincerity of Pomponazzi's submission to ecclesiastical authorities has long been questioned. Yet in 1520 just after the conclusion of the controversies on the immortality of the soul, the priors of the church of San Giovanni found Pomponazzi respectable enough to request his expertise. If he was secretly hiding subversive views, he was to a degree successful. In any case, it seems certain that his view that Aristotle deviated from the Christian dogma was sincere.

CONCLUSION

Although it is impossible to determine what Pomponazzi really thought, his written words suggest a coherent approach to philosophy that fit within some of the philosophical traditions of the time.[130] He admitted that his works on the soul caused scandal, yet he was not legally a heretic.[131] Determining that Aristotle's "mind" did not correspond with Christian theology had long roots as well. Bessarion and other Platonists had argued that Aristotle's philosophy deviated from Christianity on a number of points. During Pomponazzi's career, Tommaso de Vio suggested the same. A number of powerful ecclesiastical and secular authorities, such as Bembo, Grimani, and the Gonzaga family protected Pomponazzi. It is not, as one scholar recently argued, "always obvious when Pomponazzi is lying."[132] Rather his confessions of doubt might reflect genuine perplexity and his submissions to ecclesiastical authorities on matters of faith might confirm his conviction that philosophy should not be mixed with theology. The prohibition of *metabasis* forbids theology from interfering with physics and also prevents the study of nature from overturning doctrines of faith. Pomponazzi, like most of his colleagues at Padua and Bologna, was a layman and therefore removed from the decision-making hierarchy of the Church. Moreover, philosophy at the universities at Padua and Bologna was taught foremost to students who would then study medicine or law. Therefore at these universities it was not necessary to consider natural philosophy as ancillary to theology.

The immediate reception of Pomponazzi was mixed. Some attacked him as impious, still others followed a similar path by teaching natural philosophy according to natural principles. The Fifth Lateran Council and the controversies surrounding the soul did not eliminate the influence of Averroes' or Alexander's writings, as the next generation of intellectuals continued to consult and draw from them.[133] In one of the stranger twists of the immortality controversy, Luca Prassicio attacked both Pomponazzi and Nifo for abandoning Averroes, because he believed that Averroes was the best source of arguments for immortality of the intellect and gave the most accurate interpretation of Aristotle.[134] The controversies over the soul shook attempts to forge an orthodox natural philosophy and broadcasted the difficulty of showing that the mind of Aristotle conformed to the truths of the Christian religion.

CHAPTER FOUR

Italian Aristotelianism after Pomponazzi

Despite the Fifth Lateran Council and the controversy surrounding Pietro Pomponazzi, throughout the sixteenth century, Italian natural philosophers continued to attempt to uncover Aristotle's true positions, regardless of their agreement or disagreement with Christian doctrines. Often appealing to Averroes' or Alexander of Aphrodisias's authority, they justified their method using similar terms that Pomponazzi, Alessandro Achillini, and others had used in the first decades of the sixteenth century. Averroes' theory of a single rational soul for all humans convinced fewer in the sixteenth century than it had in the 1490s; in spite of this, Averroes remained a significant and, at times, controversial influence.

Already Agostino Nifo, and Nicoletto Vernia had abandoned the theory of the unicity of the intellect by the first decade of the sixteenth century. After Luca Prassicio, few prominent philosophers explicitly argued for it. The shift away from Averroes' theory of the soul did not depend on ecclesiastical degrees or prosecutions.[1] On this issue the Church's influence was limited. A number of thinkers, influenced by Alexander or arguments such as Pomponazzi's, interpreted Aristotle's soul as being corruptible, tied to sensations, and thereby part of the material world, a position that was also condemned at the Fifth Lateran Council. Just as earlier thinkers had done, they did not argue that this position was unambiguously true but only that this view was Aristotle's or that according to principles of Peripatetic philosophy the materiality of the soul was probable.

Even though Averroes' theory of a single intellectual soul had limited appeal, his works held wide influence among university professors in Italy during the sixteenth century. For many scholars, he was still considered the best guide to finding Aristotle's true opinion. For others he remained emblematic of mistaken methods of natural philosophy, a synecdoche for the view that philosophy is distinct from theology. It was not until the establishment of the Jesuit order and the beginning of the Council of Trent that ecclesiastical authorities became more concerned with controlling the instruction of philosophy and restraining the influence of Pomponazzi and Averroes.

Both those who valued Averroes and those who attacked his views were affected by humanism. Those who advocated reading Averroes saw their project as involving determining Aristotle's true view. As a result, philological and historical reasoning counted among their motivations. Those who condemned Averroes emphasized a different strain of humanism, one that enlisted philosophy and erudition for the cultivation of virtue. For them, the study of nature was subalterned to both religion and ethics.

AVERROES IN THE SIXTEENTH CENTURY

During the Renaissance, interest in Averroes grew. His complete works were printed over ten times before 1570s.[2] During this time, Averroes' writings were edited, translated, and retranslated.[3] Jacobo Mantino, Giovanni Francesco Burana, Abraham de Balmes, and Elia del Medigo, among others, translated works previously not extant in Latin.[4] Some of the most prominent philosophers of Italy wrote commentaries specifically on Averroes' writings. The growing interest resulted from the continuing desire to uncover Aristotle's views and the belief in the correspondence between Averroes and the Greek commentators. The scrutiny and interest in his work meant that philosophers followed his view for numerous issues, not just those surrounding the intellect.

Scholars frequently described, and presumably prized, Averroes' positions because they were "Peripatetic," a term that suggests that they conformed to Aristotle's actual view. For example, siding with Averroes and Achillini, Girolamo Fracastoro and Giovanni Battista Amico, both former students at Padua, revived Aristotle's cosmology and attempted to build homocentric models of the orbs that were capable of accounting for the trajectories of planets.[5] Since Ptolemaic innovations in astronomy such as equants and nonuniform motion could not be found in Aristotle's De caelo, Fracastoro and Amico hoped to establish a model that was consistent with the principles of natural philosophy, rather than the dictates of a mathematics or the innovations of astronomers. In 1537, Amico, crediting his teachers at Padua, Marcantonio Genua and Vincenzo Maggio, for inspiration, wrote that he was greatly persuaded by the Peripatetic opinion that required homocentric orbs for the planets and was aware that Averroes had interpreted Aristotle in a similar fashion.[6]

Interest in Averroes' solutions to natural philosophical problems led a number of scholars to write commentaries on his works, a tradition that corresponded to the needs of university education. The 1405 statutes of the Studio of Bologna specify that lectures should be given on two of Averroes' writings in the faculty of arts: *On the Substance of the Orbs* and the prologue and parts of the first, second, and fifth books of *Colliget*, Averroes' treatise on medicine.[7] Although these are not the most famed or notorious portions of Averroes' writings, they fit within the needs of the university curriculum. *On the Substance of the Orbs* bridges terrestrial physics with astronomy, and the *Colliget* added another voice to medical teachings, especially about the relation between composite and simple medicines.

Only a few commentaries on the *Colliget* are extant, despite the work's wide readership. Pietro Mainardi, a professor of medicine at Ferrara, who, in 1500, gave explanations for the fifth, sixth, and seventh books of the *Colliget*, in which he extrapolated on the differences between food and medicine.[8] Matteo Corti's 1527 comments on the seventh book of the *Colliget*, which examined cures for fevers, among other issues, are extant in a unique manuscript.[9] Corti's work appears to have been a relatively minor affair for this famed physician and professor at Pisa and Padua, who influenced by revivals of Hellenism, published numerous works on anatomy, dietetics, phlebotomy, and remedies.

The number of commentaries on *On the Substance of the Orbs* greatly increased after 1500. Although at least one late-medieval scholar, namely John of Jandun, wrote a commentary on this work, it was not a standard work for commentaries. When Nifo wrote his commentary, printed in 1508, he contended that he had found just one exposition on this work, that of John of Jandun's.[10] Subsequently, a number of Nifo's contemporaries and successors took interest in *On the Substance of the Orbs*. Throughout the sixteenth century, a steady trickle of treatises analyzed this work, which was later printed as part of the widely circulated Giuntine editions of Aristotle. Pietro Pomponazzi wrote a question commentary in 1507. Giovanni Battista Confalonieri wrote a commentary printed in 1525. Mainetto Mainetti, a professor at Pisa, wrote a commentary published in 1570; and the Ragusan philosopher, Nicolò Vito di Gozze, composed a commentary that was published in Bologna in 1580.[11]

By this beginning of the sixteenth century, treatises, commentaries, and *quaestiones* that addressed specific books of Averroes became more common than in earlier centuries. Even though the motivations for commenting on Averroes were multiple, the idea that Averroes was following the Greeks and had preserved their texts was a major rationale. Pomponazzi puzzled over what kind of book *On the*

Substance of the Orbs was exactly and why it merited further comment. According to Averroes, Aristotle wrote a treatise on the substance of the orbs, which did not survive antiquity. Therefore, Averroes' treatise was an attempt to replace this missing title. Pomponazzi admired Averroes' purpose as well as his method. He believed that the "Commentator" collected "all the roots and foundations" for this book from statements sprinkled throughout the extant Aristotelian corpus; as a result, *On the Substance of the Orbs* formed a necessary and worthy part of natural philosophy.[12] Pomponazzi's views echoed in the works of later commentators on this work. Confalonieri wrote that "our Averroes was so steady that whoever should use him as a guide . . . will perceive the strength of Aristotelian doctrine . . . just as he [Averroes] had drawn out the truth of Aristotle's mind, even when often having a corrupt text."[13] Confalonieri, as a result, was content to put forth the "true mind of Averroes" and then solve any difficulties and problems that the text displayed.[14]

Interest in commentaries on Averroes' works was not confined to sixteenth-century treatises. Printing houses made efforts to connect John of Jandun's works to those of Averroes'. An edition of Jandun's questions on *Parva naturalia* printed in 1589 included Marcantonio Zimara's question on motion and the mover, which was explained according to the "intentions of Averroes and Aristotle." Jandun's questions on *De caelo* were accompanied by *On the Substance of the Orbs* in 1552 and 1564 printings; and his questions on the *Physics* were printed with annotations and further questions written by Elia del Medigo, well known for his Latin translations of Averroes from Hebrew that Giovanni Pico della Mirandola had commissioned.[15]

Smaller tracts, typically composed of a single *quaestio* that analyzed the work of Averroes, proliferated during the sixteenth century. Nifo defended Averroes' views about mixtures of the elements.[16] Vittore Trincavelli wrote a *quaestio* on reactions according to the doctrines of Aristotle and Averroes, published as an addendum to a 1520 edition of Richard Swineshead's *Calculations* that he had edited.[17] Giovanni Francesco Beati, a professor of metaphysics at Padua from 1543 to 1546, used the seventh chapter of *On the Substance of the Orbs* to frame a *quaestio* on the finitude of the celestial region. Beati, noting the similarities among Themistius, Alexander, and Averroes, contended that "Averroes appears to have perfected the position of Themistius."[18]

Just as in early part of the sixteenth century, scholars noted the similarities between Averroes and the Greek commentators. Marcantonio Genua (1491–1563), a professor at Padua, attempted to reconcile Averroes' psychology with Simplicius's.[19] Girolamo Balduini, a professor at Padua during the middle of the sixteenth century who was knowledgeable enough about the late-antique Greek philosophy to write on Porphyry's logic, noted in his commentary on Aristotle's *Physics* that "[w]hen following Averroes we follow also the Greeks," who in turn conform to Aristotle.[20] Konrad Gesner, the Swiss physician and encyclopedist, wrote a brief biography of Averroes, contending, "In his commentaries on Aristotle he most greatly imitated the Greeks, such as Alexander and Themistius."[21]

Some linked Averroes to Greek commentators by concentrating on Averroes' opening chapters of his commentaries as material for discussing the purpose of various philosophical subjects. Simone Porzio (1496–1554) in a brief treatise on Averroes' prologue to the long commentary on *Physics* explained the proximity between the Greeks and the Commentator. According to Porzio, Averroes "wished to imitate the Greeks," and thereby wrote general introductions to Aristotle's works before moving on to interpretations of particular sentences and words, just as in fact the late-antique Greek commentators did.[22] Similarly, one of Porzio's students, Girolamo Balduini, used a lengthy discussion of Averroes' prologue to the long commentary on *Physics* as his own introduction to a commentary on that work, where he linked Averroes' positions to both the Greek commentators and Aristotle.[23] In his 1551 exposition on the prologue to *Posterior Analytics*, Giovanni Bernardino Longo recycled the commonplace "no one is Aristotelian unless an Averroist" , attributing inspiration for the phrase from Averroes' comments on *De caelo* where he wrote "no one is Peripatetic unless an Alexandrian."[24] Interest in Averroes' logical positions was not unique to Longo, Annibale Balsamo, for example, wrote a brief treatise in which he attempted to solve obscure points in the *Posterior Analytics* "according to the mind of Averroes" (ad mentem Averrois). Thus, understanding what Averroes truly believed became a goal of some sixteenth-century scholars.[25]

CRITIQUES OF AVERROES AS GUIDE TO ANTIQUITY

For some, Averroes' concordance and reliance on the Greek commentators had recommended at least some of his works, yet other scholars emphasized the negative aspects of their agreement. In a 1485 letter describing his reliance on a variety of Greek, Latin, and Arabic commentators for his Aristotelian paraphrases, Ermolao Barbaro (1453–1493) alleged that Averroes' inferiority stemmed from the fact that all his words were stolen from Alexander, Themistius, and Simplicius.[26] Barbaro had by this time apparently softened his still harsh view. Two years earlier, in 1483, in a letter to Vernia, Barbaro tried to persuade Vernia to "condemn, hate, and avoid this most wicked genre of philosophizing."[27] Some twenty years later, Symphorien Champier (1472–1539) used Averroes' similarities to the Greek commentators as a means to denigrate him for lack of originality. He wrote, "Averroes took pleasure in following them . . . and did not so much draw from them but copied them. Which is only what he was: the name commentator suited him."[28] The Bolognese professor and student of Achillini, Lodovico Boccadiferro (1482–1545) believed Averroes' supposed lack of originality could help categorize the good and bad doctrines found in his work. "Whatever is good that he has, Averroes took from the Greeks," Boccadiferro wrote, "Nothing that he got from himself is good and everything he said on his own or took from his fellow Arabs, and, I mean everything, is fatuous and confused."[29] Boccadiferro, thus, justified his occasional reliance on Averroes but distanced himself from any controversial, erroneous, or condemned doctrines. When he agreed with Averroes, he was innocently agreeing with the ancients. When he disagreed, he opposed the supposed barbarism of the Arabs.

Not all scholars saw such a tight connection between Averroes and the Greek commentators. Girolamo Borro (1512–1592), a contentious and at times controversial professor at the University of Pisa during the years 1553 to 1559 and 1575 to 1586, rejected the idea that the Greek commentators could help understand Aristotle. Borro attacked numerous aspects of humanism, deriding those who concentrated on texts, claiming that emending errors in manuscripts was both simple and of little value. Averroes, for Borro, was useful as a tool, however, in fighting those who tried to combine Platonism and Aristotelianism. Averroes' attacks on Avicenna and Avempace became a model for Borro's own disputes with Francesco de' Vieri II (1524–1591), his rival at Pisa.

In the short treatise, *The Causes of Our Ignorance Are Many*, Borro named the mixing of doctrines as one of the causes of ignorance. Avicenna was one of Borro's prime examples. The Greek commentators are the others. He wrote, "All the Greek expositors stick in this same mud of those, who mixed Aristotle's doctrine with Plato, and who wanted them to be in agreement, but who while they lived wanted there to be disputes [among each other]."[30] Calling for Aristotelian purism, he wrote that out of the Greek commentators "no doctrine is born, but rather some mixture of doctrines arises, which is neither Academic [i.e., Platonic] nor Peripatetic."[31] Borro had a dim view of the Greek commentators but higher esteem for the unadulterated positions of Aristotle. Borro, citing Averroes' criticism of Avicenna's lack of interest in the natural world, extended the critique to Plato himself, thereby denying the applicability of mathematics to discussions of nature and promoting an experiential approach, at least in theory.

The prologue of Borro's *On the Motion of the Heavy and Light* includes wildly lavish praise for Averroes, "who when he digresses, brings Aristotle with him."[32] Contrary to the judgments of numerous humanists, Averroes' method and writing style is of extreme merit, according to Borro. He wrote: "Nothing is richer, graver, more vigorous, more distinguished, and more splendid" than Averroes' expositions.[33] Consequently a number of the sections of Borro's book contain explanations of how Averroes had diligently elucidated the true opinion of Aristotle and fought against the Platonizing views of Themistius and Avempace. Borro's appropriation of Averroes, and his opposition to those who combined Plato and Aristotle, gave him authoritative support for his polemics against mathematical approaches to explaining heaviness and lightness. Borro accepted the ideals of literal exposition and keeping doctrines pure, even if he rejected humanists' concerns with language and texts.

Borro's hostility toward philology stood apart from a number of other sixteenth-century readers of Averroes, who attempted to forge more precise understandings of his work. Their goal of uncovering Averroes' intent corresponded to the related desire of finding Aristotle's view. Averroes' positions were not taken to be necessarily the truth, just as Aristotelians disagreed at times with what they determined to be the true opinion of Aristotle. Rather, trying to understand what Averroes really thought accompanied the process of making new translations. During the fifteenth and sixteenth centuries, Jacobo Mantino, Elia del Medigo, Abraham de Balmes, and Calonymos ben David, using Hebrew manuscripts, made Latin translations of Averroes that conformed to humanists' standards of Latin prose.[34]

Determining Averroes' intent was tied to attempts to establish accurate versions of his texts. Francesco Storella, a professor at Naples from 1561 to 1575, wrote two brief treatises dedicated to analyzing the new translations of Averroes' logical and natural philosophical works. Storella's observations and annotations are filled with alternative translations and comparison of the medieval translation with the new translation of Mantino. Storella used observations taken from Ambrogio Leone (1459–1525), Gersonides (1288–1344), and Tiberio Baccilieri (1461–1511) as evidence for proposed emendations.[35] In this manner, Storella partially integrated the Latin and Hebrew Averroistic commentary traditions, the latter dating from the early fourteenth century, when Gersonides wrote supercommentaries, or commentaries on Averroes' commentaries.[36] The result of this integration was that Averroes became the subject of philological scrutiny rather than just a source for philosophical or doctrinal issues. Storella's commentaries reveal a transformed Averroes, an author whose works were the subject of linguistic analysis not just as a storehouse for philosophical arguments and fragments of the Greeks' doctrine.

THE PERIPATETIC PATH

During the time when interpretations of Averroes were becoming more philologically sophisticated, he remained an authority for natural philosophy and logic.
Most Aristotelian thought had little or no direct relation to theology or ecclesiastical concerns—for example, all the logic. Still in the years after Pomponazzi's
death in 1525 scholars who sought Aristotle's opinion noted its lack of correspondence to Christianity, much in the same way that Achillini, Pomponazzi, and
their medieval predecessors had. This approach, although perhaps remaining
controversial, was nevertheless widespread, particularly in Italian universities,
but also in other locales of learning. For example, the scholar Benedetto Varchi
(1503–1565), lecturing on the rational soul at the Florentine Academy in 1543,
wrote that it was possible to speak about the soul in "two guises," one according
to "human reason . . . as the pagan philosophers had," the other according to "the
divine light," used by Christian theologians.[37] Although Varchi disagreed with
Alexander of Aphrodisias's mortalist interpretation of Aristotle, he nevertheless
noted that in some respects Aristotle's theory ran counter to Christian dogma.[38]

Perhaps the most significant of those who followed this "path of Aristotle"
in the middle of the sixteenth century was Simone Porzio, a Neapolitan who
had been Nifo's student and then became a well-paid and famous professor at
Pisa. Whereas Pomponazzi knew no Greek and integrated a number of non-
Aristotelian sources into his discussions, Porzio was a skilled reader of Greek
and his treatises betray a more narrowly philological approach to determining
the truth according to Aristotelian principles.[39] Alexander of Aphrodisias greatly
influenced Porzio's psychology, and even if true followers of Alexander's mortalist
position on the soul were hard to identify by name in Ficino's time, Porzio, along
with Pomponazzi, became the best-known proponents of the theory during the
sixteenth century.

Like Pomponazzi, Porzio did not maintain that the soul was mortal *tout court*
but that it was the best interpretation of Aristotle and most probable according to
Aristotelian principles. In the preface to his 1551 *De humana mente*, Porzio wrote
that he considered the authority of followers of Averroes, Simplicius, and Themistius because "they support his [i.e., Aristotle's] position," whereas he would
pass over the Latin commentators because their endless studies were designed so
"they could contradict and always disagree with each other," thereby borrowing
the trope of anti-Scholastic humanists to justify reading Averroes.[40] After having
noted that Aristotle was not only a man of admirable learning but also extremely
pious, Porzio explained that the Peripatetic method depended on seeking out nat-

ural principles that ultimately are based on sense experience.[41] After concluding that Aristotle held that the intellect depends on sensation and therefore cannot be immortal, Porzio explained his purpose, "We think that these matters on the human mind are to be proven out of Aristotle himself . . . not so you can disagree with the Latins on this matter, but so that with greater education you will be able to engage with Greek philosophers."[42] Accordingly, Porzio presented himself as interpreter of texts and pathway to ancient learning rather than to the truth.

In his earlier lectures, which were unpublished during his lifetime, Porzio ironically described his method as twofold. The first part of his exploration of the immortality of the soul, he suggested, should proceed so that "we demonstrate while fibbing, and in the second part we will declare the truth of our faith and what is handed down from God." The method of proceeding by lying means using only "the principles of nature," but the second method depends on the "Lord's wisdom which gives us the immortality of the soul."[43] For some this passage might suggest a hidden attempt to subvert the Catholic Church, yet Porzio presented this view in public with even Cosimo I de' Medici, the Duke of Florence, present at the lecture. Porzio's mortalist view gained detractors, but, for the most part, they arose only after Porzio's death. During his lifetime he was rewarded without apparent penalty for his investigations into Aristotle's views on the soul.[44]

Separating Aristotle from Christianity was common to many of Porzio's contemporaries. For example, Francesco Vimercati directly questioned the need to assimilate Aristotle with Christian thought. Vimercati, a professor at the Collège Royal in Paris, originally came from Milan and studied at Pavia, Bologna, and Padua before moving to France.[45] Like Porzio, he was a skilled reader of Greek. In a 1543 treatise on the soul, he asked why it was necessary for believing Christians to confirm that Aristotle believed in an immortal soul, "As if he [Aristotle] did not hold and bitterly defend many other dogmas against our faith."[46] For Vimercati, forcing pagan thought into Christianity diluted the faith, suggesting that Christianity was so weak that it could be damaged by "vain and ridiculous" propositions, such as the unicity of the intellect.[47] Still, Vimercati thought that Aristotle's views on the divine were generally relevant even if not in total agreement with Christianity. In a 1551 treatise on *Metaphysics* Lambda, he wrote, "This book here will show the treatment of God and divine minds; these are not all the same as what we believe by faith but that which is capable of being treated by the light of nature by a most excellent philosopher."[48] Aristotle's conception of divinity was not that which was found in the *De mundo*, which Vimercati thought was inauthentic and similar to Christian works, but rather his views of God stemmed from philosophical investigation that relied on reason and the consideration of nature.

A decade after Vimercati's treatise on *Metaphysics* Lambda was published, Antonio Bernardi di Mirandola wrote *The Overturning of Solitary Struggles*, a work that considered a number of Pomponazzi's views. Bernardi was the bishop of Caserta and dedicated the book to Alessandro Farnese il Giovane, a cardinal and patron of arts and letters. The supposed larger purpose of *Overturnings* was to discuss the legality and legitimacy of dueling, but the numerous digressions led to discussions of the soul, spontaneous generation, and the existence of demons. At times Bernardi disagreed with Pomponazzi, believing that he had misinterpreted Aristotle by attributing to him the view that humans could be produced by spontaneous generation, a position that Avicenna, Tiberio Russiliano, and Pomponazzi endorsed to explain how the world could be repopulated after universal floods.[49] He considered Pomponazzi's explanations in *On Incantations* as being naive, not worthy of a philosopher or a Christian.[50] For other issues, however, he sided, with Pomponazzi, contending that "Aristotle speaking on the foundations of nature never posits demons distinct from God."[51] Using *On Divination through Dreams*, a short treatise contained in Aristotle's *Parva naturalia*, Bernardi concluded that Aristotle not only denied the presence of demons in dreams and the efficacy of oracles but also that, "Aristotle according to natural reason did not consider forms except that are embodied."[52] Consequently, Bernardi interpreted Aristotle as denying the existence of earthly immaterial divine substances that are separated from bodies, such as spirits or demons. Although a number of later ecclesiastical authorities scorned these ideas as Bernardi's views, he posited them not as his own but Aristotle's.

Concerns about Aristotle and Christianity were not confined to theologians and philosophers but extended to the medical profession as well. The universities at Bologna and Padua taught natural philosophy as an introduction to medicine, among other motivations.[53] A number of professors began their careers teaching philosophy before moving on to medicine.[54] As a result, the interests and outlooks of philosophers and physicians overlapped. Sixteenth-century professors of medicine, influenced by humanism and the desire to use the oldest and, therefore, according to some, the most authoritative ancient texts, attempted to establish Aristotle as a medical authority because he predated Galen and Avicenna, the authors of standard texts used in medical instruction. The motivations for this reevaluation of Aristotle among physicians varied, coming from a desire to raise the status of medicine to that of philosophy as well as a from the realization that there were correspondences between Hippocratic writings and some texts attributed to Aristotle, such as the *Problemata* and the fourth book of the *Meteorology*.[55]

The result of the growing importance of Aristotle for medical instruction meant the overlapping of attitudes toward instructing medicine and natural philosophy at Padua and other Italian universities, including interest in Averroes' interpretation of Aristotle and the desire for more accurate interpretations of Aristotle among professors of medicine. For example, Girolamo Mercuriale, a professor of medicine at Padua, believed Averroes' adherence to Aristotle explains why he held position that the rational soul is based in the heart and its activities take place in the brain. Mercuriale wrote, "Because it is most well known that Averroes never departed from the great master Aristotle. I have always thought that he arrived to this view because he thought Aristotle judged the matter in this way."[56]

The writings of Giovanni Battista da Monte, a professor of medicine at Padua in the 1540s, further illustrate the common ground between physicians and Aristotelian commentators. In his commentary on the first fen, or section, of the first book of Avicenna's *Canon*, which was widely read as an introduction to the principles of medicine, Da Monte considered the nature of mixtures of the elements.[57] The question of mixtures was key to the study of Renaissance medicine, which depended on the concept of bodily complexions or temperaments that were the underlying cause of health or sickness. These bodily temperaments were conceptually similar to mixtures of the elements, a topic that was discussed in Aristotle's *De generatione et corruptione* and the fourth book of the *Meteorology*. Da Monte contended that the efficient cause of these mixtures, that is, what creates the form of these new beings, is conceived in different ways by theologians, philosophers, and physicians. The difference in his mind between these last two categories was apparently minimal. Da Monte wrote that he would "speak as a Peripatetic and as physician."[58] Although medicine and Peripatetic philosophy might correspond, Da Monte thought theology and philosophy should be separate. He wrote, "I do not think anything worse could happen in philosophy than to mix it with theology;" those who have made this error are Avicenna and "all the Scotists and Thomists," who have "when they combined everything, most often confused everything, wandering from the true path of the Peripatetics."[59]

On the question of generation of forms for mixtures, Da Monte wrote that there were two central views: Avicenna's *dator formarum* and Averroes' view that the form comes from an idea activated by an intelligence related to the species of mixtures rather than from a divine artificer. Averroes earned Da Monte's praise: "[H]e alone thought with the Greeks, and he is the one who always turns toward the truth." Therefore, for this issue all the Greeks and Averroes are in agreement and their "conclusion is the best and the truly Peripatetic one."[60] Da Monte's preference for Averroes over Avicenna in a work that ostensibly possesses the goal of explaining Avicenna borders on the paradoxical, but this preference is consistent throughout the work. Although he thought that Averroes errs at times, he considered Averroes to be Aristotle's "most faithful expositor," while Plato's theory of forms misled Avicenna.[61]

At Padua, throughout the sixteenth and into the seventeenth century, philosophers continued to understand the study of nature as distinct from theology. Francesco Piccolomini, a professor there from 1564 to 1598, explained that a philosopher cannot dispute with a theologian, because they disagree on starting points. Piccolomini posited that the conclusions of a philosopher will be true only suppositionally, but that the theologian's conclusions are true absolutely. He proposed proceeding in the field of physics by "perceiving the powers of nature, yet we should firmly believe in those powers which come from the mouth of God, in order to elevate ourselves above nature and be linked to God."[62] Following this approach, Piccolomini considered God the assisting form of the world and argued that, for Aristotle, providence derives from the necessity of nature and God's goodness stems from his being a mover of the world, rather than from an "influence distinct from him."[63] This view, which he believed was Aristotle's, however, results "only from proceeding through the works of nature [physica opera], but not in agreement with the exact truth, which our theologians explain most broadly."[64] Knowledge of God derived from nature differs from that of faith.

Piccolomini's colleague and at times rival, Giacomo Zabarella, allying himself with Alexander of Aphrodisias, held that Aristotle believed that all parts of the soul are mortal, although he explained that this view was false according to theology.[65] In his discussion of the prime mover, Zabarella separated approaches using natural philosophy and metaphysics. He judged a single proof that mixed these two fields to be "a monster and a hermaphrodite and most alien from Aristotelian doctrine and Averroes himself."[66] Therefore, he concluded that, while natural philosophy can establish the existence of an eternal mover, it cannot prove that this mover is distinct from matter.[67] The arguments of Zabarella and Piccolomini, both famed and well-paid professors, were published and most likely accurately correspond to their public lectures. Their views fell within the acceptable range of discourse. Ecclesiastical authorities failed to investigate or admonish them.

Nevertheless, the approach to philosophy that emphasized interpreting Aristotle using natural principles without regard to the conclusions' correspondence to religious doctrine was controversial. Some university professors dismissed Pomponazzi's view of the human soul. Boccadiferro wondered at Pomponazzi's mortalist stance, calling it "fatuous."[68] The Paduan professor Marcantonio Genua wrote a treatise dedicated to showing that Pomponazzi failed to interpret correctly Aristotle's position on the mortality of the soul.[69]

Girolamo Cardano (1501–1576) is one of the most difficult critics of Pomponazzi to categorize. A professor of medicine at Pavia and Padua and author of numerous writings on mathematics, medicine, natural philosophy, and astrology, Cardano innovated despite being inspired by Hippocratic and other ancient texts. Cardano had doubts about a number of Aristotelians. He considered Pietro d'Abano impious and dishonest because he presented himself as a magician, when, according to Cardano, he merely exploited his knowledge of natural causes.[70] Cardano's assessment of Pomponazzi was equally negative. Notwithstanding similarities in their thought and the fact that Cardano and Pomponazzi would eventually be seen as irreligious brethren by seventeenth-century thinkers, Cardano wrote that Pomponazzi's views were "not only false and impious, but also very alien from the Peripatetics, and even ridiculous."[71] Stating that "our [religious] law is the truth," he speculated that Pomponazzi's rejection of miracles had "rendered our religion false."[72] Consequently, Cardano thought that Pomponazzi had not just been impious but had also poorly interpreted Aristotle. For example, when Pomponazzi explained that a woman's eyes, rather than her breath, could cloud a mirror, he relied on his faith in ancient authorities and neither on experience nor Aristotle's true thought, according to Cardano. He wrote, "Pomponazzi cannot say this even according to the Peripatetic path [in via Peripateticorum]," as Aristotle did not believe that eyes emitted rays but held the theory of visual intromission.[73]

Yet there were similarities between Cardano's and Pomponazzi's thought. For example, Cardano used occult causes to explain wondrous and hidden effects.[74] At the end of his treatise *On the Immortality of the Soul*, Cardano digressed, considering the question of immortality according to nature (naturaliter loquentes). There he posited that the opinion of the Peripatetics and Aristotle was the "soul that is the form of each body is mortal," because its existence depends on there being a compound of form and matter, even if Cardano believed the human soul is in fact immortal.[75] His judgment on the soul's mortality corresponded at least minimally to Pomponazzi's.

Although Cardano, in his old age, was prosecuted and stripped of his right to teach and publish, the charges of heresy brought against him had little or nothing to do with his interpretations of Aristotle. Cardano's trial resulted in part from his inability to recognize the conservative shifts in Church doctrine as well as the waning power of his ecclesiastical protectors.[76] Cardano was perhaps the most frequently censured author during the sixteenth century, largely because there were so many requests to read him. The large number of censures mostly addressed his astrology, his dedication to occult causation, and his reduction of religion to natural causes.[77]

Cardano's detractors went beyond these numerous inquisitors. Julius Caesar Scaliger (1484–1558) wrote one the most influential books of natural philosophy of the sixteenth century in response to what he considered to be the misinterpretations and false accusations Cardano leveled against Aristotle. The work went through ten editions and made Scaliger one of the most respected and frequently read Aristotelian natural philosophers not just during his lifetime but also well into the seventeenth century.[78] As an Aristotelian Scaliger cut a curious figure. Influenced by the strands of Aristotelianism taught in northern Italy, he portrayed himself as a defender of Aristotle and praised Aristotle's philosophy in the highest terms. But, at the same time, many of his views deviated from common understandings of Peripatetic natural philosophy, creating the impression that his identification with Aristotle might have been just one of his many rhetorical strategies.[79] His attack on Cardano demonized forms of Aristotelianism, such as Achillini's, Pomponazzi's, and Porzio's, that openly recognized discrepancies between Aristotle's thought and Christianity.

Some of Scaliger's most profound deviations from traditional interpretations of Aristotle regarded the reconciliation of his natural philosophy with Christianity. Scaliger surpassed the efforts of even Thomas Aquinas in his attempt at linking Aristotle to Christian dogma.[80] Most notably, contrary not just to his contemporary Neoplatonist polemicists but Thomas as well, he contended that Aristotle believed in the trinity. He also argued that Aristotle believed God created the world and the heavens. Although this conclusion might seem removed from a literal exposition on Aristotle, Scaliger based himself not just on Aristotle's text but Averroes' as well. Calling the words of Averroes divine, he twisted them to produce the opposite interpretation that was commonly made at the time, contending that Averroes did not deny the heavens were created but only that their generation was of different in kind to sublunary generation. He posed the rhetorical question, "Would a Christian say something different [from Averroes]?"[81] Similarly, using Averroes' assertion that eternal beings have efficient, formal, and final causes that emanate from the first principle, Scaliger concluded that God is both administrator and creator of the world.[82] Linking God's ordinary power to nature, Scaliger concluded by asking, "What else is the providence of nature but the perpetual presence of God himself?"[83] Thus for Scaliger, Averroes' principles lead to the assertion of God's providence over sublunary particulars.

Scaliger remained one of the most influential Aristotelians well into the seventeenth century, although some who valued his work questioned his interpretations of Aristotle even if they accepted his philosophical conclusions. For example, Paganino Gaudenzio (1595–1649), even though he believed that ancient philosophy was inspired by traditions of wisdom linked to scripture, thought that Scaliger had erred in making Aristotle's thought correspond to Christianity. Despite being sympathetic to Scaliger, Gaudenzio believed Scaliger incorrectly interpreted Aristotle's views on the prime mover, the trinity, the creation of the world, and spontaneous generation.[84] For Gaudenzio, Scaliger "had triumphed," over Cardano, but had, "not always worried about the sentence of Aristotle."[85] That is, he believed Scaliger correctly understood the nature of providence, creation, and the trinity but that this understanding deviated from Aristotle's.

CONCLUSION

Many scholars of the late Renaissance recognized discrepancies between Peripatetic thought and Christian dogma, or, in the case of the Neapolitan physician Donato Antonio Altomare (1520–1566), between Galenic and Christian dogma.[86] This recognition did not necessarily entail an attempt to undermine Christianity. As Vimercati pointed out, true Christians should have enough faith that they are not worried by the philosophical views of a few ancients. Rinaldo Odoni's vernacular treatise on the soul, printed in Venice in 1557, concluded that Christians have stronger proofs than Aristotle possessed that the soul is immortal, namely in the example of Christ's life.[87]

Not only philosophers adopted these attitudes, but sixteenth-century literary figures also considered and admired Pomponazzi and Porzio. Sperone Speroni (1500–1588), the Paduan author of dialogues on numerous literary subjects, has his characters praise Pomponazzi's writings for being "full of new things worthy of consideration."[88] This praise is in spite of Speroni's low opinion of Averroes, whom he believed was neither Muslim nor Christian yet gullible enough to believe women could become pregnant from semen lingering in the bath, as was stated, as Coluccio Salutati had already noted, in Averroes' *Colliget*.[89] Torquato Tasso (1544–1595) the famed poet and courtier of the House of Este, wrote positively of Porzio. In Tasso's dialogue *Porzio, or, On Virtue*, Porzio's interlocutor describes him as "the best and most famous philosopher not only of Naples but of all of Italy."[90] Tasso, aware of the content of Porzio's writings, had his character describe his approach as following the Greek philosophers as opposed to the Latins or Barbarians, that is, the Arabic philosophers. Moreover, Tasso paraphrased Porzio's doctrine of the soul in approving terms as Porzio's interlocutor concluded in agreement with "Alexander, Simplicius, and Averroes that the moral virtues are born from the contemplation of natural and celestial matters."[91]

In Italian courts and universities, philosophers, physicians, and poets described Aristotle and his followers as legitimate sources of knowledge, even if his philosophy led to conclusions that differed from Church doctrine. Despite the spread of these views, or perhaps because of it, in the second half of the sixteenth century, religious authorities attacked these interpretations as they forged a new version of Aristotelianism that harkened back to Thomas and other medieval Scholastics who wanted natural philosophy to be ancillary to theology.

CHAPTER FIVE

Religious Reform and the Reassessment of Aristotelianism

A number of Italian natural philosophers prized Averroes or attempted to follow Aristotle's principles without regard for their correspondence to Christian doctrine. Sixteenth-century Catholic authorities reacted by trying to establish orthodox ways of considering Aristotle and Aristotelian natural philosophy. The rapid successes of the Protestant Reformation signaled the ineffectiveness of the Fifth Lateran Council's attempt at reforming the Catholic Church. In response to the Reformation, the Council of Trent, first convened in 1542, spent nearly the next twenty years setting forth ways of avoiding corruption among the clergy and its flocks. Simultaneously, the Council established new dogma and standards of orthodoxy. These attempts at reforming the Catholic Church included the reaffirmation of Aristotelian understandings of causation and reconsiderations of how philosophy should be taught.[1] Paolo Sarpi (1552–1623), the Venetian historian and naturalist, observed that Aristotelian thought was connected to so many decisions of the Council of Trent that if Aristotle "had not been adopted, we [Catholics] would lack many articles of faith."[2]

Among the most significant ways in which philosophical education changed during these years was the result of the establishment of the Jesuit order. Charged with educating youth both in Catholic lands and in missions, Jesuits established schools around the world designed to educate students to be able to use the most sophisticated intellectual tools, both philosophical and humanistic, to defend the faith. In the case of natural philosophy, Jesuits adopted Thomas's synthesis of Aristotelianism and faith and conceived philosophy as subordinate to theology. This choice entailed their affirming the "Apostolici regiminis" of the eight session of the Fifth Lateran Council, which had largely been ineffective or ignored in the years immediately after its promulgation. Consequently, Jesuit philosophers sought to minimize the influence of Averroes and eliminate that of Pomponazzi by promoting a version of Aristotelianism that privileged metaphysics and reconciled philosophy and theology. For specific doctrines, such as his views on the soul, Jesuits considered Averroes' views not only mistaken but also perverse and depraved, just as humanists had. To counter those influenced by Averroes and Pomponazzi, they relied on philosophical arguments, considerations of linguistic issues, and the authority of the Catholic Church.

Like earlier humanists, Jesuits pointed to the linguistic barriers that hindered Averroes from accurately interpreting Aristotle. Unlike these humanists, they did not adopt a strategy of invective. Rather they attempted to expunge Averroes' view of the soul from philosophical education. Their strategy called for unifying Aristotelianism by deemphasizing the existence of "sects" of philosophers. Such a strategy emerged from some of the earliest Jesuits. In the 1550s, Jeronimo Nadal, an associate of the founder of the Jesuit order, Ignatius of Loyola, envisioned a series of handbooks that could harmoniously collect the important parts of Scholastic thought.[3] In this fashion, he hoped to build a version of Aristotelianism that conformed to Catholic doctrine, fending off the alternatives, namely, materialist and Platonist philosophies.

The development of the Jesuit position toward Aristotle arose in the 1550s and was eventually codified beginning in 1589 in the *Ratio atque istitutio studiorum Societatis Iesu*, shortened to *Ratio studiorum*. This handbook established guidelines for Jesuit teaching. The Jesuits' international diffusion, prominence in education, and relatively high internal discipline fostered the diffusion of their Catholic version of Aristotelianism that emphasized the concordance of faith and reason as one of the predominant versions of philosophy in Europe and its colonies. In spite of the univocal character of the *Ratio studiorum*, many Jesuits were philosophers and therefore disagreed with each other at times. Furthermore, their positions evolved over time. Yet the internal control of the publishing of their works meant that their printed writings shared a high degree of agreement that corresponded to doctrinal orthodoxy. Jesuits frequently evoked the Fifth Lateran Council and the "Apostolici regiminis" in the formation of their own rules for conducting philosophy and in identifying heretical or unorthodox positions. Therefore, their techniques went beyond the boundaries of philosophical argument, as they used inquisitorial procedures to battle against opposing versions of natural philosophy, perhaps most notably in the failed prosecutions of the Paduan philosopher Cesare Cremonini but also in denunciations of numerous writings that led to their placement on the index of prohibited books.

THE DEVELOPMENT OF JESUIT ARISTOTELIANISM

One of the primary goals of Jesuit instruction in philosophy was the inculcation of good moral habits among its students. This goal was shared with earlier opponents to Averroes, and not just with humanists; Barozzi, for example, cited fears about morality as his rationale for banning the teaching of Averroes' unicity thesis in Padua. In order to create these good moral habits, Jesuit leaders demanded that positions that could lead to heterodoxy or impiety were to be avoided by teachers. Instruction ideally emphasized consensus about correct positions by presenting the views of philosophers who were in agreement with the faith. Jesuit teaching largely ignored those who deviated from what was considered orthodox. Such attempts at forging orthodoxy are visible in Jacobus Ledesma's instructions to professors of philosophy at the Collegio Romano in 1564–65 in which he demanded the unity of doctrine among teachers at the college, forbidding them from introducing novel opinions into their teachings without permission from a superior and from contradicting Aristotle "without an authority from the past."[4] For Ledesma, these authorities from the past had different statuses. He asked teachers to refrain from disagreeing with Aristotle and he also forbade them from praising "Averroes or other impious interpreters," but Christian and pious authorities, such as Thomas and Albertus, were to be praised.[5]

Thomas became an authority above all others for theology and philosophy from the foundation of the Jesuit order. Ignatius was familiar with Thomas from his education with Dominicans in Paris; and the Thomistic synthesis fit well with Ignatius's desire for his followers to contemplate God's creation.[6] As a result, he directed Jesuits to follow scripture and Thomas in theology. For philosophy, Ignatius's recommendation of Aristotle as an authority gave prominence to Thomistic interpretations.[7] Ledesma continued in this vein in his instructions for teachers. He wrote that the Jesuit teachers should not state that other Latin Scholastics disagreed with Thomas. Moreover, his rules prohibited "showing that Averroists, or a faction of Greeks or Arabs, oppose the Latins or theologians."[8] Even if it might be necessary to mention Averroes' or others' heterodox views, the professors should not digress and potentially give students the opportunity to become attracted to their views. Rather, their heterodox "opinions should be simply pronounced in an indifferent manner."[9] Ledesma contrasted his pedagogy with the way philosophy was instructed in Italian universities, where there was, in his eyes, "too much liberty, which harms the faith;" and, the lack of a single authority rendered their teachings "odious and contemptible."[10] For him, freedom of thought might undermine morality.

The lack of specificity in Ledesma's statement obscures which specific Italian philosophers were odious. Nevertheless, his belief that teaching philosophy should serve theology ran counter to the views of a number of the lay Italian professors of natural philosophy and medicine discussed in the previous chapters. Ledesma believed that in order to aid theology, the professors of philosophy at the Collegio Romano should "note which opinions that concern matters of faith must not be held, and they must defend with all their power [totis viribus] those which must be defended."[11] One of the tools was to say that these positions "are expressly required to be believed, also according to the opinion of Aristotle," specifically referring to the question of the immortality of the soul.[12] Ledesma's strategy for inculcating orthodoxy was to use philosophical argument as well as Aristotle's authority to defend positions crucial to Catholic dogma.

The immortality of the soul was just one of these positions for which he hoped to show that Aristotle conformed to Christian belief. In the same year he listed the doctrines that "must be taught and defended in philosophy," including that "God has providence over lower bodies, including single things and human affairs" and that creation is possible and is true according to "the principles of Aristotle and follows necessarily from his doctrine."[13] He made an exception for the eternity of the world, for which he believed that Aristotle held the universe existed from eternity. But he thought that Aristotle's view was incorrect both "according to the truth and according to true philosophy," believing that its creation in time could be proven through "natural principles" (rationibus naturalibus).[14] Ledesma, therefore, deviated from Thomas's view that the eternity of the world was a neutral question, positing that its creation could be determined not just from faith but from evidence taken from natural philosophy.

Ledesma's rules that were designed to undermine Averroes' authority in philosophy were mild compared to the polemics of some of his Jesuit contemporaries. Peter Canisius, the provincial superior for Germany, wrote in 1567 that he wished to eradicate Averroes' philosophy because he thought that it created "atheists as much as heretics."[15] According to him, anyone with too great a familiarity with Averroes must "be believed to be depraved."[16] The threat was pressing for Canisius; he contended that there were Jesuits in Munich who were teaching Averroes and "dared to call him divine."[17] Although Canisius might have overstated his case, among some of the earlier Jesuits philosophical charity extended not just to Scotists, Thomists, and nominalists but also to Averroes. Benedictus Perera, a Jesuit professor, in his 1564 instructions for professors at the Collegio Romano on how to lecture on philosophy, suggested teachers should read a number of authors, whom he described as followers (seguaci) of Averroes. These authors included Themistius, Francesco Vimercati, John of Jandun, Walter Burley, Paul of Venice, and Agostino Nifo. This reading was not merely intended to familiarize instructors with the enemy. Perera believed that reading Averroes was "very useful as much for his doctrine as for his fame that he has in Italy."[18] Despite not wanting to divide the commentators into sects, he described Averroes as "a singular splendor, glory and guard of the Lyceum" to whom Peripatetic philosophy owes more than to anyone else.[19]

Perera's views troubled others in the order, and he eventually changed his printed views as the result of pressure from superiors. In 1576 and 1577 a committee, which included Pedro de Fonseca and Ledesma, judged Perera as having followed Averroes to too great a degree.[20] In the 1579 printing of *On the Common Principles of All Natural Beings*, Perera found middle ground, writing that he would not "praise those who condemn Averroes maliciously, tear him apart with insults, and shout, as if he was a plague to all genius, that all his opinions must be abandoned."[21] Yet those who say that "Averroes was of sharp wit, grave judgment, and singular zeal and diligence"—words similar to Perera's own past judgments— also err.[22] Repeating the opinions common to numerous humanists, he noted that Averroes interpreted Aristotle "through ignorance of the Greek language, faulty manuscripts, weakness of good translations, and numerous delusions" and as a result held many opinions that were not just false but "impious and absurd."[23] Perera's changed position reflected the solidifying consensus among Jesuits that Averroes' philosophy was dangerous.

Rules for instruction at Jesuit colleges adopted positions similar to Ledesma's regarding the teaching of philosophy. The decrees of the Third General Congregation in 1573 stipulated that, because philosophy is meant to serve and be a handmaiden to Scholastic theology (theologia scholastica), Jesuit teachers should avoid reading those who "write impiously against Christian dogma."[24] The 1586 *Ratio studiorum* repeated the decree from 1573 and recommended that professors should not read Averroes "or other philosophers of his grain" in a way that would lead to a division of philosophy into "sects of Averroists, Greeks, and Latins." The 1586 *Ratio studiorum* repeated Ledesma's restriction on praising Averroes or considering his digressions from Aristotle's text.[25] Two subsequent sixteenth-century sets of rules for Jesuit professors similarly prohibit praising Averroes, following Averroes' digressions, reading impious interpreters of Aristotle, and adhering to any "sect of philosophy." These rules cited the Fifth Lateran Council to support the position that philosophical argument should be used to bolster orthodoxy.[26] The unwillingness to identify disagreements in interpretations of Aristotle hid the potential ambiguity of his writings, a trait that had long been used to undermine Aristotle and explain dissent among his followers. Hiding the difficulty of interpreting Aristotle, in turn, made the Jesuit interpretations appear more solid since challenges to their authoritativeness largely went unrecognized.

Despite the lack of details in the various versions of the *Ratio studiorum* about the identity of these "impious interpreters of Aristotle," Jesuit commentaries, written at the same time, provide more specific considerations of Aristotelians. By the time Franciscus Toletus's commentary on Aristotle's *Physics* was first printed in 1576, he had already taught at the Collegio Romano and was now preaching in the papal court. The commentary's unsigned preface warns the readers to avoid following "impious interpreters of Aristotle, whether Greek or Arab," because "nearly all these impious ones were heathens, idolaters, and some even Saracens, or Mohammedans."[27] These warnings were considered necessary because "certain ones in some academies have abandoned themselves [addictos] to impious authors."[28] According to the author of the preface, these academics wish to be called "Averroists" and consider his words to be "like an oracle," despite his works' being "full of numerous errors harmful to faith and piety."[29] Among these alleged errors were: denying providence for the sublunary world, the immortality and multiplicity of the human soul, and human freedom (libertas). For some Jesuits, Averroes was an acceptable authority but for only a few subjects that would not corrupt morals. The Jesuit Andreas Schott's 1603 list of the best commentaries on Aristotle includes Averroes' work on the *Topics* and other logical works but not his commentaries on *De anima* or other works that might threaten piety.[30]

The Jesuit order promoted humanistic studies as well as philosophical ones. The attacks in Toletus's preface on Averroes and the academies that embraced him resonate with the insults found in earlier humanist invective. The preface asserted that as a Muslim, Averroes "always possessed hatred toward the Christian religion."[31] Moreover, he relied upon a "corrupt book of Aristotle" and did not have access to all the Greek commentators and any of the Latin ones.[32] The links to humanism in this diatribe are overshadowed by his concerns with faith. The followers of Averroes are dangerous because they lead to impiety. To defend the faith, Aristotle's correspondence to Christianity must also be defended. The preface proclaimed, "Many when they hear that something is true according to Aristotle, they understand it as being true according to philosophy: even according to the best philosophy. And because Aristotle thought this, they believe that it agrees with natural reason and light."[33] This approach, however, causes these disputants to "overturn that decree of the Council [i.e., "Apostolici regiminis"] out of carelessness [imprudentia]," because they argue "with the highest zeal" that it seems that Aristotle believed some issues in a way contrary to Christianity when they could have adduced arguments that made Aristotle conform to "faith and truth."[34]According to this view, these unnamed academics who follow Averroes missed an opportunity to demonstrate the compatibility of Aristotelianism to Christianity and as a result disobeyed the Fifth Lateran Council.

Although the preface did not specifically name its opponents, it seems likely that the targets were lay professors at Italian universities. The tactics to discredit them were twofold, highlighting the alleged moral consequences of not following the Fifth Lateran Council and associating these philosophers with Averroes, whose reputation for impiety was almost unrivalled. The psychological positions of Averroes continued to create concerns, and he served as a stand in for those who wished to separate natural philosophy from theology. Consequently, Toletus's work promoted the Thomistic path of philosophy and his demonization of Averroes lay along that path.

By emphasizing the moral consequences of teaching that Aristotle held impious positions, the Jesuits pushed aside the discussion of whether Aristotle actually held those views and promoted the study of philosophy in order to promote orthodoxy. Among ecclesiastics in the Catholic world after the Council of Trent, concerns with proper religious practice and belief took on a greater weight, which in turn made the philosophical practices of Pomponazzi appear even more dangerous than they had in the 1520s.

THE SPREAD OF ORTHODOXY

Ecclesiastical orders besides the Jesuits were also concerned with Pomponazzi's legacy and the versions of Aristotelianism taught in Italian universities during these years. Alfonso de la Vera Cruz, an Augustinian hermit stationed in New Spain wrote, in his 1569 *Speculations on Physics*, that maintaining that the immortality of the rational soul cannot be proven by natural reason is erroneous because of the decisions made at the Fifth Lateran Council.[35] Franciscans also reacted to naturalistic versions of Aristotelianism in both writing and sermons. Frans Titelmans, a Franciscan originally from the Low Countries but active in Bologna, wrote an elementary textbook on natural philosophy that mixed summaries of Aristotle's works with religious poetry that Titelmans designated as "psalms." His work went through numerous editions, providing religious schools with an alternative to the more advanced and less-religiously minded works based on university lectures as well as being a counterweight to the Lutheran textbooks that were similar in terms of their combining discussions of natural philosophy with religious instruction.[36] The Franciscan preacher Francesco Panigarola, in sermons he gave at Saint Peter's Basilica in 1577, attacked unnamed philosophers whose conclusions contradicted theology and who denied the world's absolute dependence on God as creator. Panigarola contended the Pythagorean, Epicurean, and Averroistic views of the soul derived from the assertion of an eternal universe and the refusal to admit an actualized infinite.[37]

Other sermons addressed the teachings of the universities in even more precise terms. Cornelio Musso (1511–1574), a Franciscan who became the bishop of Bitonto, used his philosophical and theological training at Padua, Pavia, Bologna, and Rome to address the immortality of the soul in his sermons. Musso, a highly regarded humanist preacher, gave opening sermons at for a session at the Council of Trent in 1545.[38] In later sermons, arguing that the foundation of the Christian faith is based on the immortality of the soul and that divine providence, hell, paradise, and the commandments depend on such a concept, he lamented those who doubted its demonstrability. To those who say that "one cannot prove in philosophy this conclusion, and that it is not in Aristotle," Musso answered that immortality was a "truth proven by Socrates, Plato, Pythagoras, Homer, . . . and preached by Moses; all the prophets by the greatest philosopher of all philosophers, Jesus Christ; by Saint Paul; by all the Apostles, and by numerous Peripatetics."[39] Musso's sermon revealed his specific knowledge of philosophical discussions of the soul. He condemned Averroes' position that the soul is the assisting form of the body, "like a sailor in a boat," Galen's position that the soul arises from matter, and Alexander's view, repeated by Pomponazzi, that the intellect cannot exist without sensation or phantasms.[40]

Dominicans had been among the first to attack Pomponazzi during his lifetime and their attacks persisted in the sixteenth century. Giacomo Nacchiante, the bishop of Chioggia, in his 1557 work dedicated to correcting philosophical errors, argued that Aristotle believed in the creation of the world.[41] Furthermore, he thought it was possible to convince according to "the principles of natural knowledge and especially according to the mind of Aristotle that the human soul is incorruptible and immortal."[42] Consequently, according Nacchiante, Pomponazzi's views were "not just impious, rather greatly impious, but also do not truly derive from the light of nature and satisfy Aristotle even less."[43]

The Salamancan Dominican, Domingo de Soto (1494–1560), who influenced Toletus among others, also took up the issue of creation and accused Averroes of being "that Saracen," who held that "nothing can be generated out of nothing" and ridiculed "our faith."[44] The Dominican theologian Melchor Cano (circa 1509–1560) offered a different view of Aristotle than the Jesuits, yet still considered the Aristotelians of the Italian universities threatening. On a number of issues, such as the immortality of the soul, providence, creation, and punishments in the afterlife, Plato's doctrine is more clear and constant that Aristotle's, according to Cano. He cited Tommaso de Vio's interpretation as further evidence that Aristotle had erred on the question of the soul. Accusing Aristotle of ambiguity and obscurity, Cano warned against considering him an oracle, "We have heard that certain Italians give just as much of their time to Aristotle and Averroes as they do to sacred letters."[45] As a result, "Pestilential dogmas on the mortality of the soul and lack of divine providence in human affairs were born in Italy."[46] The version of Aristotelianism that was most dangerous, therefore, was tied to reading Averroes.

In order to strengthen his assertions, Cano described four points where Aristotle's philosophy deviated from Christianity. Aristotle denied that God created the heavens and the possibility that God causes dreams . In addition, Cano believed that Aristotle understood God as just one intelligence above other intelligences, which were the movers of the heavenly orbs. Consequently, Aristotle denied divine providence for human affairs.[47] Despite these concerns, Cano maintained the necessity of Aristotelianism. He contended that Thomas had corrected Aristotle on these points thereby creating a pious version of his philosophy. Nevertheless, Cano's assessments of Aristotle roused Jesuits. The Jesuit Antonio Possevino argued that Plato was for many topics wrong "not only according to the Catholic faith but also according to natural reason," and then went through a point-by-point refutation of Cano's objections to Aristotle.[48]

Although not all Catholics ecclesiastics agreed about ancient philosophy and its relation to Christianity, a number of Jesuits maintained a loyalty to Aristotelianism and a hostility to Averroes and Pomponazzi well into the seventeenth century. Fonseca, for example, described his theologically minded philosophy as following Thomas more than any other philosopher or theologian and believed that Aristotelian doctrine corrected the theological errors of all other schools of philosophy.[49] Another example is found in the Coimbrans' commentaries on Aristotle, which were perhaps the most widely proliferated Jesuit works of natural philosophy, repeatedly printed, used as textbooks in numerous colleges, and read even in Protestant lands such as England.[50] Fonseca assigned the Portuguese Emmanuel de Goes to write many of these commentaries. Goes completed his work on De anima 1598, when it was printed for the first of 18 times.[51] In this commentary, he touched on points key to Jesuit interpretations of the role of natural philosophy. For example, Goes referred to the Fifth Lateran Council and the Council of Vienne, which determined that the intellective soul is the form of the body.[52] He criticized Pomponazzi for holding that the indivisibility of the rational soul can only be proven by faith, and he briefly dismissed Averroes' position on the unicity of the intellect.[53] Although in the *De anima* commentary Averroes was dismissed, his rejection was not universal. In other places, the Coimbrans' commentaries repeated assessments of Averroes' fidelity to Aristotle. The commentary on *De caelo*, addressing the question of whether the heavens are composed of matter and form, after noting how rarely Averroes is praised, described him as having "followed the mind of Aristotle" better than all other Peripatetics.[54]

Pomponazzi remained an object of criticism for Jesuits for years. In 1603, the Jesuit Martin Delrio, in a book that condemned natural magic, expressed surprise that it had taken so long for *On Incantations* to be placed on the index of prohibited books. Delrio believed that Pomponazzi's contention that celestial influences could have caused miracles and established religion proved he was a poor philosopher and a worse Christian.[55] Girolamo Dandini wrote in his massive 1611 *On the Ensouled Body* that not only John of Jandun's, Pomponazzi's, and Porzio's views were contrary to the decisions of the Fifth Lateran Council but so was Scotus's position that denied the immortality of the soul could be proven through philosophy.[56] Antonio Rubio, a Jesuit missionary who spent more than two decades in New Spain, wrote about Pomponazzi and Porzio in a similar light, concluding that Aristotle thought that the soul does not arise simply from matter but is created by a primary cause that can be traced back to the divine. Accordingly, Rubio concluded the dictates of natural philosophy should agree with the Catholic faith as defined at the Fifth Lateran Council that the soul is immortal.[57]

Despite the broad consensus among Catholic authorities about the nature of Aristotelian thought, there was not complete agreement that the Thomistic synthesis should be endorsed at the Council of Trent. Participants objected to forcing Aristotle's thought to agree with Catholic doctrine, just as Tommaso de Vio had at the Fifth Lateran Council. Primo del Conte, a theologian originally from Milan, who attended a number of the sessions at Trent, openly questioned the desire to make Aristotle into a Christian. Marcantonio Maioragio, a scholar who wrote primarily on ancient rhetoric, expressed doubts about whether he should publish a paraphrase of Aristotle's *De caelo* on the grounds that his attempt to reveal Aristotle's true thought also revealed Aristotle's impiety. Conte lauded Maioragio's method that was tempered by humanism. Conte contended that when explaining Aristotle we should explain his view, not ours, because "in no way is it possible to reconcile" Aristotle with Christianity.[58] As a result, he advised Maioragio that, "[w]hen you interpret Aristotle, do it so that Aristotle speaks and I will not hear a Christian," because, Conte concluded, "We all know Aristotle was not a Christian."[59]

TAURELLUS AND CESALPINO

Both Protestants and Catholics endorsed Aristotelian philosophy during the sixteenth century. Although Martin Luther had demonized Scholastic philosophy,
subsequent generations of Lutherans remade Aristotelian thought for their own
ends.[60] Luther's associate Philip Melanchthon, who reformed Lutheran university education, delivered an oration in 1537 that recounted Aristotle's biography,
emphasizing his eloquence, much as Bruni had.[61] Melanchthon believed Aristotle most likely held the human soul to be immortal despite his not satisfactorily
demonstrating this position.[62] In any case, Aristotle's systematic account of the
natural world was useful for him and subsequent generations of Lutheran scholars who taught in universities such as Wittenberg and Tübingen. For Protestants
just as for Catholics, Aristotelian thought was fundamental for creating knowledge that conformed to religion.

By the end of the sixteenth century, the divisions that divided Europe spiritu-
ally differentiated approaches to Peripatetic thought. The Lutheran professor of
medicine Nicolaus Taurellus (1547–1606) developed a method that made natural
philosophy dependent on metaphysics and in turned demonized natural philosophers living in Catholic lands. Accordingly, Taurellus believed his philosophy
successfully combined theology and the study of nature. Even though this goal
was similar to that of Jesuits and other Catholic orders, Taurellus distinguished
his philosophical views on God from the Coimbrans' commentators and other
theologically minded philosophers. His attacks, however, were most vehement
and relentless against lay Italian philosophers who held that natural philosophy
should be independent from considerations of the divine. He found two Italians
to be the most egregious: Francesco Piccolomini and Andrea Cesalpino, both of
whom he saw as producing a naturalistic version of philosophy that overlooked
key questions of faith. For Taurellus, the dangers of naturalism were remedied by
a consideration of metaphysics that corresponded to theology.

The most extensive analysis of Taurellus's work depicts him as shifting German philosophy from being dominated by studies of logic to studies of metaphysics, a change that became fully realized in the work of Leibniz, who indeed
was familiar with Taurellus's writings.[63] After his studies in Tübingen with Jacob
Schegk and Samuel Heiland, Taurellus taught in Basel. His first published work,
The Triumph of Philosophy, That Is, the Metaphysical Method of Philosophizing,
1573, established Taurellus's philosophical outlook. His defense of metaphysics
possesses similarities to medieval and early modern criticisms of versions of Aristotelian natural philosophy as being freed from theology. He described how he

considered and rejected the possibility of a double truth whereby something "that is false theologically could be true for philosophers."[64] He concluded that when philosophers' conclusions contradict theological truth, the fault is not in science or logic but in "human weakness."[65] For Taurellus, opposing reason to faith is a moral failure.

To him, this human weakness was found not just in his Catholic contemporaries but in Aristotle himself. Here Taurellus differed from his Jesuit contemporaries who wished to minimize or remain silent about Aristotle's deviations from Christianity. Taurellus was confident that, "[i]f Aristotle would have been educated by the light of Christ, through contemplating, he would have retracted his errors."[66] Taurellus, despite having great esteem for Aristotle, whom he considered "the father" of philosophy, pointed to Aristotle's doctrine of the eternity of the world as an example of where Aristotle's demonstrations are false, injurious to God, contrary to faith, and repugnant to theology.[67] For him, such a position cannot be held, even hypothetically or as a supposition.

The interconnectedness of theology and philosophy was a consistent theme in Taurellus's writing throughout his life. In a work on medical diagnostics and prognostics, printed in 1581, he equated the belief that there is "double truth" with Satan.[68] In the last book that was published during his life, a metaphysical consideration of the eternity of the world, he specified the differences between philosophy and theology. It is a mistake, Taurellus believed, to suppose that scripture and Aristotle respectively represent revelation and philosophy, "[b]ecause many things contained in the Bible are philosophical, and there are many things held in the books of Aristotle, which are not true, and certainly are not philosophical."[69] His belief that there was only one truth, did not mean that he did not distinguish religious and philosophical truth. Some truths could only be known through theology, namely the nature "of Christ, human redemption, the graceful remission of sins, faith, the hope of future happiness." Everything else—"God, laws, humankind, the world"—was part of the domain of philosophy.[70] This philosophy, however, should be tied to metaphysics. Unlike the Jesuits, who endorsed the possibility of physical proofs for the eternity of the world, Taurellus thought they should be metaphysical.[71]

Perhaps Taurellus's most famous work was a diatribe against the Aretine philosopher and physician Cesalpino. Cesalpino served as personal physician to Pope Clement VIII and wrote a number of works on medicine, mineralogy, botany, demonology, and natural philosophy. Notwithstanding Taurellus's interpretation of his writings, Cesalpino applied metaphysics to the natural world. For example, he famously attempted to reform botanical classification so that the categories would be based on a consideration of plants' substance.[72] He limited, however, the category of substance to living things; inanimate mixtures of the elements were not truly substances, in his view. This meant that these mixtures and their transformations, such as in meteorological phenomena were without their own formal or final causes. As a result, they should not be considered in relation to God's providence.[73] Ultimately the lack of providence in nature depended on metaphysics, namely on Cesalpino's interpretation of the prime mover, which he considered to be a "speculative intelligence" rather than an "active intelligence" that directs and intervenes in nature.[74]

These views, among others, upset Taurellus. In a book-length polemic combining philosophical argument with invectives, he attacked Cesalpino and his sympathizers for espousing materialistic teachings about the soul and spontaneous generation, for holding that the question of whether the world was created in time to be a *problema neutrum,* and for denying God's providence in the sublunary realm. His polemics also question the efficacy or willingness of Catholic ecclesiastical authorities to enforce orthodoxy. While writing about Cesalpino's view of the human intellect, which he equated to Averroes', he asked rhetorically, "If once Averroes' sentence on the oneness of the intellect was condemned in the Lateran Council, why is it permissible for Cesalpino to revive it?"[75] More vitriol fell upon Cesalpino's equating of God with universal soul. Taurellus wrote, "I do not know if Cesalpino was a Christian. Nevertheless, I would not dare to affirm duly that it is possible to find a greater impiety in the writings of pagans."[76] According to Taurellus, Cesalpino's error is found partly in his poor philological skills. He and his followers ignore Christian piety and are misled by Averroes, Taurellus wrote, so "[i]n order to understand the meaning of the Greeks they do not consult the Greeks but rather barbarous interpreters."[77] Taurellus proclaimed philosophical independence, maintaining that "[w]e are attached to no sect; neither the authority of Averroes, nor Alexander, nor even Aristotle himself moves us."[78] For Taurellus, "[i]f you are a Christian, it is necessary to admit that Aristotle often erred" and therefore to recognize that blindly following his authority leads away from Christian philosophy.[79]

The differing views of Taurellus and Cesalpino partly stemmed from different visions of what natural philosophy should do. For Taurellus, natural reason was paired with religious truth. In his work on the eternity of the universe, Taurellus attacked not just Cesalpino but a number of thinkers coming from Catholic lands: Francesco Piccolomini, Francisco Vallès, and the Coimbrans. In a separate diatribe he dismissed Piccolomini's view that God is an "assisting form."[80] Taurellus contended that reason, when properly applied, should take into account metaphysical and physical accounts in order to demonstrate truths about the nature of God and his creation. In his eyes, the God of religion must be identical to the God of philosophy, and part of philosophy's purpose should be to investigate providence, something he found missing in Cesalpino's writings. The confessional divide perhaps heightened Taurellus's rhetoric, as separating philosophy from theology meant endorsing the "Satanic" belief in "double truth," which supposedly circulated among Catholics.

Views similar to Taurellus's on the dangers of impious philosophy can be found in the works of some of his Protestant contemporaries. For example, the Calvinist philosopher Bartholomew Keckermann (1571–1609) believed in the harmony of theology and philosophy and described Pomponazzi as impious.[81] Eliminating "double truth" was one of the primary goals of Otto Casmann's natural philosophy, which he envisioned as a means to obtain a correct understanding of God's creation. Casmann (1562–1607), a Calvinist who taught in Steinfurt, shared Taurellus's misgivings about Aristotle and his unidentified followers who held "that the world and its sea are eternal is true in philosophy but false in theology."[82] In his 1596 treatise on the sea and tides, Casmann expounded on the various interpretations of philosophy that put it at odds with theology. He maintained that the unity of the two fields is binding not just for Christians but also for pagans, such as Aristotle. He wrote, "If Aristotle or some other pagan babbled pronunciations contrary to the theosophical oracles, these would not be philosophical but rather warts and manure."[83] For Casmann, Aristotelian warts and dung exist: numerous followers of Aristotle deviate from true philosophy based on divine wisdom.

CONCLUSION

By the end of the sixteenth century, Aristotelianism had divided into two main groups. A group, composed mainly of lay university professors and physicians in Italy, followed a path similar to Pomponazzi's, arguing that they were using natural principles or following Aristotle's opinion. Others, however, were worried about the moral implications of natural philosophy. The Reformation and the Catholic Church's reaction to the growing sentiment that it needed to reform shaped philosophical paths. For some Protestants, Italians provided an example of impious natural philosophy. For Catholic ecclesiastical authorities, Thomas's synthesis and the idea that philosophy should be a handmaiden to theology directed the study of philosophy so that interpretations of Aristotle conformed as much as possible to Catholic dogma. Jesuits adopted strategies to minimize the influence of Averroes and the versions of natural philosophy that were independent of theology. For them, the goal of philosophical instruction was moral. Arguments that made the mortality of the soul seem probable or convincing weakened the ethical and religious ends of their pedagogy. With these justifications, Jesuits gave new life to the "Apostolici regiminis" of the Fifth Lateran Council. Before the Council of Trent, this decree had little effect, but, during the second half of the sixteenth century, numerous writers cited the decree and even pointed to authors who violated it.

Catholic ecclesiastical authorities, however, had less concern over natural philosophy than they did for Lutheranism, for other forms of Protestantism, and for sorcery. The Holy Office, newly established to root out heresy, led few investigations into natural philosophers. Between the processes against Pomponazzi and Cremonini, it is possible that only one natural philosopher, Girolamo Borro, was investigated for teaching the mortality of the soul. In the inquisitorial proceedings, he cited Tommaso de Vio as evidence that Thomas's view was incorrect. This charge was one of many against Borro. Inquisitors charged him with Lutheranism and the possession of forbidden books as well. In the end, Borro's powerful patrons intervened. First Duke Francesco de' Medici and then Pope Gregory XIII obtained his release.[84] In another case of prosecution, this time against a suspected Calvinist in Venice named Teofilo Panarelli, the accused's supporters maintained that he denied the existence of purgatory "as a philosopher," thereby suggesting that philosophical considerations of erroneous positions were tolerated but Protestantism was not.[85] Pomponazzi's works escaped being placed on the index of prohibited books for many years. Only in 1576, years after their publication, did Catholic authorities censure his *On Incantations* and *On Fate* by placing them on the index. His *On the Immortality of the Soul* never appeared on the index.[86]

The freedom of these Aristotelians should not be attributed to Venetian liberty that extended to Padua.[87] Although Venice maintained privileges and demanded independence from Rome, the state curtailed intellectual freedom. Venice issued the earliest index of prohibited books in 1543.[88] Inquisitors investigated and executed suspected practitioners of black magic and suspected Protestants.[89] Targets of these investigations included famed physicians such as Gabriele Falloppia and Ulisse Aldrovandi, both of whom were absolved after being accused of being Protestants.[90] Perhaps the two most famous intellectuals who were victims of late sixteenth-century inquisitions, Tommaso Campanella and Giordano Bruno, were denounced in Padua and Venice.[91] Moreover, following the "via Aristotelis" was not an approach contained by the borders of Venice's empire but rather natural philosophers throughout the Italian peninsula adhered to it.

By 1600, both dominant forms of Aristotelianism—those of lay professors and those of Jesuits and other religiously minded writers—were legitimate and licit, even if they were at odds. That these were seen as the best or only interpretations of Aristotle appeased many, as commentators continued to write and lecture on Aristotle in secular universities and Jesuit colleges. Yet for others they were unsatisfying. Aristotelianism's impious character or supposed lack of freedom of thought prompted a number of attacks from those who wished to supplant Aristotelian thought with a more pious version of natural philosophy.

CHAPTER SIX

Learned Anti-Aristotelianism

Instruction in natural philosophy during the sixteenth century was overwhelmingly Aristotelian, in universities, in newly founded Jesuit colleges, and in other schools and religious institutions throughout Catholic and Protestant Europe. More commentaries on Aristotle were written between 1500 and 1650 than in the previous millennium.[1] During these years, printing houses published hundreds of editions of Aristotle's philosophical works in both Greek and Latin.[2] Yet alternative versions of natural philosophy increasingly competed with the traditional. These new philosophies questioned fundamental propositions of Aristotelianism. Some reformulated matter theory. For example Bernardino Telesio and Girolamo Cardano revised the number of elements or prime qualities that explained the physical world. Others questioned the application of syllogistic logic, following Petrus Ramus; or, basing themselves on Platonic or Hermetic visions of the universe, as Francesco Patrizi did, they rejected Aristotle's conception of the divine.

These promoters of alternatives to Aristotelianism presented themselves in opposition to traditional ways of conceiving nature or of teaching philosophy. The needs of university teaching, the values of Renaissance humanism, and the ideal of *prisca theologia* shaped the character of their opposition. Their concerns with Aristotle's natural philosophy went beyond disagreements about matter theory, dialectic, and philosophic authority. Religion counted among their motivations for dismissing Aristotle. To many, Pomponazzi's and Porzio's interpretation of Aristotle suggested the incompatibility of Christianity with Peripatetic thought. Simultaneously the promotion of ancient rhetorical traditions and the persistence of Neoplatonism offered material for devising new explanations of the universe and for attacking Aristotle's.

In the years that Italian university professors and Jesuits were honing their versions of Aristotelianism, authors, who promoted the field of rhetoric and endorsed Cicero, attacked Aristotelians much as they had done in the previous centuries. Despite continuity with earlier polemics, the specifics of their critiques changed just as Aristotelianism changed. Sixteenth-century polemicists confronted Pomponazzi and others who separated Aristotle's views from theology because of their supposed effect on the morals of students and youths, just as Jesuits had done. For example, in 1546 Paolo Giovio, who had studied with Pomponazzi at Bologna, described the attempt to prove the mortality of the soul according to Aristotle's view as being so nefarious that "nothing more pestilential could be induced that would corrupt youth and dissolve the discipline of Christian life."[3] The vehemence of Giovio's words is particularly striking given that they were written by a Catholic only decades after the advent of Protestantism. The danger Pomponazzi posed apparently outweighed Luther, for Giovio.

Giovio pinpointed Pomponazzi as the cause of immorality, but others criticized Aristotelians more generally. Averroes and Averroists remained a target, as did the alleged indiscriminate application of logic and poor Latin prose found in the work of Aristotelians. The writers of anti-Scholastic invectives of the previous centuries provided much of the inspiration for attacks on Aristotelianism.

HELLENIZING AND MEDICINE

New contexts for humanistic invectives arose in the sixteenth century. As the ideals of humanism permeated universities, several scholars within universities castigated what they saw as impure or outdated versions of Aristotelianism. The Hellenizing tendency of university professors inspired critiques of curricula based on Arabic-writing authors and promoted ancient writings as the greatest source of both wisdom and technical knowledge. As a result, these scholars attacked Averroes and those influenced by him, not just on linguistic grounds but also for deviating from religious orthodoxy as well.

During the sixteenth century, attitudes toward the teaching of medicine reflected growing interest in Greek texts and the privileging of the authors of antiquity. Throughout the late Middle Ages and Renaissance, Italian universities used Avicenna's *Canon* as a central instructional text for medicine.[4] The *Canon* was translated from Arabic into Latin in the twelfth century and so suffered from the perceived linguistic and conceptual flaws that some humanists also had attributed to Averroes' work. In Ferrara, Nicolò Leoniceno (1428–1525) reacted against the traditional curricula based largely on Avicenna and contended that the Arabico-Latin tradition should be entirely replaced by Greek authorities. To remedy what he considered to be the dire state of medical instruction, Leoniceno collected manuscripts and made new translations of Galen. Others at Ferrara, such as Giovanni Manardi (1462–1536), followed this tradition, advocating the use of Greek among physicians to avoid terminological confusion. Similarly, Antonio Musa Brasavola (1500–1555) embraced Galen as an authority, making an index of the Galenic corpus and promoting Galen's commentaries on Hippocrates' works such as *Regimen in Acute Diseases, Epidemics,* and the *Aphorisms.*[5] A large part of the appeal of Hippocrates and Galen was based on their age of their writings. Older texts, temporally closer to the world's creation, presumably reflected purer and more accurate understandings of nature and the human condition.

The old age of the texts was not the only argument in favor of the Greek medical writings. The French physician Symphorien Champier argued that physicians should prescribe European *materia medica.* He maintained that the medicines found in Arabic authors' books were unsuitable for the French. Even though his argument could have been made on the basis that medicines should be local in order to match the climatically conditioned temperaments of the patients, Champier employed religion and ethnicity as key distinguishing points. He wrote, "We are Christians, not followers of Muhammad; French, not Arabs."[6] He continued, "Barbarians write for barbarians, and Arabs for Arabs; French therefore for French."[7] His chauvinistic language reveals his distaste not just for the Arabic-writing herbalists but for all followers of Islam, which he called "the dirtiest and most nefarious religion."[8] His low opinion of the use of Arabic medical texts reflected an ideology that saw French Christians as medically and culturally distinct from Arab Muslims.

These medical humanists mostly weighed the values of Galen, Hippocrates, and Avicenna, but their attitudes affected conceptions of Aristotle and Averroes, partly because of the importance of Aristotle to the foundations of physiology and partly because of Averroes' medical writings. Champier criticized Averroes, whom he saw as impious as well as a poor interpreter of Aristotle. Averroes, in his eyes, however, was not a Muslim strictly speaking, because "he denied the teachings of Moses and Christ and deserted his Muhammad."[9] Rather, according to Champier, "the impious Averroes" did not follow "the law of the infidels" but was a "pagan" and a "boor" whose theory of the soul did not follow Aristotle's mind and deviated from Christianity because it eliminated the possibility of resurrection.[10]

Even though Champier attacked Avicenna's reduction of miracles to natural causes, which Pomponazzi appropriated, Averroes fared worse than Avicenna in the minds of many Renaissance medical authors.[11] For example, Gesner pointed to Averroes' allegedly poor moral character, marked by envy for Avicenna.[12] Many at this time believed that Avicenna and Averroes were contemporaries and that both had lived on the Iberian Peninsula, despite Avicenna's having lived a century before Averroes and in the eastern part of the Islamic world rather than in al-Andalus. A mistranslation of the Arabic falsely made Avicenna a prince in the minds of many Renaissance authors and artists. That is, they saw him as an actual crowned secular ruler not just a prince of philosophers.[13] The most extreme versions recounted Averroes' envy as being so great that he murdered Avicenna by poisoning him.[14] The inconsistent situation emerged in which minor errors about antiquity and Latin usage engendered controversy and reproach, but gross misunderstandings about Arabic writing and its past were either embraced or went unchecked.

CRITIQUES OF HUMANISM

During the sixteenth century, the instantiated positions of humanists engendered hostility, as their critics equated their inflexible rules on language with pedantry. Much as Erasmus had ironically praised the folly of priests, Scholastics, and rulers, satirical works mocked the pompous learnedness of the ascendant class of intellectuals. Ortensio Lando, for example, lampooned erudition in his 1544 *Paradoxes, That Is, Sentences beyond Common Sense*, a work that argued that it was better to be ugly than beautiful, ignorant than learned, imprisoned than free, and poor than rich. In some of these paradoxes, Lando undermined the authority of ancients, for example, by calling Cicero ignorant of philosophy and rhetoric. As an equal opportunist, Lando chipped away not just at the most esteemed ancient authority for eloquence and rhetoric but for natural philosophy as well, dedicating two separate paradoxes to dismantling Aristotle.[15]

Lando's approach to mocking Aristotle was twofold: he cast doubt on the authenticity of his writings and called into question Aristotle's personal morality. Using a tactic that derived from Gianfrancesco Pico della Mirandola and that would later be used by Petrus Ramus, Francesco Patrizi, and Pierre Gassendi, Lando maintained that the books that are now ascribed to Aristotle were in fact not his. The self-mocking nature of the *Paradoxes* renders it difficult to assess Lando's sincerity——he later wrote a rebuttal of his own work. Regardless of the book's joking nature, he put forth textual evidence for the spuriousness of Aristotle's writings. Citing Strabo and Plutarch, he repeated the story that Aristotle's works were lost in the years after his death. Strabo wrote that after Theophrastus's death, Aristotle's books were largely unavailable to his followers. After the corpus had fallen apart from neglect, Apellicon of Teos, a man more interested in book collecting than philosophy, copied Aristotle's works, but poorly. As a result, the remnants of the Aristotelian corpus abound with errors. Plutarch offered a briefer version of Strabo's story, which became prime evidence that Aristotle's writings were not genuine and therefore lack authority.[16] Lando also cited Cicero and Simplicius. Both wrote that Aristotle penned dialogues and therefore provided evidence that the extant Aristotelian corpus differed greatly from his actual production.[17]

Lando's second anti-Aristotelian paradox attacked Aristotle's doctrine, person, and his adherents. Lando set forth the immodest thesis that, "Aristotle was not only ignorant but was also the most evil man of his age."[18] In this paradox, he maligned not just Aristotle but others closely linked to the Aristotelian tradition. He called Averroes a liar and barbarian. Lando then evoked Simone Porzio, whom he rhetorically asked, "With your most beautiful genius have you not penetrated so far that you have known that this so familiar Aristotle was a steer?"[19] Moving from the bizarre to the philosophical, Lando listed Aristotle's supposedly immoral doctrines, including the mortality of the soul and the possibility of blessedness for the living.[20]

A number of Lando's contemporaries concluded that Aristotle denied the immortality of the soul, yet his attempt at biography stands out even among the hostile discussions of his era. Lando maintained that Aristotle held expertise in poisons and used this knowledge to help murder Alexander the Great. Although seemingly preposterous, Plutarch and Arrian both wrote of this plot, albeit without endorsing its truth.[21] More evidence of this crime, according to Lando, is found in the fact that Aristotle's disciple Callisthenes killed himself by leaping out of a window.[22] To complete the circle, Lando, repeated the story, found in Diogenes Laertius, that Aristotle killed himself by jumping into the Euripus Strait, after desperately fleeing from Athens.[23] Either ignoring or rejecting Bruni's biography, Lando mustered bits of Diogenes' biography to demonstrate further Aristotle's immorality, as evidenced by his falling in love with a "shameless whore."[24] Even Diogenes' physical description of Aristotle suggested, to Lando, effeminacy and therefore somehow a "lack of care about divine matters."[25]

Lando's work was not meant to be serious. As a result of the satirical nature of his writings, the careful consideration of ancient sources, as is found in say Bruni's biography, is absent. Yet most of the material of his polemic derived from ancient sources and resonated with others who wished to subvert Aristotelianism. Lando, however, was uncharacteristic for his time, as many who wished to pull down Aristotle endorsed other ancient authorities such as Plato or Cicero. Although perhaps effective, the satirical tract was exceptional among opponents of Aristotle.

INVECTIVE AND PLATONISM

Lando used satire combined with a selective reading of ancient sources to attack Aristotle during the same years when more traditional humanistic castigations of Aristotle also circulated. The tradition of attacking Aristotle among promoters of rhetoric and dialect shows continuity with earlier polemicists such as Juan Luis Vives and Lorenzo Valla. Mario Nizolio (1498–1566), a proponent of ancient oratorical traditions and author of works on Cicero's rhetoric, divided Aristotelians into sects of "Albertists, Thomists, Scotists, Averroists, and Avicennists," finding little positive in any of these thinkers, "whom no learned person considers worthy of either the title Philosopher or learned man."[26] A critic of the use of syllogism, Nizolio found special rage for Averroes, who allegedly understood Porphyry's logic so poorly that he supposedly nonsensically thought, "Adam is the genus of man."[27] Echoing Vives and citing Averroes' prologue to his commentary on *Physics*, Nizolio used Averroes as an emblem of "pseudo-philosophers," who are unwilling to deny Aristotelian principles. Even though Averroes' alleged slavishness, barbarisms, and faulty use of dialectic condemned his thought, Nizolio agreed with his general understanding of Aristotelian philosophy, believing that Aristotle had opposed "the authority of Plato and the Old Greeks" by separating theology from the study of nature.[28] To be sure, Nizolio believed this separation was flawed, writing, "Aristotle had greatly erred by removing theology from physics," as had Peripatetics in general.[29] Accordingly Nizolio saw Aristotle's philosophy as distinct from theology and contrary to Plato's.

Contrary to Nizolio's strong preference for Plato over Aristotle, the desire to reconcile Plato and Aristotle had not died with Giovanni Pico della Mirandola but remained central to a number of sixteenth-century humanists. The Spanish humanist Sebastián Fox Morcillo (circa 1526–1560) argued that on many issues Plato and Aristotle agreed with each other as well as with Christian dogma.[30] The physician Jean Fernel (1497–1558), who posited seeds as the cause of disease, believed that Plato influenced Aristotle's tracing of the origin of the universe back to God, despite differences in their proofs of God's creation of the world. Aristotle and Plato agreed in other ways. According to Fernel, Aristotle believed in a double world, much as Plato did. One world is bodily and formed of the elements; the other, incorporeal and divine, is "the pattern of this lower transitory world."[31] Similar to Morcillo and Fernel, Francesco de' Vieri wrote a treatise in the Italian vernacular in which he approvingly discussed Plato's theology. He found that Aristotle and Plato agreed on fifteen key points, including that the soul is immortal, that humans experience beatitude or punishment after death, and that philosophy is a gift of God that brings knowledge of both the divine and human, even if our soul in relation to the divine "is like the eye of a bat is to the light of the sun."[32] Like Fernel, de' Vieri thought Aristotle believed that the universe is real in two ways: as material particulars and as a "copy of God."[33] Therefore, de' Vieri interpreted Aristotle as pious, differing greatly from Porzio and Pomponazzi, among others.

In his *On the Comparison of Plato and Aristotle*, Jacques Charpentier, a professor at Paris, saw Aristotle and Plato not necessarily in agreement but complementary. His analysis is reminiscent of ancient Greek Neoplatonists. Aristotle was superior to Plato in physics, but Plato held the advantage in metaphysics.[34] Yet Charpentier, basing himself on *De mundo,* a work that is now considered spurious and that he himself had doubts about, argued that Aristotle's theology "was neither vulgar nor very different from that of Platonists."[35] These attempts at reconciling Plato and Aristotle, however, met a variety of opponents and was rendered more difficult by investigations into the history of theology, as an increasing number of scholars became aware that early patristic writers not only thought Plato and Aristotle disagreed but also that Aristotle erred on important theological issues.

THE REVIVAL OF THE CHURCH FATHERS

The Hellenizing tendency of sixteenth-century scholars and theologians reshaped conceptions of Christianity, partially as a result of growing interest in early Greek patristic writings. Although some of their works had been known and studied throughout the Middle Ages, late medieval scholars for the most part shifted their studies away from the exegesis of scripture toward formal theological investigations that depended on creating systematic theological renderings of Church doctrine. During this time period, Peter Lombard's *Sentences* became the central text for theological commentary rather than the Bible itself.[36] This shift, coupled with the fact that few scholars read Greek in the Latin West before the fifteenth century, meant that Greek patristic writings had few direct influences on late-medieval conceptions of Christianity.

By the first decades of the fifteenth century, as part of the rise in interest in antiquity, scholars began to give greater attention to the early Christian writers. For example, the Florentine humanist, Ambrogio Traversari, who translated Diogenes Laertius into Latin and therefore was key to developments in the biographical tradition of Aristotle, turned his attention to the study of the early Latin Church Fathers Tertullian, Lactantius, and Jerome and translated Basil and Chrysostom from Greek into Latin.[37] The arrival of Greek scholars, such as Pletho, Trapezuntius, and Bessarion, at the ecumenical councils further highlighted the importance of not just literature and philosophy written in the Greek language in general but also theology more specifically. Trapezuntius, for example, translated the patristic authors Eusebius and Chrysostom into Latin.

Increasingly, these early Christian writers gained authority because of their chronological proximity to the establishment of Christianity. They appealed to humanists who looked negatively upon medieval theology because of its reliance on Aristotelian logic; and, their antiquity recommended them to those who sought out the *prisca theologia*. Greek Church Fathers offered solutions to interpreting scripture and, as both Bessarion and Pomponazzi knew, considered at least parts of Aristotle's philosophy alien to Christian belief. By the early years of the sixteenth century, the scholarship of Erasmus and Lefèvre proliferated translations and interpretations of early patristic writing. Church reformers, both Lutheran, such as Philip Melanchthon, and Catholic, such as Gasparo Contarini, endorsed the study of these early Christians.[38]

By the sixteenth century it became common to cite patristic authors as evidence for Aristotle's impiety. In particular, just as Bonaventure and Bessarion had known, some of these patristic authors interpreted Aristotle as denying God's providence for sublunary particulars, that is, individual substances whose place is between the center of the earth and the moon. The issue of providence in natural philosophy grew in importance during the sixteenth century. Melanchthon's reformulation of Aristotelian natural philosophy attempted to make the concept of providence a foundation for investigations into nature. Accordingly, Lutheran natural philosophers saw providence as playing a significant role in explanations of a variety of natural phenomena, including the nature of living things, meteorological phenomena, and astronomy.[39] Professors at Wittenberg and other Lutheran universities wrote textbooks on natural philosophy that integrated religious concepts with discussions of nature.[40] Calvinist and Catholic theologians and philosophers soon followed similar paths, emphasizing the importance of divine providence. By the beginning of the seventeenth century, some of the most widely available works on natural philosophy, such as Jean Bodin's *Theater of Universal Nature*, repeatedly invoked divine providence.[41]

During the middle of the sixteenth century, a revived interest in the Church Fathers' views of Greek philosophy undermined attempts to reconcile Aristotle with the Christian doctrine of special divine providence. In the 1550s, Guillaume

Postel, accused of impiety by both Calvinists and Catholics, brought forth the teachings of Justin Martyr in an effort to discredit Aristotle's followers or in his words "in order to overturn the authority of Aristotle, wherever it is contrary either to divine authority or reason."[42] According to Postel, his work (and Justin Martyr's as well) was needed because since antiquity "cohorts of atheists" have used Aristotle's ambiguity to prove that there is no providence for particulars.[43] One member of this group was Averroes, whom Postel described as "the greatest enemy of providence."[44] Postel believed developers of Scholastic doctrine in Paris, following Averroes, "strip these [divine] truths from Aristotle" while "he [i.e., Aristotle] himself lies completely neglected."[45] In his eyes, Scholastics ignore the real historical Aristotle, distorting his writings and siding with Averroes.

Postel's unorthodoxy made him unattractive for many, yet the association of Aristotle and Averroes with the denial of providence became more widespread during the second half of the sixteenth century. Pirro Ligorio's reaction to the 1570 earthquake in Ferrara perhaps helps illustrate the shifting weight placed on concerns over Averroes' position on providence. At the time of the earthquake Ligorio was an antiquarian employed in the House of Este's court. Dismissing papal accusations that the cause of the earthquake was divine punishment against the Ferrarese for accepting the Jews, whom Pius V had recently expelled from the Papal States, Ligorio identified another cause for these temblors. The divine punishment was not for protecting Jewish refugees but rather aimed at "those who are so bold that . . . they deny God's providence, having been deceived by Aristotle, Galen, Averroes, Alexander of Aphrodisias, and other Peripatetics."[46] According to Ligorio's scenario, Aristotle and his followers denied providence by dismissing the role of divine intervention in the sublunary world. The scholars, who followed the Peripatetic view that there was no divine providence in the sublunary sphere, angered God to such a degree that, in order to demonstrate his control of natural particulars, God destroyed Ferrara.

Ligorio did not name his source for this interpretation of Aristotle and Averroes as denying providence, but it is likely that it stemmed from those influenced by interpretations of the Church Fathers. For example, Jacopo Mazzoni, in his 1576 comparison of Plato and Aristotle, after arguing that Plato's thought conformed to Catholicism based on the authority of patristic writings, concluded that Plato's views on providence agree with scripture, but Averroes' belief that matter is uncreated leads to potential ambiguities on the subject.[47] In a similar vein, de' Vieri noted that Plato's doctrines were well received by the "Greek Doctors" of the Church.[48] Among Plato's doctrines that de' Vieri believed conformed to Christianity but not to Aristotle's thought was that "God has providence over

all things, in particular over man."[49] Related doctrines, allegedly both Platonic
and Christian but not Aristotelian, included that God answers prayers, that God
uses angels to take care of humans, and that God protects the weak and punishes
the proud.[50] De' Vieri differed from Postel in that he thought that Aristotle himself put forth numerous doctrines that were contrary to Christianity, rather than
blaming Scholastics who allegedly capitalized on his ambiguity and distorted his
views.

PETRUS RAMUS AND REFORMS OF ARISTOTLE

De' Vieri was not the first to put forth a list of the doctrines where Aristotle deviated from Christian dogma. Gianfrancesco Pico della Mirandola, for example,
attempted to undermine the legitimacy of Aristotle's thought and its conformity
to religious truth in his skeptically inspired defense of revealed truth. Other reformers of natural philosophy devised replacements for Aristotelianism, instead
of promoting skepticism as Gianfrancesco Pico had done. They often cited the
impious nature of Aristotle's philosophy as a motivation, shifting the epithet of
"impious" from Pomponazzi and Averroes to Aristotle himself. Even when Averroes remained a target, these reformers at times implicitly agreed with the interpretations of Aristotle made by Averroes or sixteenth-century Italian natural
philosophers, such as Pomponazzi, Simone Porzio, or Francesco Vimercati. In
their eyes, Pomponazzi had correctly interpreted that Aristotle believed in the
mortality of the soul and Averroes accurately understood Aristotle's limitation
of God's direct influence to celestial bodies. These reformers, however, thought
that Pomponazzi and Averroes incorrectly held that natural philosophy should
be independent of theology or considerations of the divine. Consequently, they
concluded that Aristotle should be thrown out, because his thought was irreconcilable with the higher truths of religion.

Petrus Ramus was a convert to Protestantism who spent much of his career
devising systems of dialectic and rhetoric designed to replace Aristotelian logic.
These innovations provoked polemics between him and other scholars working in
France, including Charpentier and Vimercati.[51] These polemics were heated, yet
his life ended even more violently as a Huguenot victim in the St. Bartholomew's
Day Massacre in 1572.

Ramus's motivation for devising a new system of dialectic was largely peda-
gogical. His proposed curriculum shortened standard ones of his time, because
he came up with innovations that would allow a greater number of students to
learn the rudiments of higher education that ideally would propel them to higher
social positions. In order to shorten the curriculum, Ramus chose to eliminate

teaching Aristotle's texts and to focus on dialectic and rhetoric.[52] This decision provoked much controversy and eventually led François I to ban Ramus from teaching philosophy. As a result of this ban, Ramus began to teach mathematics. After François I's death, Ramus's right to teach philosophy was restored, and in 1551 he took up the appointment of royal lecturer in eloquence and philosophy.[53]

A hostility to Aristotelians marks Ramus's early writings. In *Aristotelicae animadversiones* he questioned, on the basis of the Strabo's and Plutarch's authority, the integrity of the Aristotelian corpus, just as Lando had done in his satirical paradoxes. According to Ramus, the texts of Aristotle's logical works are corrupt yet still held a monopoly over the practice of philosophy.[54] Ramus's associate and collaborator, Omer Talon, repeated Ramus's view in his commentary on Porphyry's dialectic.[55] Ramus's hostility toward Aristotelian logic softened over the course of his career or was at least less apparent.[56] Nevertheless, in his posthumous *Commentaries on the Christian Religion*, first printed in 1576, he condemned Aristotle as impious on a number of grounds: Aristotle's logic denies that future contingents can be true, thereby limiting God's knowledge; and, despite his broad endorsement of teleology, exemplified by the dictum "nature does nothing in vain," Aristotle "composed his physics and metaphysics against the divine creation and administration of the world."[57]

The subsequent influence of Aristotle steered awry both ancient and contemporary figures. According to Ramus, "Epicurus, Mani, and Celestius, stained by this philosophy of Aristotle, maligned the foreknowledge and total providence of the divine ordering."[58] Ramus's assessment of Aristotle's influence on these figures is unlikely, but his selection was significant. Augustine had represented Manicheans and Celestius as heretics. Yet Epicurus was perhaps more important as a symbol in the sixteenth century. Epicurus had long been associated with the denial of providence and hence impiety.[59] Lutherans used his name as a synecdoche for this position.[60] Thus Ramus's linking of Epicurus to Aristotle evoked a series of longstanding denunciations of illegitimate philosophy. More specifically, Ramus described Epicurus and Aristotle as "impious and even atheist" because of their belief that contemplation does not continue beyond life.[61] The use of the word "atheist" at the time typically specified a range of heretical positions, which could include the explicit denial of the existence of God.[62] The early modern term overlaps with the modern one. Ramus understood Aristotle's denial of providence and God's foreknowledge of future contingents as denying the existence of an omnipotent divinity. For Ramus, a God whose power was limited was not truly God.

These positions were not the only reasons for identifying Aristotle as impious, according to Ramus. Implicitly siding with Pomponazzi and Porzio, Ramus contended that Aristotle "mocked the immortality of the soul," and therefore implicitly also derided such doctrines as justice in the afterlife, the remission of sins, and bodily resurrection.[63] Aristotle's belief that human happiness emerges from living well rather than from rewards in the afterlife further defined him as impious.[64] Ramus ignored the numerous interpretations of Aristotle as believing in the immortality of the soul and sided with those who tried to follow the "Peripatetic path" by determining Aristotle's true view without regard to Christian religion. By ignoring the tradition that reconciled Aristotle with Christianity, Ramus not only tried to damage the Scholastic teachings based on Aristotle but also implicitly accepted the readings of secular Italians.

FRANCESCO PATRIZI

During the years that Ramus tried to reform dialectical pedagogy, Francesco Patrizi found in Platonism and Hermeticism the bases for understanding nature and the universe. Patrizi also put forth one of the most sustained critiques of Aristotelian natural philosophy of the sixteenth century and one of the most philologically sophisticated of human history. From Dalmatia, then part of the Venetian Empire, Patrizi studied at Padua before becoming attached to the Ferrarese court and eventually the court of Pope Clement VIII. Like Ramus, Patrizi offered a textual interrogation of the authenticity of the Aristotelian corpus. His *Peripatetic Discussions* (1571) elaborately discussed numerous ancient texts, casting doubt on whether Aristotle actually wrote the works now attributed to him.

In addition to the skeptical take on the provenance of the Aristotelian corpus, Patrizi put forth an equally elaborate history of Aristotelianism. Starting with the time of Aristotle, Patrizi divided Aristotelianism into ten distinct periods. The final period was Patrizi's era, when scholars, including Achillini, Nifo, Pomponazzi, Porzio, and Vimercati, began reading Greek and "took up the task of explaining Aristotle."[65] He contrasted those of his age to medieval scholars who "yoked Aristotelian philosophy to Christian philosophy;" during the Middle Ages, "nothing could be thought to stand in theology, unless it was established on Aristotelian foundations."[66] According to Patrizi's historical analysis, because Aristotle's writings contradict themselves so often, discord arose among Latin medieval philosophers and they "divided into various and multiple sects of nominalists and realists, the number of which is uncountable."[67] Therefore, he implicitly described the Latin medieval philosophers as not accurately following Aristotle but rather combining him with theology. The ambiguity and textual corruption of Aristotle's corpus and the impossibility of their task of creating consistency within Aristotle's thought divided them into numerous squabbling groups.

Compared with his assessment of the Latin philosophers, Patrizi's consideration of Averroes was favorable. He noted that "[m]any think Averroes' view is the most Aristotelian of all the interpreters," because "Averroes embraced Alexander of Aphrodisias's view in nearly all cases."[68] For that reason, since the introduction of Greek studies in Italy at the beginning of the sixteenth century, the Italian "philosophers of the schools mixed the Greek interpreters to Averroes and the Latins."[69] Despite the continuing habit of writing "eternal disputations," as a result of their greater adherence to Greek texts, Boccadiferro, Porzio, Genua, Maggio, and Vimercati were able to "discover a purer exposition of Aristotelian words."[70] This purer exposition, however, meant only a more accurate interpretation of Aristotelian texts, not a better understanding of nature and God.

For understanding the truths of the universe, Patrizi thought Platonic and Hermetic texts offered greater insight. He believed they were older and confirmed Christian truths. Stating that Aristotelianism and Christianity were combined only in the late Middle Ages, Patrizi attempted to show the impiety of Aristotelianism in the 1591 *New Philosophy of the Universe*. In his habitually exhaustive manner, Patrizi cited a number of patristic sources in his discussion of Aristotle's doctrinal missteps. Like Bessarion and Pletho during the previous century, he argued that Platonic philosophy corresponded better to Catholic theology than Aristotelian thought did. In the preface dedicated to the "future" Pope Gregory XIV, he began with the rhetorical question: "Why are only those parts of Aristotle's philosophy read, which are most injurious to God and the Church?"[71] Contending that for the previous four hundred years "Scholastic theologians" have used "Aristotelian impieties for the foundations of faith," he listed Dionysius the Areopagite, Justin Martyr, Clement of Alexandria, Arnobius, Lactantius, Cyril, Basil, Eusebius, Theodoretus, Augustine, and Ambrosius as proponents of Platonism and opponents of Aristotle.[72]

One of the doctrines Patrizi found most problematic was the absence of providence in the sublunary realm. Patrizi noted that Origen held that "Aristotle was worse than Epicurus, because he was impious in divine providence."[73] Clement of Alexandria complained that for Aristotle "providence extended only to the moon."[74] Accordingly, Patrizi placed the Church Fathers in agreement with Averroes' interpretation that Aristotle limited providence, yet he disagreed with that Aristotelian position. Those in "the Parisian school," that is, medieval Scholastics, he wrote, "explained the universe, imitating Averroes' commentaries," thereby mixing "the most sordid" into the Catholic faith.[75] For Patrizi, the alternative was Plato's thought, which he believed corresponded to Christian theology. Ecclesiastical authorities had other ideas, bringing Patrizi to the Inquisition, where the censor determined his Platonism undermined the Scholastic principles that formed the basis for Catholic doctrine.[76] Patrizi abjured, revised his writings, and stopped publishing on Platonic philosophy.[77]

Since the time of Bessarion and Ficino, if not before, it had been the aim of a number of scholars to use Plato's philosophy as a basis for Christian dogma. In the age of the Council of Trent and the reaffirmation of Aristotelian philosophy by Jesuits and Dominicans, two orders intent on enforcing orthodoxy, Platonism became just one more source for heresy. In the eyes of some, because of the apparent similarities between his thought and Christianity, Plato was actually more dangerous than Aristotle, because the correspondences were deceptively similar.

Roberto Bellarmino (1542–1621), a Jesuit leader whom Clement VIII (1536–1605) named cardinal, is best known for his 1616 warnings to Galileo not to teach heliocentric cosmology.[78] His defense of Aristotelianism was long standing. He explained to Clement VIII that "a more certain danger emanates from Plato than from Aristotle," because "Plato is closer to Catholic doctrine than Aristotle is."[79] From this perceived affinity, Plato seduces believers into vice and heresy. Fearing "heretic authors more than pagan writers," Bellarmino gave the example of Origen, who was "overcome by the sweetness of Plato and corrupted by his eloquent allure."[80] As a result, Origen wrote books full of errors, and both he and Plato were condemned during Justinian I's rule.[81] For Bellarmino, a consideration of ancient history showed not that the Church Fathers preferred Plato but rather demonstrated the dangers of Platonism and its close relation to heresy, revealing that some patristic authors were heretics.

The study of the past is a tool that can provide evidence for multiple camps. Aristotelians mustered historical arguments that countered Patrizi's and others' assertions that the Church Fathers found Aristotle impious. The Jesuit Pedro de Fonseca contended that, if we possessed all of Aristotle's books listed by Diogenes Laertius, the Platonists would not think Plato's philosophy was more pious than Aristotle's. In his view, Augustine and Cicero prized Plato over Aristotle because of Plato's eloquence rather than his doctrines.[82] Pedro Juan Nuñez echoed Bellarmino's assessment in the comments to his erudite 1594 edition of the biography of Aristotle often attributed to Ammonius. Conceding that some prefer Plato's theology, he wrote that, in *Metaphysics* Lambda, Aristotle "disputed accurately on God and on other minds, and even with clearer argument and direction than Plato, without wrapping them in fables and Pythagorean symbols."[83]

Girolamo Pontano, a professor at Rome with connections to Bellarmino and the papal curia, argued that the Church Fathers had erred in their understanding of Aristotle.[84] As Bruni and Petrarca had pointed out, few read Aristotle in the centuries after Jesus. Those who read him did not necessarily understand him well. After Aristotle's death, Pontano argued, all Peripatetics were followers of Alexander of Aphrodisias. As a result, Gregory of Nazianzus, Justin Martyr, Gregory of Nyssa, and others, who were persuaded that Aristotle believed in the mortality of the soul, were misled by the Alexandrian interpretation that, according to Pontano, dominated the ancient world during their lifetimes. In a work on the immortality of the soul printed in 1597, Pontano wrote that Alexander was wrong in his interpretation, thereby dismissing not only Pomponazzi, Porzio, and Galen but also Platonists and other anti-Aristotelian polemicists.[85] Pontano extended historical analysis, which humanist polemicists used for reinterpreting Aristotle, to the very sources that the humanists and Platonists used in their critiques. Historicism is faced with ambiguities just as philosophy is. Pontano's views held some influence, resurfacing in the Giulio Cesare La Galla's lectures given at the Gymnasium Romanum in 1618, where he asserted that true Aristotelian teachings of the soul correspond to the decrees of the Council of Vienne and the Fifth Lateran Council.[86]

The Holy Office's concerns went beyond Platonic threats to Aristotelianism. They included novel natural philosophies as well. Jesuits, among others, were wary of the introduction of new theories that were not backed by ancient ones, largely because unexpected implications about Christian doctrine might arise. The promoter of a novel natural philosophy based on the actions of the qualities hot and cold, Bernardino Telesio (1509–1588) perhaps unexpectedly created more controversy than he anticipated. By making these two sensible qualities universal, Telesio theorized that the heavens are corruptible, a position that, by the end of his life, was increasingly problematic because of Copernicanism's challenge to Aristotelian natural philosophy. Yet what attracted the attention of inquisitors was his justification for creating a new natural philosophy. Telesio argued that Aristotle's philosophy was contrary to scripture because of the absence of his acknowledging God's "knowledge and administration of human affairs."[87] Telesio, like so many others, thought Aristotle denied God's providence in the sublunary realm.

Telesio's work provoked defenses of Aristotle. In 1587 Giacomo Antonio Marta wrote a treatise against Telesio. Earlier he had written polemics against Porzio's *De humana mente*, where he tried to defend Aristotle from the accusations of being a materialist.[88] Turning to Telesio, Marta defended Aristotle for his nobility, virtue, and wisdom. Perhaps strangely for someone who attacked Porzio and maintained Aristotle was pious, his high esteem of Aristotle echoed Averroes', as he called Aristotle "a miracle of nature" and "a contemplator of the highest creator."[89] Marta attacked Telesio not just for his use of sensible principles to explain nature but also on account of his conviction that women contribute seed in reproduction. Telesio's belief in female seed was not novel; Galen endorsed this view, for example. Yet Marta objected because it supposedly cast doubt on Adam's responsibility for the transmission of humankind's original sin.[90]

Despite Marta's objections, inquisitors apparently ignored the question of female seed and its correspondence to the account of creation in Genesis. Yet similarities remain between Marta's ardent defense of Aristotle and ecclesiastical authorities' condemnation of Telesio. Girolamo Pallantieri, who corrected Telesio's work for censors in Padua in 1600 more than a decade after Telesio's death, upheld the utility of Aristotle for religion. Pallantieri cited Clement of Alexander as evidence that Aristotle knew about the Old Testament and reaffirmed that "Holy councils and so many of the greatest popes made the doctrine of Aristotle" orthodox by eliminating repugnant doctrines, such as the unicity or mortality of the human intellect.[91] Telesio's demand that natural philosophy should be based on sensible qualities was incorrect, according to Pallantieri, because both Aristotle and scripture demonstrate the existence of immaterial and insensible substances.[92]

CESARE CREMONINI

Bellarmino adamantly endorsed Aristotelian natural philosophy, contending that its Platonist rivals were riskier. Jesuits, however, accepted only certain versions of Aristotelianism. Whereas earlier in the century those who argued for the separation of natural philosophy from theological considerations were left largely untouched, in the 1590s and 1600s Jesuits attempted to prosecute Cesare Cremonini for his literal interpretations of Aristotle, or for teaching "in via Aristotelis" in Padua. Cremonini's encounters with Jesuits paralleled Patrizi's. They knew each other from their time in Ferrara, and Patrizi even dedicated a treatise to him.[93] Although their distant philosophical poles made them unlikely companions, their approaches to Aristotle were similar. Both wished to understand the true thought of Aristotle, rather than incorporate it into theology. Patrizi thought such reconciliation was impossible for Aristotle and listed forty-three separate points where Aristotle disagreed with the Christian religion. Whereas Cremonini considered Aristotle's philosophy separate from theology, and in a sense separate from the truth, even if it provided the standard paradigm for theories about nature.

In his *Apology for Aristotle's Words on the Quintessence of the Heavens*, 1616,

Cremonini described his method as suppositional. Although admitting that his arguments might be repugnant to the Christian faith, he believed they were proven from Aristotle's principles. To illustrate his point, he wrote that it is possible to make a Democritean demonstration, explaining effects based on the principles of atoms and void. Such a demonstration, in spite of internal coherency, will be false, according to Cremonini. The same is true for Aristotle. Using Aristotelian principles, it is possible make demonstrations that are valid according to the initial suppositions but not necessarily true. In this manner, Cremonini presented his hedging approach to Aristotle. He wrote, "I do not say that he simply reached the truth, since he erred whenever he opines against the faith."[94] Accordingly, religious dogma trumps Aristotle's arguments, contended Cremonini.

Interpretations of Cremonini have wildly differed, even during his own lifetime. Because of his unwillingness to look into the telescope of his friend and colleague Galileo Galilei on the grounds it gave him a headache, Cremonini has taken the appearance of an unbending defender of Aristotle. His self-defense from the Inquisition and the charges themself have made him emerge as proponent of libertinism. His rigorous philosophical writings paint a picture of him as staunch rationalist.[95] For some, Cremonini's unwillingness to define Aristotelianism as the unadulterated truth suggests insincerity or masked impiety, but it need not. Tommaso de Vio, Bessarion, and Patrizi, among others, have not received the same accusations for arguing Aristotle is incompatible with faith.

Whether or not Cremonini was sincere, he applied his suppositional method to the question of providence. He wrote that Aristotle had "fallen into error, since it is stated by faith and by the decrees of holy theology that God's providence is administered rightly and truly over all singulars."[96] Still, Averroes was in a sense correct, not dogmatically, but as an interpreter of Aristotle. According to Cremonini, Averroes had rendered the doctrine according to "good Aristotelian sense," although scarcely according to the truth. Yet as textual interpretation it makes good sense because Aristotle "was ignorant of true religion" and its rewards, believing that philosophy itself was its own prize.[97]

Jesuits, who were establishing a college in Padua, objected to Cremonini's literal interpretations of Aristotle, which they believed corrupted the youth. A significant part of his clash with Jesuits and the Holy Office surrounded jurisdiction and the traditional rights of Padua's university to have a local monopoly on higher education.[98] In an oration defending the rights of the Studio of Padua, presented in December of 1591, Cremonini cited Venetian laws that established the university and argued that the establishment of the Jesuit school would create discord among scholars.[99] The Jesuits, in turn, accused Cremonini of teaching that the soul is mortal and holding secret meetings with private students in which they exchanged blasphemous sentiments.[100] The Venetian government, however, saw the Jesuits as encroaching on their independence from Rome and supported Cremonini. The Venetians declared that Cremonini was a good Christian and sent a priest to testify that Cremonini did not teach the mortality of the soul.[101] Although it is impossible to know what he might have said secretly, or successfully hid from the public, his lectures promoted the position that the intellect is immortal, although not necessarily according to Aristotle.[102] In a short treatise dated to 1598 he argued against Galen's materialism, maintaining that the soul must be immaterial and cannot derive from the underlying matter.[103] Furthermore, although his relations with the Jesuits were strained, he was on good terms with members of other orders, such as the Benedictines, to whom he left a significant part of his possessions in his will.[104]

In the places where Cremonini believed Aristotle deviated from Christianity, such as the eternity of the world, he defended his teaching by noting that he was explaining Aristotle's thought. Although this proviso might appear a mere excuse, Cremonini cited statutes of the university that demanded professors "clearly explain and reveal [declarare] the text of the authors" whose texts were the subject of the lectures.[105] Cremonini's declared practice conformed to university's statutes as well to techniques that Patrizi described as belonging to the latest school of Aristotelianism, which was intent on explaining Aristotle accurately without regard to the medieval tradition of incorporating natural philosophy into theology.

In spite of Cremonini successfully defending himself from the Inquisition and preventing the Jesuits from teaching philosophy at Padua, he marks the end of that tradition, in part because of the waning influence of Aristotle, in part because of the growing unacceptability of his approach.[106] Even his contemporaries at Padua attempted to integrate the medieval theological tradition into their lectures on Aristotle. Giorgio Raguseo, for example, rued that his predecessors at Padua, such as Zabarella, Federico Pendasio, and Piccolomini neglected Latin commentators.[107] Giovanni Cottunio followed Thomas to a far greater extent than his predecessors did.[108] Many of Cremonini's successors either rejected Aristotle or Cremonini's approach to reading him.

CONCLUSION

Criticism of Aristotle and Averroes came from numerous sources throughout the sixteenth century. Humanists, purveyors of new natural philosophies, and Platonists aligned and attacked Aristotle, Averroes, and other Peripatetics for their alleged impiety. These attacks created more reaction than concession by Catholic

authorities. By the 1590s, the Holy Office adopted a more aggressive stance for determining and maintaining orthodoxy. The best known aspect of this increased
level of censorship was the Church's reaction to Copernicanism. Concerns with
heliocentricism arose from multiple causes—its apparent contradiction of biblical passages or its implicit overturning of the epistemological hierarchy of natural
philosophy and mathematics.[109] The Catholic Church's rejection of the earth's
motion also calcified the orthodoxy of Aristotelianism because of its philosophical underpinnings for geocentric cosmology. The form of Aristotelianism endorsed by ecclesiastical authorities confirmed the idea of natural philosophy as a
handmaiden to theology and strengthened the institutional support for medieval
forms of Aristotelianism, most significantly those inspired by Thomas and, to a
lesser degree, Scotus.

The instantiation of Catholic Church's position, besieged by the multiple
threats of astronomical innovation and novelties in natural philosophy, created
an environment that encouraged more precise controls over the interpretation of
Aristotle, especially in Catholic countries. Those controls, some more successful than others, censored those who held a range of positions, including Neoplatonists, such as Patrizi, who thought Plato surpassed Aristotle; inventors of
novel approaches to nature, such as Telesio, who thought that innovation could
replace Aristotle's allegedly impious understanding of God's relation to nature;
and Aristotelians, such as Cremonini, who believed that the accurate exposition
of Aristotle demanded explaining the ways his philosophy deviated from Christianity. Despite the fundamental differences in the philosophical goals of Patrizi,
Telesio, and Cremonini, their interpretations of Aristotle were in essential agreement. For them, Aristotle could not be reconciled with religious truth. For them,
the demands of textual exposition and consistent philosophy made Aristotle a
figure of impiety.

CHAPTER SEVEN

History, Erudition, and Aristotle's Past

By the late Middle Ages, some interpretations of Aristotle's philosophy and its relation to Christianity depended on knowledge of the past and on conceptions of historical practice. Readings of Augustine and Cicero shaped Petrarch's, Bruni's, Ficino's, and Valla's depictions of Peripatetic philosophy. The discovery of the Hermetic corpus and the ideal of *prisca theologia* led Giovanni Pico della Mirandola, Augustinian Platonists, and Francesco Patrizi either to try to reconcile Aristotelianism to Christianity or to forsake it in the attempt to discover ancient wisdom. The desire to find Aristotle's intent by relying on Greek texts and the views of the Greek commentators led a number of sixteenth-century Italian university professors to conclude that Aristotle deviated from key tenets of Christianity.

Knowledge of Aristotle's past reached new levels of sophistication at the turn of the seventeenth century. The Valencian humanist Pedro Juan Nuñez (1529–1602) edited a more accurate version of the biography of Aristotle attributed to Ammonius. Nuñez provided not just a commentary but a chronology of his life, which demonstrated the impossibility of Aristotle's having studied with Socrates.[1] Isaac Casaubon (1559–1614) created a critical edition and translation of the Greek text of Aristotle's writings in 1590; and, versions emended by Giulio Pace came out in the following decades.[2] Increased knowledge of Aristotle's life made his pagan roots difficult to ignore. Guillaume Du Val, in his summary of the *Metaphysics* that accompanied Causabon's and Pace's 1619 edition, invited the readers to "wonder at the theology of this pagan man," because of Aristotle's belief in an eternal God.[3] Yet Du Val did not see Aristotle as conforming completely to Christianity, interpreting him as believing that God does not see human actions, thereby rendering his philosophy impious because he "denied divine providence and justice."[4]

Awareness of Aristotle's biography even led a few Catholic authorities to ponder Aristotle's impiety. The Spanish Jesuit Francisco Arias contended that, since Aristotle was raised in "paganism without the faith of Christ, he becomes useless for the master of virtue."[5] Arias noted that, even though Aristotle's moral doctrine was the "most reconcilable and discerning," of all the pagans, he misunderstood the nature of God and thought abortion and infanticide were permissible.[6] Emanuel do Valle de Moura, the bishop and general inquisitor in Evora, Portugal, maintained that Aristotle "frequently sacrificed to demons," thereby rendering it plausible that he obtained diabolic aid in his discoveries about nature.[7] Valle de Moura also recounted Aristotle's supposed wickedness, which included not just Aristotle's beliefs regarding providence but also his alleged indulgence in sodomy in his old age. Nevertheless, he left open the possibility that Aristotle, after confessing on his deathbed, recognized the immortality of the soul.[8]

Despite Arias's and Valle de Moura's limited influence on the Catholic Church's endorsement of Aristotle, historical interpretation, nevertheless, guided many seventeenth-century evaluations of Aristotle and Aristotelians. These interpretations often built on, or criticized, the works of sixteenth-century authors. Those interested in Aristotle's past frequently expanded their scope, looking not just at Aristotle but at the larger Aristotelian tradition. Seventeenth-century histories of philosophy examined ancient schools in addition to medieval and Renaissance thinkers. Patrizi pioneered this approach when he divided Aristotelians into ten historical groups, the last being those who defined the field during his lifetime.[9] Although reasonable, Patrizi's historical analysis of schools Aristotelianism was partisan, assisting in a larger polemic over Aristotelian philosophy's inferiority to Platonism and Hermetic thought. Similarly, the seventeenth-century historical evaluations of Aristotelians emerged from polemics and debates over religion and philosophical orthodoxy that emerged in the aftermath of the Council of Trent.

In France during the first half of the seventeenth century, investigations in Aristotelianism's past circled around two main issues, one institutional, the other intellectual. The institutional question regarded whether university instruction must be based on Aristotle's thought. The rise of Jesuits imposed Aristotelian thought on French universities and colleges during the first decades of the 1600s. As Jesuits took over existing colleges and founded new ones that competed with already established universities such as at Paris, deviations from Aristotle became less accepted in universities. In 1611, the Sorbonne reformulated the statutes surrounding instruction and made explaining Aristotle obligatory for professors.[10]

Similar statutes had been passed elsewhere in Europe: in Oxford in 1586 and at Padua in 1607.[11] In 1624 the Sorbonne surpassed these universities in their insistence that professors follow Aristotle when it banned the teaching of an alternative matter theory and asked the Parlement of Paris to issue an arrest warrant for the neoterics, Antoine Villon, Jean Bitaud, and Etienne de Clave, who proposed these innovative theories. In this case, university officials cited religious doctrine for mandating Aristotelianism.[12]

Those who questioned the Sorbonne's decision pointed to the humanist trope of the liberty of philosophizing (*libertas philosophandi*) and the historical circumstances by which Aristotle's thought became the basis for university curricula and ancillary to theology. Historical investigations into Aristotle's piety and the tumultuous *fortuna* of his thought became a tool plied to subvert the renewed ascent of Scholasticism.

The major intellectual issue surrounded the question of whether natural philosophy should be a handmaiden to theology. From the time of Michel Montaigne, a number of French thinkers used skeptical arguments, genealogically linked to Sextus Empiricus and Gianfrancesco Pico della Mirandola, to show that the truths of religion were unknowable and distinct from natural reason. The sincerity of these skeptics was, and continues to be, doubted. Even though Montaigne's follower Pierre Charron (1541–1603) argued that skepticism was not only compatible with Christianity but also bolstered it, some polemicists equated skepticism with deism, atheism, and other forms of heresy.[13]

By the 1620s, a number of French thinkers, many tied to the circle of Cardinal Richelieu, endorsed a version of Christian skepticism. They argued that some of the truths of religion were mysteries that cannot be proven through philosophy. These *érudits* used their broad historical reading of past philosophy to show that the Thomistic synthesis ran counter to various judgments made by patristic authors and ecclesiastical authorities. Their historical analyses underlined the vicissitudes of the idea that philosophy should be ancillary to theology, thereby potentially undermining the Jesuits' adoption of the handmaiden position and questioning the validity of the Council of Trent's implicit endorsement of Aristotelianism.

The antiquarianism of these seventeenth-century *érudits*, and their opponents, altered conceptions of Aristotelianism. During the sixteenth century, French scholars investigated the Middle Ages in order to settle questions of legal interpretation.[14] Similarly, seventeenth-century historical research into medieval universities shaped arguments about curricula and the subalternation of subjects of learning. Examinations of the past included new interpretations of the previous century. Sixteenth-century figures, such as Pomponazzi, Nifo, and Porzio, whose philosophical influences were limited in seventeenth-century universities, were subject to the needs of polemicists who at times overlooked the actual historical context of sixteenth-century Italy and at other times used the historical context to contrast seventeenth-century restrictions on teaching and publishing with an allegedly more tolerant past.

GIULIO CESARE VANINI

In 1277, Etienne Tempier characterized the separation of philosophy and theology as "double truth," listing it among propositions that could lead to error and heresy. Without using the rhetorical strategy of Tempier, Jesuits as early as the 1560s complained that the teaching of Aristotelian philosophy in an improper way, as presumably they thought was done in Padua and other Italian universities, could lead to moral corruption. Pomponazzi and Porzio led lives that conformed to social and religious mores, making the Jesuits fears seem overstated, although rumors had circulated in the sixteenth century that Pomponazzi was an atheist or had renounced immortality on his deathbed.[15] Believing Aristotle held the soul to be mortal and endorsing a separation between the methods of natural philosophy and theology need not undermine religious faith. Still, the Jesuits' fears appear justified, at least slightly, in the first decades of the seventeenth century, when Pomponazzi's writings served as a source for Giulio Cesare Vanini, who was convicted and executed for atheism in Toulouse in 1619.[16]

Vanini had been a Carmelite and student at Padua before fleeing first to England in 1612 and then to France two years later. In France, Vanini published two works that presented arguments against the plausibility of numerous matters of faith. In his *Amphitheater*, printed in 1615, he cited Pomponazzi repeatedly. Vanini transformed Pomponazzi by grouping him with Niccolò Machiavelli and Cardano, a cohort that in his eyes believed that religious laws were merely tools for political control rather than divine truths. The assimilation of these three thinkers distorted the diversity of their views. Cardano polemicized against Pomponazzi; Pomponazzi shows no signs of having read Machiavelli, or vice versa; and, even though the structure of Cardano's presentation of evidence at times mirrors Pomponazzi's, his astrological explanations for changing political regimes differ from Machiavelli's virtue and fortune.[17]

A frequent charge against Pomponazzi is, and has been, that his assertions for the veracity of religious truth seem weak when compared with the elaborate naturalistic accounts given for marvelous and biblical events. There is no consensus, however, if or where Pomponazzi was disingenuous. To the contrary, despite numerous ambiguities in his work, Vanini appears, to nearly all commentators, to have tried to undermine arguments for the existence of a providential God, the creation of the world, the immortality of the soul, and the resurrection of the dead. His work, which purports to be a polemic against atheism, has been read, with few exceptions, by both his contemporaries and present-day scholars as a thinly veiled defense of atheism because of its lengthy presentation of arguments that oppose Christian tenets.[18] Recognizing the unorthodoxy of these arguments, Vanini added provisos to his rejections of Catholic dogma, such as "only through religion am I persuaded that demons exist."[19] The overall effect of his works suggests little respect for theology. Nevertheless, his works do not deny the existence of God and his contemporaries' labeling him an atheist reveal their fear of heresy and blasphemy rather than his open rejection of the existence of the divine.[20]

Vanini's printed works often cite Pomponazzi and Averroes. In a chapter that considers possible natural explanations for oracles, he praised Pomponazzi as being a "very acute philosopher, into whose body Pythagoras would have judged the soul of Averroes to have transmigrated," and, in a punning reference to the marvelous phenomena considered in the work, he judged On Incantations to be "admirable," despite that by this time the book was on the index of prohibited books.[21] What precisely Vanini intended by his praise of Pomponazzi and by his suggestion that the Mantuan philosopher and Averroes shared a soul becomes less clear, considering he categorized Pomponazzi as an atheist together with Machiavelli and Cardano. In an argument that supposedly aimed to show the existence of God's providence from the miraculous nature of religious law, Vanini described in detail the views of alleged atheists who opposed this position: Machiavelli suggested that rulers and power-hungry priests invented these religious laws, fabricating miracles to give them credence, and Pomponazzi argued that the stars were the efficient cause not just of the establishment of each religion but also of the miracles that convinced the population of the veracity of new religions by endowing animal and herbs with wondrous properties. Vanini's consolidation of these thinkers defies their actual views.[22] Pomponazzi's position is not one of imposture, as is Machiavelli's. Although Pomponazzi argued that the establishment of religion is in accordance with nature, he did not argue against its truth. His naturalistic account, which is presented as, at best, a probable explanation, rules out the possibility that religion has resulted from human deception because its source is nature.[23]

Unlike Pomponazzi, Vanini was unable to escape his prosecutors. In Toulouse, he was charged with heresy for holding anti-Trinitarian views. Convicted and sentenced to death, Vanini, instead of maintaining his innocence, turned against his prosecutors and in a vindictive fit publicly announced his unbelief before being executed.[24] His death increased his notoriety, and he became the prime example of an atheist inspired by Averroes and sixteenth-century Italian thought. Despite the near universal disapproval of his views, Vanini's grouping of Pomponazzi with Cardano and Machiavelli convinced his distractors. Polemicists who wrote anti-atheist treatises expanded the group to include Vanini. Thus the four along with Averroes became emblems for atheistic arguments that posited that religion either stemmed from imposture or directly from the working of nature without God's intervention.

In the thirteenth century, Giles of Rome had criticized Averroes for allegedly holding "[t]hat no law is true, although it can be useful," a statement not found in but most likely derived from an interpretation of comments on the *Metaphysics* or of the *Destruction of the Destruction*.[25] Yet the category of "political Averroism" was absent from medieval discourse.[26] After Vanini, it became common to link Averroes, Pomponazzi, Cardano, and Machiavelli to the idea that religion is an imposture that leaders have created to control the populace. The possible influence of Pomponazzi on the Venetian scholar Paolo Sarpi's materialist views and on the *Theophrastus Reborn*, albeit a treatise that had a limited circulation and was largely unknown at the time, made linking Pomponazzi to unorthodoxy even easier later in the seventeenth century.[27]

ARISTOTELIAN LIBERTINES AND MERSENNE'S POLEMICS

During the first decades of the seventeenth century, confessional divisions gave rise to a new genre of French polemical writing that focused on attacking atheists or supposed atheists. Frequently grouped among the lot of atheists were skeptics, deists, and libertines. These writings attacked Montaigne and Charron for their attempts to shake the foundations of knowledge. They demonized unnamed libertines for doubting the veracity of conventional religion and leading lives of questionable morality. The existence of both of these targets is a matter of controversy. Tullio Gregory argues that skeptics, such as Charron, were hidden atheists, but Richard Popkin interpreted Charron's doubting as a version of fideism that asserts that religious truths can only be known through revelation and not reason.[28] Regardless of Charron's sincerity, these skeptical writings influenced numerous French writers often referred to as *libertins érudits*. The category of *libertinage érudit* dates back as early as the 1884 work of Jacques Denis but gained more currency first in 1919 from Jacques Charbonnel and then from René Pintard's 1943 work on the topic. Denis, Charbonnel, and Pintard described a circle of intellectuals who surrounded Richelieu.[29] According to Pintard, these scholars—who included Gabriel Naudé, Pierre Gassendi, Gui Patin, François La Mothe Le Vayer, and Cyrano de Bergerac—used their extensive learning to undermine traditional morality. Followers of Pintard, namely Henri Busson and Giorgio Spini, found the sources for the alleged dissimulation and impiety of these figures in sixteenth-century Italians, such as Pomponazzi, Francesco Vimercati, and Cardano.[30]

Although there remain supporters of Pintard, Busson, and Charbonnel, criticism has come from multiple fronts.[31] Many accusations of the impiety of Italians depend on the *Naudeana et Patiniana*, an anonymous treatise not printed until the 1690s, that purports to report the irreverent witticisms and questionable behavior of Pomponazzi, Girolamo Borro, Cesare Cremonini, and Agostino Nifo, among others.[32] The reliability of the *Naudeana* is doubtful and Pintard's insistence on the existence of a culture of secrecy and hidden messages likely reflects the composition of his work in occupied France more than the activities of a group of scholars in Richelieu's circle.[33] Regardless, seventeenth-century French polemicists connected impiety to figures such as Pomponazzi, transforming sixteenth-century Italian philosophers into allies of contemporary heretics by minimizing or ignoring their attempts to understand Aristotle's true thought and emphasizing their unorthodoxy. In the climate of seventeenth-century France, a number of defenders of orthodoxy reinterpreted the lengthy tradition of considering nature without recourse to theology as a clever means to weaken Christian dogma.

Soon after Vanini's death, his heterodoxy, tied to Pomponazzi and Cardano, became the subject of polemical treatises, such as the Jesuit François Garasse's 1623 treatise *The Curious Doctrine of Great Wits of This Time*. This work expressed outrage of what he considered to be the scandalous behavior of Theophile de Viau, a poet and coauthor of "Le Parnasse satyrique," which the Parlement of Paris judged impious and blasphemous.[34] In *The Curious Doctrine*, Garasse connected Viau's behavior to the philosophy of Vanini, Pomponazzi, and Cardano, positing that irreligion and hedonism stem from denying creation. According to Garasse, libertines followed Pomponazzi and Cardano by rejecting the truth of Genesis for the belief that humans originated from spontaneous generation out of pig and frog feces. Garasse pointed to Muhammad as also holding this view as well as to other "stronger atheists" who believe that humans originated from the sperm of monkeys.[35] Garasse distorted Pomponazzi's thought further by linking his insistence on the separation of natural philosophy and theology to skepticism and impiety referring to those who have after having "proposed maxims of impiety" relinquish themselves to the authority of the Church and excusing themselves by saying they are laymen and "leave this question to Theologians."[36] For Garasse, Pomponazzi's and others' deference to theologians on questions of religion was mere imposture, in spite of Pomponazzi's public defense of his views and the support given to him by powerful ecclesiastical and political authorities.

Garasse's exaggerations met opposition from ecclesiastical authorities. Jean Duvergier de Hauranne, or Saint-Cyran, a follower of Cornelius Jansen, attacked Garasse, accusing him of making numerous errors and dismissing his ability to reveal clandestine atheists. Saint-Cyran, who endorsed an antirationalist theology, that is, one that held that religious truths can only be known through faith, found Garasse's identification of all forms of skepticism with atheism to be reactionary as well as an affront to his conception of theology. Saint-Cyran, at least in the short term, was successful, as he persuaded the faculty of theology in Paris to condemn Garasse's work in 1626.[37]

Despite Garasse's overstatements and imprecision, his association of skepticism with the naturalism of Vanini and the Aristotelianism of Pomponazzi was common to more careful scholars. Marin Mersenne's 1623 *Questions on Genesis*, although ostensibly a commentary on the first six books of the Old Testament, possessed as its main goal the refutation of supposed atheists, deists, and proponents of magic who followed the Hermetic tradition. As a Catholic polemicist, Mersenne's definition of atheism was broad and included variants of Protestantism, which he saw, at least in his early years, as a heresy that demanded to be stamped out. After the lack of success with this strategy, Mersenne converted to a more tolerant form of persuasion through epistolary engagement.[38]

Mersenne, a member of the religious Order of Minims, possessed an outlook similar to those of contemporary Jesuits. He was not the first to link Pomponazzi's and Cardano's explanations of miracles to magic and heresy; for example, the Jesuit Martin Delrio had done so several decades earlier.[39] Like Garasse, Mersenne saw skepticism and Vanini's naturalism as pernicious to rational theology, which he believed should be largely in accordance with Thomas's views.[40] Mersenne rejected the views of neoterics, such as Bruno, Telesio, Galileo, and William Gilbert, who believed that Catholic theologians "only follow Aristotle and swear an oath on his word, even though the phenomena and experiences show the contrary."[41] Rather, he thought that these innovators have misunderstood Scholastics because, as Mersenne wrote, "[t]heologians assent to no authority that lacks reason."[42] In this light Mersenne proffered various proofs for the existence of God and lamented that "[a]theists ignore metaphysics."[43]

Vanini's *Amphitheater* provoked Mersenne's wrath. He associated Vanini with Pomponazzi and Cardano, as they all allegedly promoted arguments against the existence of God and for the imposture of religious laws. Mersenne pointed to the "atheism of politicians [politicorum], followers of Cardano and Pomponazzi," the politicians presumably being a reference to those influenced by Machiavelli.[44] For Mersenne, Pomponazzi's assertion that Aristotle held the soul to be mortal became assimilated to the idea that belief in immortality is a fable invented by legislators.[45] In a similar light, Mersenne rejected Pomponazzi's and Cardano's use of vapors and the force of imagination to explain purported miracles as supporting the belief that religion was a mere tool that derived from the deceptions of the powerful.[46]

In a manner reminiscent of the Plato-Aristotle debates of the fifteenth century, Mersenne pointed to the question of providence as crucial to definitions of heresy and orthodoxy, a question that dominates Vanini's *Amphitheater*. Contending that Thomas's interpretation of Aristotle's view of providence was the best, he accused atheists of following Averroes by maintaining that God's creation was necessary rather than being the result of free will. Additional mistaken views on the nature of providence arose from Averroes' denial of God's providence over "corruptible singulars," that is, sublunary particulars. Mersenne also identified as erroneous Vanini's following of Alexander of Aphrodisias's and Averroes' position that the existence of monsters shows that some imperfect beings arise from material necessity beyond the perfection of God's power. The recourse to material necessity independent of God's will pointed to limitations of divine power and providence.[47]

In the next years, Mersenne followed his *Questions on Genesis* with two similarly themed books. His *The Impiety of Deists, Atheists, and Libertines* (1624) defended what he believed was the truth of the Catholic faith from the supposedly impious philosophies of Charron, Cardano, Bruno, Machiavelli, David Gorlaeus, and Vanini among others.[48] One year later he published *The Truth of the Sciences: Against the Skeptics or Pyrrhonists*. This work attacked not just skeptics as named in the title but also unnamed alchemists, who counted among their beliefs the naturalistic view that demons are nothing but formations of air.[49]

In *The Truth of the Sciences*, Mersenne reaffirmed his high esteem for Thomas, who "perhaps had the best mind, [and was] the most solid, the steadiest, and the most judicious of all those who followed him."[50] According to Mersenne, Thomas "had not wanted to establish principles other than those of Aristotle, as much as he judged them to be true."[51] Changing his position from that found in *Questions on Genesis*, Mersenne pointed out Aristotle's failings: he believed the world to be eternal and he might have denied that God's providence extended to the smallest beings of the universe. Mersenne wrote that *De mundo*, which he admitted to not knowing whether Aristotle had truly written it, "had almost prevented him from believing that Aristotle had denied that God's providence extend to the smallest things."[52] In this regard, Mersenne's interpretation was not far from Charron's, one of the main targets of his polemics. Charron grouped Aristotle with the "Epicureans and Irreligious" in that they all denied providence. Charron believed that Aristotle held that providence extends to "only the celestial, lofty and incorruptible" and that providence is only for the "large and general and not particular or detailed."[53]

By endorsing Thomas, Mersenne's theological views were backward looking. But he pointed to a relativistic strategy that could lead to evolutions in interpretations of the acceptability of writings. According to Mersenne, the Catholic Church could condemn any book that is used to support heresy. The list of condemned books, therefore, should not be static, as the interpretation of books changes over time in concert with the arrival of new heresies. Novel heretical readings of older works alter the status of their acceptability and justify creative interpretations that condemn figures such as Pomponazzi.[54] Thus Mersenne believed that enemies and dangers of the Catholic Church evolved over time and that past decisions could be reviewed and overturned because of new circumstances. Present dangers were more important than the strict adherence to past decisions.

TOMMASO CAMPANELLA

Tommaso Campanella agreed with Mersenne's assessment of Pomponazzi and the threatening rise of Machiavellian imposture but disagreed with him on the importance of Thomas. Mersenne attacked Italian Aristotelians as part of polemics against alleged deists, skeptics, and atheists; Campanella attacked Aristotelianism, using historical analysis to show that for centuries numerous thinkers found Aristotle incompatible with Christianity. Campanella wrote *On Gentilism That Must Not be Followed* in 1609 while imprisoned in Naples , where he was held by Dominican inquisitors for rebellion and heresy. His questionable views included a defense of Telesian natural philosophy and heliocentric cosmology.[55] After his release, he made his way to France, where he was received by the king and participated in Parisian literary culture.[56] It was only in 1636, at the end of his life, that *On Gentilism* was printed. Despite the twenty-five-year delay in publication, this work was still relevant, unlike much of Campanella's thoughts on natural philosophy, which had become obsolete as the result of innovations by Bacon, Kepler, and Galileo.

The premise of *On Gentilism* was that the Catholic Church's recent acceptance of ancient philosophy and condemnation of novelties in natural philosophy was mistaken, contrary to the Church's traditions, and the result of mistaken interpretations of pagan thought generally and Aristotelian thought specifically. According to Campanella, scripture provides a basis for accepting new philosophical discoveries, namely a passage from the book of Daniel, which was prized by Bacon and became a motto for the Royal Society and reads, "Many will travel the earth, and knowledge will multiply."[57] For Campanella, this passage linked the discovery of America with the natural philosophies of Giovanni Pico della Mirandola, Telesio, Francisco Vallès, and Paracelsus; with Galileo's discovery of new stars; and with Copernicus's discovery of the motion of the earth. Although some of these new philosophies err at times, "nevertheless they indicate that the entirety of philosophy must be renewed," according to Campanella.[58]

In order for philosophy to be renewed, the philosophy of the Greeks must topple. Thus Campanella argued not just that the acceptance of novelties conformed to the religious truth of scripture but also that "the philosophy of the Greeks possesses infected roots and false dogma," which lead to error, heresy, and immorality.[59] Even after centuries of anti-Aristotelian polemics, Campanella's treatise displays an originality in terms of argumentation and novel application of historical arguments. For example, since the 1560s Jesuits had cited the eighth session of the Fifth Lateran Council to justify the necessity of teaching Aristotle held the soul to be immortal. To the contrary, Campanella interpreted the decrees of "Apostolici regiminis" as "affirming pagan philosophy and poetry have infected roots, that is, false principles."[60] As a result, Campanella contended that "a holy council should choose a new philosophy that conforms to holy theology and sacred canons."[61]

For Campanella, Aristotle represented the worst of pagan philosophy. His interpretation of Aristotelianism simultaneously demonized the Greek commentators, Averroes, and Pomponazzi but still judged their interpretations of Aristotle to be correct. Aristotle put forth an impious philosophy, and these commentators' attentiveness to the texts correctly revealed his errors. Campanella, distinguishing medieval theologians from other commentators who sought to uncover Aristotle's intent, wrote, "Even though Thomas and Albertus [Magnus] and others tried to redirect Aristotle into a good reading and they oppose him where he openly counters faith, nevertheless, not only Greeks and Arabs but Christians say that St. Thomas did not understand or distorted the Greek Aristotle."[62] Although John of Jandun, Pomponazzi, Walter Burley, and Cremonini might have understood Aristotle better than Thomas. Based on their readings of Theophrastus, Alexander, Themistius, Simplicius, and Averroes, their judgment that Thomas twisted Aristotle compounded the evils found in Aristotle's texts. Campanella believed that "from suspicion against Divine Thomas coming out of Aristotle and Averroes, and others, Machiavellianism was born, that root of evils that makes religion a political tool [ragion di stato]."[63] Therefore, according to Campanella, the versions of Aristotelianism that maintained that Thomas had incorrectly combined theology with hermeneutics was the source of cynical attitudes about religion and power.

Campanella associated the views of Averroes and Aristotle with those believed to be contained in *On the Three Impostors*, a fictitious book that supposedly maintained Jesus, Moses, and Muhammad invented their religions as a means of political control. When he was under investigation in the 1590s, Campanella had been accused of being the author of this nonexistent book.[64] According to Campanella, recent Aristotelians, such as John of Jandun, Nifo, Cremonini, and Antonio Bernardi did not specifically argue for the imposture theory but nevertheless compromised the strength of faith by "mumbling" their censures of Aristotle's positions on the eternity of the world and mortality of the soul. They simultaneously write that "[a]s Christians they believe in the immortality of the soul, but the contrary must be pronounced according to philosophy."[65] Campanella's rebuking of these Aristotelians both condemns them by placing them on the same level of the "Antichristians" who allegedly wrote *On the Three Imposters* and implicitly accepts the interpretation of Pomponazzi, Porzio, and others who held that Aristotle denied the personal immortality of the soul.

Just as Giles of Rome, Gianfrancesco Pico, and Patrizi had done, Campanella composed a list of Aristotle's errors. The list contained the familiar errors of eternity of the world; the mortality of the soul; limitations on God's power, knowledge, and providence. It also included the statement "Religion is the art of ruling and keeping the population dutiful and obedient," based on a loose interpretation of the *Politics*.[66] Following Melchor Cano's assessment that Machiavellianism— that is, the belief that religion is an imposture used by governments to control the population—grew from Italian Aristotelian schools that taught the mortality of the soul and the absence of providence in the sublunary sphere, Campanella maintained that a sect of followers of this view, of "politicians and libertines," had grown in northern Europe. In a fashion characteristic of early modern religious polemics, Campanella suggested that these libertines should be grouped with "the Calvinists, Lutherans, and Puritans, because the aristocrats of these sects are all Machiavellian and atheists, just like Calvin and Averroes," thereby contending that Protestantism developed as a means for the powerful to control the population as a whole.[67]

Campanella, however, was occupied by mistakes within Catholic doctrine more than he was with Protestants. *On Gentilism* attempted to demonstrate that the reconciliation of Aristotelian thought with theology has no basis in Catholic tradition. Campanella believed the introduction of Aristotle into Christian schools during the Middle Ages was merely a matter of "chance."[68] Echoing many earlier humanists, Campanella pointed to the adoption of Platonic thought by Augustine and other Church Fathers. He used this evidence to argue against Bellarmino, who had maintained that Platonic doctrine was more dangerous since it appeared to be closer to Christianity.[69] In contrast to Plato's piety, Aristotelians opposed revealed religion from its beginnings. In the second century BC, Antiochus IV founded a gymnasium, which Campanella maintained taught Aristotelian philosophy, in Jerusalem in order to eradicate Judaism.[70] Later, Plato's works were unavailable in the "rude centuries" after Charlemagne established schools, so Aristotle and his "Arab" followers entered the curriculum, "because the manuscripts of other philosophers had not been discovered."[71] Citing Roger Bacon and Thomas's treatise *On the Unicity of the Intellect*, Campanella believed that by the fourteenth century all Aristotelians were followers of Alexander or Averroes, despite the thirteenth-century condemnations at Paris and Thomas's attempt to check the promulgation of Averroes' views on the intellect.[72]

Campanella's interpretation of Thomas peculiarly proposes that Thomas did not follow Aristotle but merely, as a commentator, revealed Aristotle's thought. He wrote Thomas "never swore that Aristotle is true but only revealed him just as if someone should expound on Virgil or Muhammad."[73] Such a reading permitted Campanella to separate Thomistic thought from those who "swear on the words of pagan philosophies"; he concluded that such trust in heathens "is heresy, perjury, and the greatest impiety."[74] By removing Thomas from the camp of Aristotelians, he could defend Thomas and still hold that the Church should accept new natural philosophies and that the staunch defense of Aristotle demanded after the Council of Trent was in error.

Mersenne and Campanella condemned the works of Pomponazzi, Averroes, and other Peripatetics because of their allegedly pernicious effects, which potentially led to immorality, atheism, and the belief that religion was merely a political tool. Mersenne proffered a relativistic judgment: that the meaning of texts changed with contexts and what determined their acceptability was their influence rather than the authors' intent. Mersenne's reading of Pomponazzi that linked him to Vanini, Cardano, and Machiavelli ignored the setting of Pomponazzi's writing. As a result he inferred a danger that results not from the text but from subsequent uses of its authority. In a similar fashion Campanella found the faults of Averroes and Aristotle in their application, faulting them for supposedly having provided material for *On the Three Impostors*. His investigations into the history of medieval thought provided additional evidence for the historical contingency of Catholicism's adoption of Aristotle. Moreover, he showed that ecclesiastical hostility toward Aristotle was common to Church Fathers and thirteenth-century theologians, suggesting that the hardening of Catholic endorsements of Aristotelianism at the expense of recent discoveries was contrary to ecclesiastical tradition.

GABRIEL NAUDÉ'S HISTORICIZING
OF ARISTOTELIANISM

Campanella's historical analysis revealed a sophisticated, if at times erratic, knowledge of the history of philosophy. Its polemical origins, however, blinded him to the historical contexts that allowed for the toleration of Pomponazzi and his followers. Linking the condemnations of the thirteenth century to the Fifth Lateran Council, Campanella saw a continuity in ecclesiastical hostility to Aristotle that did not exist. Furthermore, he overestimated the extent to which these condemnations of Aristotle affected philosophy and ignored the existence of the toleration for a wide range of philosophical positions, including those that purported philosophy to be distinct from theology. Gabriel Naudé responded to Mersenne's interpretations by providing historical analyses of Pomponazzi and his circle that took into account the changing standards of heresy. Naudé's historical approach mirrored Campanella's, demonstrating the degree to which the monopoly of Aristotelianism among seventeenth-century clergy defied ecclesiastical tradition.

Naudé could have very well been one of the libertines and politicians that Mersenne and Campanella demonized.[75] In the 1620s, Naudé travelled to Italy to study medicine, where he met Paduan professors of philosophy and medicine, including Cremonini. Despite the fact that he rarely cited Cremonini in his printed works, in a 1626 letter to the physician René Moreau, Naudé described Cremonini as an imposing intellect who dominated the university. According to Naudé, Cremonini was Pomponazzi reborn; he "securely and fearlessly revealed, taught, cultivated, saw, and defended the mind of Aristotle."[76] In a later letter, Naudé wrote that Cremonini uncovered the true and genuine doctrine of the Peripatetics in his writings, likening him to Alexander of Aphrodisias and Averroes.[77] When he wrote those letters, Naudé's most significant writing was *An Apology for All the Great Men Who Have Been Accused of Magic*, a work that defended historical figures who had been accused of associating with demons. Naudé contended that there was little or no philosophical evidence for the existence of demons and believed that most past assertions of the existence of magic were doubtful. Natural causes could account for what others attributed to supernatural causes. The work was significant not just for its naturalistic outlook but also for raising doubts about the likelihood of the proliferation of witchcraft, which at the time was seen as a major political, religious, and social threat throughout western Europe.[78]

Naudé's *Apology* is largely a study of history, although for him, as for many during this time, history was intertwined with philosophy. For many scholars, experience, taken from historical accounts found in ancient writing and medieval chronicles, provided evidence for the presence of the supernatural on earth. Evaluating evidence from the past, plying the tools of the historian, therefore, could establish the probability of the efficacy of magic and the likelihood of the existence of demons. By questioning the coherency of past testimony and examining the motivations of ancient and more recent authors, Naudé tried to exonerate famous thinkers from antiquity to the seventeenth century who had been accused of practicing magic.[79] Naudé believed their perceived guilt arose only from the "lies of charlatans, the vanity of alchemists, the foolishness of magicians, the enigmas of cabalists, and the combinations of the Lullians."[80] His unveiling of this dishonesty and deception depended on combining history with natural philosophy.

Aristotle was among those whom Naudé defended, exonerating him from the charge that his intelligence was the result of communications with a demon. In Plato's *Symposium*, Socrates is described as being inspired by a demon or genie.[81] After dismissing this assertion, arguing that Socrates' "familiar demon" was "nothing but the good habits of his life, the wise conduct of his actions, and experience," Naudé addressed those who believed Aristotle must have had a demon as well since his wisdom exceeded Socrates'.[82] If one accepted Naudé's contention that Socrates' demon referred to character traits rather than supernatural intelligences, then there is no need to posit that Aristotle had one as well. Nevertheless, he used additional evidence based on Aristotelians principles that suggest Aristotle had no truck with demons. According to Naudé, "[t]here is nothing so certain in the doctrine of Aristotle and so constant among all his interpreters that he never admitted the existence of intelligences other than those that he gave to each of the orbs . . . rejecting all other kinds of Demons and Angels." Naudé argued that Aristotle believed that forms could exist only in bodies and that seemingly supernatural events could be attributed "to Nature, that is to say, to the properties of natural things, to humors and temperaments of animals, to the conditions of places, and to their vapors and exhalations."[83] Therefore if a

demon had affected Aristotle, he would not have denied the presence of supernatural forces in the sublunary world.

Naudé analogously dismissed Aristotle's possible recognition of the trinity.
Noting that Bessarion had mocked Trapezuntius for attributing Aristotle with
belief in a triune God, Naudé objected to the imaginative interpretation of a
passage from the *De caelo*, in which Aristotle wrote about sacrificing to the gods
three times, that some believed suggested correspondence to Catholic theology.[84]
To the contrary, because Thomas rejected the possibility of establishing the trinity through natural reasons, Naudé argued that attempts to do so were impious.
He wrote, "To want to make Aristotle and Plato so clairvoyant and well aware of
the mysteries of our religion is to totally reverse the philosophy of Jesus Christ."[85]
Turning the belief in the compatibility of Aristotle and Christianity on its head,
Naudé concluded the mysteries of the faith defied philosophical explanation.

For Naudé, those mysteries include the existence of demons and magic.
Therefore, Platonists' belief in them "cannot be proven by reason and experience." In this manner he rehabilitated Pomponazzi, Cardano, and Antonio Bernardi, whom he interpreted as "show[ing] very pertinently that it is better to have
recourse to the proofs of our religion in order to believe in angels and demons
than to the mass of all experiences that can be explained by the principles of natural philosophy."[86] Unlike Mersenne and Campanella, Naudé presented these
sixteenth-century Italian thinkers, not as the source of atheism, but, as having
correctly distinguished the mysteries of religion from rational and experiential
discourses on nature.

In the *Apology*, Naudé suggested that Aristotelians' naturalism conformed to
Christianity; elsewhere he provided historical explanations for recent hostility
toward Pomponazzi, Nifo, and other sixteenth-century natural philosophers. In a
biographical preface to a collection of Nifo's treatises on moral and political philosophy, Naudé recounted Pietro Barozzi's condemnation of Averroist psychology
and the controversies over Pomponazzi's *On the Immortality of the Soul*, assimi-
lating Pomponazzi to the ideals of French erudition by suggesting he "escape[d]
censures through some witticisms."[87] Displaying deep knowledge of natural philosophers of the early sixteenth century, Naudé attributed the relative intellectual
freedom of Nifo, Pomponazzi, Nicoletto Vernia, and Tiberio Rusiliano to different standards for heresy. Before the Council of Trent, he wrote, "Philosophers
were accustomed to speak freely and write about all things" and thereby sustained
propositions "to which now no one could assent."[88] Naudé thereby maintained
that interpretations of these philosophers as heretics, as well as famed medieval
astrologers, such as Pierre d'Ailly and Pietro d'Abano, ignore changing definitions

of orthodoxy. The Council of Trent blurred the reception of the Peripatetic tradition making heretics of those who were faithful according to the standards of their time. In this manner, Naudé defended the naturalism of earlier generations of natural philosophers and reaffirmed the potential orthodoxy of the view that religious doctrines defy rational explanation by suggesting the Council of Trent introduced novel standards of heresy.

Naudé cast doubt on the degree to which the Council of Trent conformed to earlier Church decisions. His fellow *érudit* François de La Mothe Le Vayer questioned the degree to which the Council of Vienne and the Fifth Lateran Council made Aristotle an authority for Catholic doctrine.[89] Like Naudé, La Mothe Le Vayer, in *A Short Christian Discourse on the Immortality of the Soul*, 1637, pointed to medieval condemnations and patristic authors' dislike for Aristotle to show the relative novelty of Aristotle's "rul[ing] with so great an empire in all the universities, that there has been seen recently an arrest given for those in Paris who attempt to teach philosophy using principles other than those of the Lyceum."[90] Also like Naudé, La Mothe Le Vayer used skeptical strategies to argue that reason and experience cannot adequately prove the mysteries of religion, including the immortality of the soul.[91]

La Mothe Le Vayer responded to the contention that the Council of Vienne and Fifth Lateran Council determined "that the immortality of the soul can be proven by demonstration, even according to Aristotelian principles."[92] As for Aristotelian principles, they, in his eyes, "seem to conclude necessarily for mortality," even though he conceded that this type of demonstration has neither the validity of a geometric proof nor "recourse to the authority of the faith." Moreover, the decrees themselves neither explicitly endorse Aristotle's philosophy nor forbid new philosophical approaches. According to La Mothe Le Vayer, the Council of Vienne "declares only that those are heretics who sustain that the rational or intellectual soul is not in itself essentially the substantial form of the human body, without saying that it conforms to the principles of Aristotle, nor that it is demonstrable by philosophy or that it must depend on faith."[93] Although his analysis underplays the significance of the use of the term "substantial form," a concept necessarily linked to Aristotelian philosophy, he accurately read the decree's failure to endorse specifically Aristotelian philosophy or philosophical proofs in general.

La Mothe Le Vayer approached "Apostolici regiminis" in a similar manner, using a close reading of the decree of the Fifth Lateran Council to suggest that those who cite it as maintaining the orthodoxy of Aristotelianism have misread the bull. He contended that the decree condemned as heretical only those who "positively taught the mortality of the soul," or who argued for Averroes' unicity of the intellect. According to La Mothe Le Vayer, those at the council held that the unicity of the intellect can be refuted by philosophical arguments. Yet refuting Averroes' psychology does not lead to "the consequent that the immortality is an issue that can be demonstrated mathematically."[94] The historical circumstances surrounding the Fifth Lateran Council provided La Mothe Le Vayer with further evidence that "Apostolici regiminis" was neither intended to give additional authority to Aristotle nor maintained the demonstrability of the immortality of the soul. First the Council took place "in a time when the opinions of Aristotle and Averroes were sustained with so great stubbornness . . . that the intention of the priests was nothing but to oppose these so dangerous maxims and not to attribute to our reasoning a certitude that pertains only to faith."[95] His analysis twisted the events of the Fifth Lateran Council in the interest of showing the respectability of his skeptical or fideistic views toward religious doctrine.

Although it is true that the "Apostolici regiminis" targeted Averroistic psychology, by equating mortality with Averroes, La Mothe Le Vayer ignored that many of these "priests" at the council believed Aristotle held the soul was immortal. La Mothe Le Vayer's interpretation that this decree is compatible with the view that the immortality of the soul cannot be proven but accepted only by faith defies the actual motivations of those who enacted the decree. It was likely aimed at those, such as Alessandro Achillini and Pomponazzi, who argued that according to Aristotle or the principles of philosophy the soul appears to be mortal even if we must accept immortality as a matter of faith. Nevertheless, La Mothe Le Vayer correctly saw that the decree neither specifically endorsed Aristotelian philosophy nor argued that immortality of the soul could be proven with philosophical certainty. He was also correct in noting that Scotus and Tommaso de Vio contended that the immortality of the soul is a matter of faith rather than subject of demonstration, thereby associating his own fideistic view with esteemed doctors of the Church.[96]

La Mothe Le Vayer's analysis offered a corrective to those who argued that deviations from Aristotelianism countered Catholic orthodoxy and that Catholic councils have proclaimed that the truths of religion are demonstrable through philosophical means. For him, Pomponazzi, instead of posing a danger, correctly understood the limits of human reason. According to La Mothe Le Vayer, Pomponazzi rightly maintained that asserting the immortality of the soul could not be proven by Aristotelian principles and agreed with both Christian orthodoxy and the views of early Church Fathers.[97]

CENSORSHIP AND GASSENDI

La Mothe Le Vayer's questioning of whether Aristotelianism was demanded by Church decrees took place in the years when opponents of traditional natural philosophy faced prosecution and banishment from universities. Ecclesiastical responses to anti-Aristotelianism began with actions against Telesio and Patrizi and continued throughout the seventeenth century. Famously in 1633 Galileo was sentenced to house arrest, causing fear among proponents of new natural philosophies. Descartes, for example, suppressed the publication of *Le Monde*.[98] French authorities censored opponents of Aristotelianism. In 1624, the Sorbonne declared fourteen theses that proposed new principles for matter theory to be "dangerous with respect to true philosophy" and "contrary to the principles of faith and religion."[99] The fourteen theses rejected Aristotelian principles such as prime matter, the four elements, the four primary qualities, and the transformation of elements. Jean-Baptiste Morin, who was responsible for the ban, described these positions as "erroneous," "scandalous," and "close to heretical," as they undermined explanations of the Eucharist that relied on substantial forms and on the Aristotelian notion of qualitative change.[100] The university's rector, probably fearing competition from Jesuit colleges and their potential judgment that the university was too permissive, agreed with Morin's assessment and asked the Parlement of Paris to authorize the ban.[101] The broad range of propositions banned because they were contrary to Aristotle meant that the Sorbonne and the Parlement had eliminated voices that dissented from Peripatetic philosophy from the university.

In 1624 Jesuits indirectly influenced the Sorbonne into making Aristotelianism a philosophical monopoly. At other schools, they had a more direct effect. For example, at the Collège d'Aix, Jesuits forced out Pierre Gassendi, himself ordained and a canon of the cathedral at Digne.[102] Gassendi, then, began associating with intellectuals in Paris, including Mersenne and Naudé, and wrote an anti-Aristotelian treatise, *Paradoxical Exercises against the Aristotelians*, 1624. Gassendi advocated abandoning traditional natural philosophy for a Christianized version of atomism, that was ultimately rooted in Epicurus's thought.[103] Gassendi took inspiration from Juan Luis Vives, Charron, Petrus Ramus, and Gianfrancesco Pico.[104] Quoting frequently from Vives, he adopted the style and arguments of earlier humanists, who had attempted to destroy the foundations of Aristotelian thought. Gassendi illustrated the textual unreliability of the corpus, the supposed meaningless of his logic, contradictory positions, the alleged blind acceptance of authority among Aristotelians, the poor prose of Aristotelians, and the impiety of Aristotle.

Earlier humanists centered their invectives on university professors who followed Averroes' views on the intellect and the eternity of the world or believed that philosophy could not be reconciled with theology. Gassendi's target, while still being Aristotelians, was vastly different: Jesuits and university officials who deemed non-Aristotelian approaches to natural philosophy to be erroneous or dangerous. The Aristotelianism of Jesuits differed greatly from that of those Vives or other earlier humanists attacked. Jesuits accepted that faith and reason were reconcilable. Moreover, like Vives, they demonized Averroes. As a result, Gassendi transformed a number of the tropes of humanists to fit the new circumstances of universities' and colleges' intolerance to innovations in natural philosophy.

Many of the techniques Gassendi used to discredit Aristotelians he shared with earlier thinkers. His attacks on dialectic and metaphysics corresponded to Vives' and Ramus's. His questioning of the authenticity of Aristotle's writings resonated with Gianfranceso Pico and Patrizi, even though Gassendi apparently was unaware of Patrizi's *Peripatetic Discussions* until after he finished the *Paradoxical Exercises*.[105] Like Valla, Gassendi accused Aristotelians of taking away the liberty of philosophizing (libertas philosophandi), using Averroes as evidence for the intolerant nature of Aristotelianism. Gassendi equated the attitudes of an unnamed professor of theology who supposedly believed that "whatever is contained in the works of Aristotle is most true," with Averroes' assessment that no one had discovered an error in the works of Aristotle in the previous fifteen hundred years.[106] Then, quoting Averroes' passage that read "Aristotle was a rule and exemplar, that nature invented," Gassendi accused his contemporary Aristotelians of eliminating the freedom to philosophize and transforming Aristotle into an object of religion.[107] Averroes, a perennial object of humanist scorn, acted as a tool Gassendi used to soften the rigidness of French universities and Jesuit colleges, which, in fact, had minimized Averroes' influence on their teachings.

In addition to associating Aristotelians with Averroes, Gassendi tried to weaken the privileged status of traditional natural philosophy by casting doubt on Aristotle's piety. His biographical musings on Aristotle influenced a number of thinkers, not just in France, but also among proponents of Epicurean thought in England, notably Walter Charleton. Gassendi emphasized Aristotle's pagan roots, stating that Aristotle "was not a Jew, not a Christian, being of the pagan persuasion" and punctuating this point by repeating the tales, found in the ancient biographical tradition, that Aristotle took up philosophy because of the Pythian oracle and later made sacrifices to the gods dedicated to his wife.[108] After quoting Lactanti-us's assertion that Aristotle "neither worshipped nor cared about God," he then listed points where he believed Aristotle differed from Christianity: that God is an animal, placed in the heavens, bound by fate and necessity, and uncaring of small matters; that the world is eternal and uncreated; that the resurrection of the dead is impossible; and that the soul is mortal.[109] Furthermore, Gassendi repeated descriptions of Aristotle behaving immorally: he helped poison Alexander the Great, was "dedicated to obscene pleasures," and committed suicide.[110]

Gassendi's assault on Aristotle's piety employed rhetorical techniques that hid the contradictory nature of his critique, such as that the Aristotle of the Jesuits was not that of Averroes. A Swiss physician living in Paris, Jean-Cécile Frey, in an equally one-sided polemic called *The Riddle of the Philosophers*, pointed out Gassendi's inconsistencies.[111] *The Riddle* sought to defend Aristotle from the on-slaughts of Patrizi, Bacon, Telesio, Ramus, and Gassendi by pointing out what Frey considered errors of interpretation. Frey defended Aristotle on matters as diverse as logic and the location of the Milky Way. Pointing to the inconsistency in believing that Aristotle did not care about the divine, yet made sacrifices to the gods on behalf of his wife, Frey accused Patrizi and Gassendi of falsely charging Aristotle with impiety.[112] According to Frey, their interpretations of Aristotle's views about God's power, his position, and form were mistaken and Gassendi's assertion that Aristotle denied the possibility of the resurrection of the dead over-looks the fact that Christian doctrine does not consider bodily resurrection to be a natural phenomenon.[113]

During the following decades, as Gassendi produced alternatives to Aristo-telianism in the form of a revived Epicurean atomism, he faced more criticism from those who embraced traditional natural philosophy. Morin continued his attack on atomist and other non-Aristotelian matter theories. He defended tra-ditional natural philosophy until his death in 1656, turning his attention away from the chemists De Clave and Villon to new perceived threats to orthodoxy. In the 1631, he published a treatise that dismissed Copernican cosmology. In 1650, he wrote a refutation of Gassendi's atomism, which in turn, provoked a response from François Bernier, a physician at Montpellier.[114] Bernier's defense of Gassendi, which primarily addressed philosophical issues, was published with a sophisticated history of Aristotelianism, written by Jean de Launoy, a Parisian theologian. Launoy's *On the Varied Fortune of Aristotle in the Parisian Academy* examined the history of censuring Aristotle. Launoy's work served the purposes of the Gassendists by revealing that the dominance of Aristotelianism in Church and university was incongruous with earlier ecclesiastical decisions.

Launoy's history bears some resemblance to Campanella's *On Gentilism*. In fact, Launoy had advised Campanella, leading to some changes between the 1636 and 1637 editions of that work.[115] Launoy, however, had a greater grasp of medieval history and provided extensive quotations that supported his interpretation that prior ecclesiastical authorities repeatedly condemned Aristotelian philosophy. Like Campanella, he collected passages form the early Church Fathers that rebuked Aristotle and recounted the condemnations of Peter Abelard and of Aristotelian works in in Paris in 1209, 1215, 1231, and 1265, apparently unaware of those of 1277.[116] Not committed to defending Thomism, Launoy rebuked Campanella for his contention that Thomas was not Aristotelian, although he confessed not knowing how Thomas and Albertus Magnus had been able to lecture on Aristotle after the ban of 1231.[117]

Launoy's treatise moved through the centuries, contrasting the views of the Church Fathers, the hostility of thirteenth-century censures, and the fifteenth-century controversies between Bessarion and Trapezuntius with the seventeenth century, when the condemnations of Villon and De Clave evinced Aristotle's renewed dominance at the Sorbonne. What Launoy called the "extreme fortune of Aristotle" was such that in locales where he was once read he is no longer and where he previously was not read he now is.[118]

Launoy modestly wrote that he "speaks historically" and did not wish to enter in battles, despite the fact that his work was published alongside Bernier's counter to Morin.[119] Nevertheless, the work's concluding pages explain the moral of his history. Launoy pointed to the historical origins of the idea that natural philosophy should be a handmaiden to theology. He wrote, "Recent and new doctors of theology named pagan philosophy a handmaiden, even if I do not find ancient Christians that spoke such."[120] Therefore, the subalternation of natural philosophy to theology does not conform to the Church's tradition. "Scripture and traditions are sufficient in themselves as the two principles of sacred doctrine," Launoy wrote, as he believed that making philosophy a handmaiden to theology should be foreign to Christianity.[121] He ended the treatise by approvingly quoting a passage from Antonio Bernardi's *Overturning*, a treatise demonized by Jesuits as well as by Mersenne, that argued that Aristotle's conclusions, following natural reasons, differ from the truths of faith.[122] Largely in agreement with Naudé and La Mothe Le Vayer, Launoy rehabilitated notorious sixteenth-century Italian natural philosophy as support for the idea that natural reason and Aristotle's philosophy were extraneous to the mysteries of the Church.

CONCLUSION

The spread of historical studies spurred reevaluations of Aristotle, based on new editions that provided more accurate access to ancient writings. Knowledge of the medieval world opened the possibility for revised conceptions of Aristotelianism. Broad reading, extreme erudition, and a nuanced understanding of historical context ushered in sophisticated if partisan understandings of the history of Aristotle's legacy. These readings emphasized Aristotle's pagan background, the controversial circumstances in which his thought combined with Christian dogma in the late Middle Ages, and the effects of the Council of Trent on freedom of philosophical discourse. Historical investigations suggested that the institutionalization of Aristotelian thought as a handmaiden to theology after the Council of Trent differed from late antiquity when Church Fathers found Aristotle a source for heresy; from the Middle Ages, when Aristotle was condemned in Paris; and, from the early sixteenth century, when, scholars openly distinguished philosophical investigations from theological ones.

Despite the use of primary sources and the invocation of historical context in the interpretation of the history of Aristotelianism, the seventeenth-century historical analyses emerged from the concerns of the time. The history of philosophy altered discussions about religion, epistemology, and natural philosophy itself. The wavering acceptability of Aristotelianism over the centuries suggested that the Council of Trent's and Jesuit's promotion of Aristotle as an authority, if not the authority, for philosophy defied the traditions of the Catholic Church. The multitude of thinkers, including Scotus and Pomponazzi, who believed philosophy could not prove the immortality of the soul, became allies to seventeenth-century skeptics, such as La Mothe Le Vayer, who questioned natural reason's ability to explain the mysteries of religion. Gassendi's historically minded critiques of Aristotle attacked those who believed that deviations from Aristotle were impious.

Besides his criticisms of Aristotle, Gassendi promoted Epicurean atomism as an alternative to traditional natural philosophies. Thus he used his knowledge of ancient philosophy to topple Aristotle, resurrecting in its place an atomistic natural philosophy, which he believed was more compatible to Christianity. His atomism grew out of historical investigations, not just through the discovery of the Epicurean explanations that served as models for his conceptions of atoms and voids but also by delineating the impiety of Aristotle. For Gassendi, the development of novel versions of natural philosophy involved looking back to the ancients in order to understand both the failings of Aristotle and the successes of Epicurus. Natural philosophy and antiquarianism were different tools used to arrive at a single goal because knowledge of ancient thought was embedded in the forging of new tools for thinking about nature. A number of seventeenth-century thinkers connected looking at the past with creating new versions of natural philosophy. The next chapter discusses how the perception that Aristotelian thought was impious influenced other aspects of the scientific revolution.

CHAPTER EIGHT

The New Sciences, Religion, and the Struggle over Aristotle

In a 1647 treatise addressed to Mersenne, a Capuchin monk from Warsaw named Valeriano Magni stated, "Atheism [atheismus] is such a crime that no other one equally touches the anger of God. Atheism is such an evil that there is nothing more dangerous to the human race."[1] Despite writing this sentence before the Peace of Westphalia, when Catholics and Protestants still waged war against each other throughout much of Europe, Magni did not equate atheism with Protestantism. Rather, this "tyrant," who was "more pernicious than any other heresiarch" from any other time or place, was praised "by pagans, by Christians, by Catholics, by heretics."[2] This "tyrant" was Aristotle.

Calling the short treatise *On the Atheism of Aristotle*, Magni wrote, "Aristotle ushers in atheism with arguments such that more effective ones are inconceivable."[3] These arguments maintain the world is eternal and uncreated, "ruled by the fatal necessity of the motions of the heaven;" remove God as an efficient cause; and attempt to establish the mortality of the rational soul.[4] Labeling Aristotle an "infidel" because he eliminated creating and ruling the world from God's duties, Magni referred to unnamed "atheists supported by the authority and doctrine of Aristotle" and to "the blameworthy Stagirite atheism." He also described others of "twisting [Aristotle] to the dogmas of the faith," implying that attempts to reconcile Aristotle with Christianity, such as Thomas's, deliberately misread Aristotle's writings.[5] Calling him a tyrant suggested that Aristotle usurped the freedom to devise alternative philosophies.

Magni's polemic lacked the detailed historical analysis of Jean de Launoy's, Pierre Gassendi's, and Tommaso Campanella's works, yet shared their conclusions. Magni's other interests recall Campanella's assertion that as a result of

Aristotle's "greatest impiety . . . it is permissible to forge a new philosophy."[6] In addition to detailing Aristotle's supposed atheism, Magni argued that the conceptual foundations of traditional natural philosophy were incorrect. For example, he argued that the belief in four material elements was wrong because experiential evidence showed water to be incorruptible and therefore not capable of transforming into air or earth.[7] Magni also tried to prove experimentally that vacuums exist in nature through demonstrations that were so similar to Evangelista Torricelli's that they garnered accusations of plagiarism.[8] Whether or not he knew of those experiments, Magni's experimentalism coincided with his view of Aristotle's impiety: both provided sufficient reasons for rejecting Aristotelianism.

Magni was by no means the first to cite Aristotle's impiety as a justification for innovations in natural philosophy or logic. In the sixteenth century both Petrus Ramus and Bernardino Telesio cited Aristotle's impiety as a motivation for their works, and, in the early part of the seventeenth century, Gassendi and Campanella did the same. Magni's condemnation of Aristotle for atheism indicates the continuation of this strategy among proponents of new or alternative philosophies during the seventeenth century.

Throughout the 1600s, universities and ecclesiastical authorities out of fear of heresy condemned teaching novel natural philosophies. The counterattacks provoked by these bans emphasized the supposed impiety or atheism of Aristotle, subverting past fusions of Aristotelian concepts and Christian dogma. They argued that using the philosophy of a pagan of dubious moral character led to gravely mistaken conclusions about God, creation, and the human soul. These assaults on Aristotle emerged not just among Polish clergymen, such as Magni, but primarily in France, the Netherlands, England, and, to a lesser extent, Italy. A number of thinkers contended Cartesianism or atomism was more compatible with Christianity than were Aristotle, Pomponazzi, or Averroes. Histories of Aristotelianism provided proof for their arguments.

ENGLAND AND ANTI-ARISTOTELIANISM

The relative weakness of early Renaissance English universities coupled with the absence of Jesuit schools made Aristotelianism far weaker in England than in much of continental Europe.[9] Nevertheless, versions of Aristotelianism influenced by humanism took root in Oxford during the last decades of the sixteenth century. Giordano Bruno sneered at Oxonian professors for being pedants obsessed with grammar and ignorant of the medieval traditions of metaphysics that had once flourished there.[10] Whether or not Bruno's characterizations were accurate, in 1586, Oxford, in accordance with humanist-influenced versions of Aristotelianism, enacted a decree that required literal expositions of Aristotle, thereby reducing the likelihood of public lectures that deviated from the ancient texts or that were bogged down by distinctions derived from medieval Scholasticism.[11]

In England, criticism of Aristotelians came not just from renegade Italians, such as Bruno, but also from within. English proponents of Platonism, Mosaic philosophy, and experimental and mechanical philosophies raised attacks against Aristotle and his supposed stranglehold on universities. Although English universities enjoyed a long tradition of questioning Aristotle's authority, competing institutions of learning, such as Gresham College and later the Royal Society, enabled critics of Aristotelianism to give an even greater voice to their views.[12] The absence of inquisitors and the emergence of latitudinarian attitudes toward religion during the Restoration fostered the proliferation of such dissent.[13]

Still, English anti-Aristotelian polemics owed much to a broader European setting. The Hermetic and Neoplatonic philosophies that had influenced Francesco Patrizi, Bruno, and Campanella took on new life in Cambridge. Pomponazzi's and other Italians' interpretations of Aristotle gained authority as accurate interpretations of an impious pagan. Controversies over university teaching echoed French and Dutch debates about the freedom of philosophizing. These English critiques of Aristotelianism, which opened a path for new natural philosophies, frequently noted the lack of correspondence between Christianity and Aristotle's natural philosophy.

England's religious landscape, in which many saw Catholicism, on the one hand, and anti-Trinitarianism, on the other, as dual threats, inspired both novelties in natural philosophy and fears over Aristotelianism.[14] Proponents of mechanical natural philosophies contended that their new systems gave a better account of God's providence than their rivals. Experimentalists believed their method avoided the metaphysical approaches that depended on dogmatism rather than direct considerations of God's works. And, the followers of Mosaic philosophies demanded understandings of nature that corresponded to scripture, instead of to the Scholastic tradition as Catholic orthodoxy demanded.

The two leading branches of Aristotelianism mapped nicely onto the main religious targets of English divines. In their eyes, the neo-Scholasticism championed by Jesuits corresponded to papist dogmatism. To the contrary, the naturalism of Pomponazzi led to denials of God's providence and the immortality of the soul, denials that English thinkers, such as Robert Boyle, identified with atheism and anti-Trinitarian Socinianism, even if it is unlikely that Pomponazzi influenced Fausto Sozzini.[15] Anabaptists and Socinians believed the soul dies or sleeps in between death and resurrection. Philosophical arguments that attempted to demonstrate the impossibility of proving the immortality of the soul might bolster the reasonableness of this doctrine.[16]

From its inception, Calvinism had been concerned with idolatry, as John Calvin condemned Catholic rituals and practices as idolatrous because of their supposed worship of images.[17] By the beginning of the seventeenth century, historical accounts of the origins of idolatry attempted to explain the source of what were considered to be the religious errors of paganism.[18] One influential and exhaustive account, that by the Dutch savant Gerardus Johannes Vossius, traced the origin of paganism to the incorrect identification of nature with God, a charged that was leveled at Aristotle.[19] For example, in 1605 Phillipe de Mornay (Du Plessis), a French Protestant with Platonist leanings, published a book, *De veritate religionis*, that traces the history of philosophy from Hermes to the end of antiquity. In it he wrote that Aristotle's and Averroes' emphasis on nature and belief in the utility of sacrifice to God was a form of impious paganism. He characterized Peripatetics as a superstitious sect that tarnishes God's glory by giving dignity to various gods and demons, by considering God a living creature, and by attributing divine powers to the heavens and stars.[20] In his view, Aristotle "investigated nature rather than God as author of nature."[21]

As a result of these religious and historical considerations, English polemics emphasized Aristotle's paganism in relation to his conception of nature as well as the uncertainty of his belief in immortality of the rational soul, his denial of the creation of the world, and removal of God's providence over humankind. Richard Bostocke wrote one of the earliest English attacks on Aristotle in his 1585 *Auncient Physics*, a book that promoted Hermeticism in place of Galenic medicine and traditional natural philosophy.[22] Whereas Scholastics dismissed particular positions of Aristotle without destroying the entire edifice, Bostocke believed the impious propositions in Aristotle's writing were interwoven throughout his entire thought, which formed a "heathnish Philosophie." According to Bostocke, the celestial orbs in Aristotle's cosmology, which are identified with divine intelligences, necessitate eternal motions and substances separated from God, who is "onely so much the more excellent then the rest [of the intelligences]."[23] It follows from Aristotle's positions, therefore, "that God medleth not under the Moone, and that he is not the maker nor the creatour of any thing but onely the mover of the heaven."[24] Thus Aristotle denied God's providence in the sublunary realm as well as God's creation.

Aristotle's misunderstanding of God and the cosmos accompanied false beliefs about humans and other earthly beings, according to Bostocke. He chastised Aristotle because he "maketh no mention of the immortalitie of the soule, neither doth he attribute any felicitie to it after the death of man" and thereby led Alexander of Aphrodisias to conclude the soul is mortal.[25] Bostocke cited a passage from the *Physics* that reads "Man and sun generates man" as evidence that Aristotle believed that both the human species and the sun are eternal and infinite as well as suggesting that heat is the cause of human generation.[26] Bostocke linked this passage to physicians who held that substantial forms can arise out of an underlying mixture rather than directly from God. Although Bostocke did not name Averroes, his view that Aristotle's "heathnish Philosophie doth not admit any Metaphisicall principle in naturall thinges," corresponded to Averroes' conception of natural philosophy. According to Bostocke, "Such naturall Philosophie is the next way to make men forget thee, O God, and to become Atheists," as it is deals with nature without "considering the creator."[27] Therefore, for Bostocke, Alexander of Aphrodisias correctly interpreted Aristotle's pagan thought, which is indelibly marked by a naturalism that restricts the powers of the divine and causes atheism.

That Aristotle was a pagan and his philosophy emphasized nature at the expense of the divine became a frequent point in English attacks on Peripatetic philosophy. Thomas Lydiat, a mathematician and astronomer, who had trained and taught at Oxford before dedicating himself to chronology, lamented that those who contended that Aristotle's positions did not conform to the scriptures and were "true according to physics but not theology." This tactic, which he presumably equated with Italian natural philosophers, "was nothing more than an attempt to expound theology with the scoff of godless men [hominum atheorum]."[28] Lydiat's 1605 work on the nature of the heavens and the elements attempted to explain the universe by employing physical causes that conformed to holy scripture even if they differed from "the minds of pagan philosophers, especially Aristotelians."[29] His theory aimed to "demonstrate the same thing is true physically and theologically," thereby providing an alternative to the approaches of Italian naturalists.[30]

Francis Bacon's polemics against Aristotle perhaps exceeded Lydiat's and Bostocke's in their vehemence. His *Refutation of the Philosophies* (1608) argued that Aristotle had uncharitably dismissed the Presocratics' philosophies. Even if Bacon's atomism differed conceptually correspond to Democritus's, he wished to reestablish Democritean thought and thereby compared Aristotle's treatment of the earlier materialist philosophers to that of "the Prince of Imposture, the Anti-Christ."[31] In the 1621 *De augmentis scientiarum*, Bacon contended that Aristotle had replaced God with the concept of nature, despite disagreeing with Bostocke's position that he had separated the study of nature from metaphysics. Rather, according to Bacon, Aristotelians misplaced the inquiry into final causes by making it part of physics rather than metaphysics. The result of this misplacement meant that natural philosophers vainly searched for teleological explanations in nature rather than "actively pressing the inquiry of those [causes] which are really and truly physical; to the great arrest and prejudice of science." Despite this faulty conception of final causes being common not just to Aristotle but to Galen and Plato as well, Aristotle's interest in teleology was more pernicious than Plato's, according to Bacon. Aristotle "left out the fountain of final causes, namely God, and substituted Nature for God; and took in final causes themselves as the lover of logic rather than theology." By shifting the source of final causes to nature, Aristotle "had no further need of a God."[32] Consequently, Bacon saw Aristotelian teleology as being doubly pernicious; it directed inquiries away from physical causation and conceived of purpose in the universe as having a natural rather than a divine source.

The reasons for Bacon's rejection of Aristotle were multifold. Many of Bacon's motivations stemmed from his desire to reform method, as is evident in his concerns with the proper application of teleology, which he associated with the impiety of Aristotle's natural philosophy. Theology affected his vision of a reformed natural philosophy as well. Just as a number of sixteenth-century thinkers had done, Bacon emphasized the importance of early Church Fathers, prizing Augustine's theology. Part of the attraction of Augustinian thought was its usefulness in crafting arguments against Calvin that maintained God's omnipotence while still allowing for the existence of evil.[33]

Ramus's and Patrizi's polemics provide additional links between Bacon and the humanists' demonization of Aristotle. In the 1580s, Ramists debated defenders of Aristotle in Cambridge, where Bacon had been a student.[34] Although these debates centered around rhetoric and demonstration, familiarity with Ramus might have fostered the growth of his religious sentiments against Aristotle, even if by 1605 Bacon rejected Ramism.[35] Seventeenth-century English scholars saw similarities between Bacon's and Patrizi's negative description of Aristotle. In 1665, Joseph Glanvill, a promoter of mitigated skepticism and of the Royal Society, which itself was partially inspired by Bacon's thought, likened Bacon's accusation that Aristotle was uncharitable to his predecessors to Patrizi's earlier critique of the same issue.[36]

English polemics that pointed to the impiety of Aristotle owed much to the judgments of Patrizi, Campanella, and Gassendi. Yet they differed as a result of their persistent hostility toward Catholicism and anti-Trinitarianism. English thinkers, following the interpretations of Pomponazzi, pointed to the heterodoxy

THE NEW SCIENCES, RELIGION, AND THE STRUGGLE OVER ARISTOTLE   151

of Aristotle's positions such as the mortality of the soul and the eternity of the world, just as Campanella and Gassendi had done. However, these English thinkers also dismissed Aristotelianism because it was too closely linked to Catholicism and because they saw it as not truly a philosophical investigation into nature but merely a branch of theology devoted to maintaining the existence of immaterial beings, for which the only evidence of their reality derived from Catholic traditions. For example, in his 1638 *The Discovery of a World in the Moone*, the polymath John Wilkins contended that the "question is much controverted by the Romish divines" of whether it is "dangerous to admit of such opinions that doe destroy those principles of Aristotle."[37] In this book, Wilkins argued, in line with Galileo and against Aristotelians, that the moon has mountains and valleys similar to those of the earth. Wilkins defended his deviations from traditional natural philosophy by contending that "more dangerous opinions have proceeded [from Aristotle]: as that the world is eternall, that God cannot have while to looke after these inferiour things, that after death there is no reward or punishment, and such like blasphemies, which strike directly at the fundamentalls of our Religion."[38] Therefore, according to Wilkins, disagreeing with Aristotle is less problematic than following him, and the study of astronomy using new observations helps demonstrate God's providence.[39]

In *Leviathan*, Thomas Hobbes attacked universities' adherence to Aristotelianism by associating it with Catholicism. Hobbes had spent some of the 1630s in Paris and therefore was likely aware of the polemics surrounding Etienne De Clave, Gassendi, and Jean-Baptiste Morin.[40] In his view, mixing scripture with the "vain and erroneous philosophy of the Greeks, especially of Aristotle" was one of the four causes of "spirituall darknesse." Hobbes incorrectly traced this mixing of pagan philosophy with scripture to the time of Charlemagne. He falsely believed Pope Leo III established a university in Paris. Because of the supposed papal origins of the university, Aristotle was altered to sustain the Catholic Church's dogma.[41] Accordingly, Hobbes wrote that even now those in universities only study philosophy "as a handmaid to the Roman Religion," although perhaps Hobbes was using the reference to Catholicism to underhandedly attack Presbyterians as well.[42] What they study at universities, according to Hobbes, "is not properly philosophy (the nature whereof dependeth not on authors,) but *Aristoteleity*."[43] Hobbes thought the mixing of pagan philosophy and theology had unfortunate consequences for not just philosophy but also religion. Doctrines derived from Aristotle surrounding separated essence, transubstantiation, and the soul were all examples of "the errors, which are brought into the Church, from the *entities* and *essences* of Aristotle."[44] Hobbes's view that scripture provided no evidence for immaterial beings may have remained unorthodox to most, but his critique of universities for tying theology to Aristotle resonated in disputes over university instruction that took place in the 1650s.

In 1654, the polemicist John Webster attacked universities' Aristotelian curricula, criticizing Aristotelian concepts, application of logic, and supposed matters of fact. He also attacked Aristotle's piety and morality, influenced, most likely, from his reading of Gassendi and the ancient biographical tradition. Webster emphasized that Aristotle was not just a "heathen," but because of his recourse to oracles also "diabolical." As a result, according to Webster, "neither were his principles and tenents [sic] any whit differing from such Diabolical directions." Webster listed such principles as his belief that God is an animal and does not concern himself with "minute and small things," his denials of creation, of the resurrection of the dead, and of the immortality of the soul. Webster's personal attack repeated some of the stock examples of Aristotle's moral failings: that he was involved in a plot to poison Alexander the Great, sacrificed "to his meretricious mistris," and was ungrateful to his teacher Plato.[45]

Seth Ward and John Wilkins, both affiliated with Oxford, responded that Webster mischaracterized the extent to which Aristotle dominated English universities. Ward and Wilkins asked rhetorically, "But where is it . . . that Aristotle is preferred before Christ? Is it at Oxford or at Cambridge? Are not the Christian Ethicks of Daneus, Scultetus, Amesius, Aquinas . . . preferred before him in the Universities?"[46] That Wilkins was a warden of a college at Oxford and the author of anti-Aristotelian treatises suggests that Webster exaggerated the degree of intellectual intolerance in English universities.[47] Nevertheless, Ward's and Wilkins's apology for the university defended Aristotle from the charge of impiety, maintaining that Aristotle acknowledged God's existence in his concept of the prime mover, a near reversal of the view found in Wilkins's 1638 treatise on the lunar landscape.[48]

Webster favored new natural philosophies and Mosaic thought as alternatives to Aristotle, citing Paracelsus, Bacon, Gassendi, Kepler, Descartes, William Harvey, and William Gilbert among others as good examples. Webster's positive assessment of Mosaic thought highlights two major sources for critiques of Aristotle's piety in England: Robert Fludd (1574–1638) and John Amos Comenius (1592–1670). In the 1630s Fludd promoted his "Mosaicall Philosophy" by tearing down Aristotle with invectives reminiscent of humanists from previous centuries. Gassendi and Fludd, despite their differences, both agreed that Aristotle's thought did not conform to Christianity. Comenius, spurred by Baconian ideas for reforming knowledge, rued the combination of "heathen philosophy with

Christian Religion."[49] The result of this combination was that "haeresie sprung out of haeresie, till . . . all things degenerated into Antichristianisme" and that this "Antichristianisme" meant that "Aristotle had an aequall share with Christ."[50] Comenius spent several years in England during the 1640s. His work on natural philosophy, originally published in Latin in 1633, was translated into English in 1651. It was perhaps the most influential work in early modern England that had its goal to base natural philosophy entirely on scripture.[51]

Similar to Comenius, Fludd contended that "[i]n this later age of the world, in which Satan, the prince of this world which is darkness, hath the upper hand." Christian philosophers adore and follow "their Master *Aristotle*, as if he were another Jesus rained down from heaven." Fludd reported that some of Aristotle's followers "become Atheists, and will acknowledge no God." Unlike humanists who aimed to reconcile pagan thought with scripture, Fludd believed "[h]eathen men were ignorant in the mysteries and abstruse operations of God." Proof of Aristotle's ignorance in these mysteries came from his naturalism, in particular that found in his meteorology. By considering "lightning and thunder, to be a common natural thing," rather than being "the immediate works of God's hand," these "pages or followers of the Ethnick Philosophers" reject the idea the God acts "but medieately, namely, by other necessary natural or supernatural means, as essential efficient causes."[52] Thus somewhat paradoxically, Fludd believed Aristotle's philosophy to be "diabolicall" because of its exclusive dependence on natural causation. Not admitting the existence of demons defined him as demonic.

Fludd returned to Moses in order to discover the foundations of natural philosophy. Other scholars in England sought out ancient pagan authorities besides Aristotle. Cambridge Platonists, such as Henry More (1614–1687) and Ralph Cudworth (1617–1688), adopted humanist views of wisdom that ultimately stemmed from Ficino, Steuco, and other Italian Renaissance Platonists. More and Cudworth both believed in the possibility of reconciling theology and philosophy, which required the acknowledgment of the existence of immaterial beings. More contrasted Aristotle's God as prime mover with not just Platonism but with "all the established religions of the civilized parts of the world."[53] In More's view, "Pomponatius, Cardan and Vaninus follow" Aristotle. Consequently, More believed Aristotle's God was limited in power and that the notorious Italians interpreted him correctly.[54] Even though Cudworth preferred Aristotle over Democritus and Descartes, he noted that Aristotle failed to proclaim "the immortality of human souls and providence over men."[55] Cudworth's interpretation of Aristotle's psychology has affinities to Pomponazzi's. He concluded that Aristotle's affirmation of the "corporeity of sensitive souls" and failure to assert the "incorporeity of the rational" meant that it was "left doubtful whether he acknowledged any thing incorporeal and immortal at all in us."[56] Accordingly, the Cambridge Platonists' critique of Aristotle's theology used Pomponazzi rather than Thomas for determining what Aristotle truly thought.

Walter Charleton (1619–1707) disagreed with Cudworth's judgment on the relative piety of atomists and Aristotle. Charleton, greatly influenced by Gassendi, promoted Epicurean philosophy as an antidote to the "darknes of atheism" that he believed afflicted his times. Following Gassendi and Diogenes Laertius, Charleton described Aristotle as "an *Ethnick*, poysoned with the *Macedonian* and *Grecian Idolatry*, nay so given over to that sottish impiety *Polytheisme*, that he could be content to make a *Goddesse* of his *Wench*."[57] Charleton pointed out that Aristotle "seems little better then wholly silent in all things that immediately concern *Theology*," although he conceded his silence was preferable to Plato's verbose but unhelpful treatment of the subject.[58] Prefacing his defense of atomism with an analysis of the history of philosophy, he used a sexual metaphor to explain why "*Junior Aristoteleans* . . . villifie and despise all doctrine, but that of the *Stagirite*." Aristotelian thought forms part of the "Female sect" of philosophy, Charleton wrote, "Because as women constantly retain their best affections for those who untied their Virgin Zone; so these will never be alienated from immoderately affecting those Authors who had the Maiden head of their minds." Charleton contrasted these adherents to the "Female sect" of philosophy with "assertors of philosophical liberty," such as Galileo, Tycho Brahe, Harvey, and Descartes, who staked out independent views on nature.[59]

Charleton was for a period of his life associated with the nascent Royal Society, where a number of its members used the impiety of Aristotelianism as an impetus for developing alternative experimentally based natural philosophies. Joseph Glanvill, a prominent defender of the Royal Society, its experimental practices, and mitigated skepticism, associated Aristotelianism with Catholic theology and the dogmatism of Jesuits, whom he dismissively thought "do but subtilly trifle."[60] In the *Vanity of Dogmatizing*, 1661, Glanvill wrote that scholastic theology is based on "steril, unsatisfying *Verbosities*" that grew out of Aristotle: "Thomas [is] but *Aristotle sainted*."[61] Yet Scholasticism misunderstood the impious nature of Aristotelian natural philosophy. Glanvill listed Aristotle's misconceptions about the possibility of bodily resurrection, God's knowledge, the eternity of the world, and God's idleness. Moreover, Glanvill pointed to the biographical tradition that linked Aristotle to paganism, "that he *philosophiz'd* by command from the Oracle."[62]

A few years later in 1668, Glanvill wrote a history of learning from antiquity to the present that had the ulterior motive of being a defense of the charge of atheism that was leveled at him by the Aristotelian Robert Crosse.[63] Following the lines of arguments found in Patrizi and Gassendi, Glanvill noted that the authorship of the Aristotelian corpus was uncertain and that it seemed probable that recent versions of his thought "made up a *Philosophy* that was *quite another thing* from" his actual writings. Glanvill repeated the views that Aristotle "was *impious* in his *Life* and *many* of his *Doctrines*," that later ancients preferred other philosophers, and that his authority was established in the Middle Ages or, as he put it, in *"Times* of *blackest Ignorance."* Therefore, followers of Christianity should not "bow down to the *Dictates* of an *Idolater* and an *Heathen*."[64] His criticisms border on incoherency by his asserting that the books attributed to Aristotle were not his, but the authority that stems from them, nevertheless, is unwarranted since Aristotle was an impious idolater. Nevertheless, in this manner, he aimed to subvert the authority that Aristotelian thought held because of its perceived proximity to Christian dogma. In Glanvill's eyes, seventeenth-century Aristotelianism developed out of medieval Catholic theologians' perverse transformation of the corrupt texts of a impious heathen into a system of vain disputation.

Samuel Parker (1640–1688), who would become bishop of Oxford in the last years of his life, attacked numerous theological enemies in his 1678 *Disputations on God and Divine Providence:* Epicurus, Vanini, and Hobbes among them. He too believed that those who followed Aristotle "built their theology on the foundations of impiety."[65] According to Parker, Aristotle was the "prince of all atheists and a more open enemy to religion than Epicurus."[66] Just as Taurellus had done nearly a century earlier, Parker pointed to Cesalpino as a prime example of an impious Aristotelian, who destroyed the concept of providence by making the prime mover a speculative rather than active intelligence and believed that humans and other animals could arise by chance through spontaneous generation. Parker cast aspersions against the Catholic Church by expressing incredulity that Cesalpino's writings seemingly had the approval of the papacy.[67]

Charleton, Glanvill, Fludd, Parker, and the Cambridge Platonists appropriated the rhetoric of Gassendi, Ramus, and earlier humanists such as Vives. Their focus on Aristotle's impiety was not a mere stratagem to lessen Aristotle's authority but reflected real concerns about natural philosophy and religion. Just as the frequent references to factual errors in Aristotle's works, such as his ignorance of the New World, reflected actual motivations for discarding his texts as the foundations of university instruction, his lack of conformity to Christianity impelled those who devised alternatives to his natural philosophy. Robert Boyle's works illustrate the sincerity of those in the Royal Society who decried Aristotle's impiety.

Before Boyle became famous for his practice of the experimental philoso-phy, his intellectual interests pointed toward religion as much as they did toward nature. Boyle associated with many of those who attacked Aristotle for impiety, including Wilkins and the Cambridge Platonists; and his library contained books by Comenius.[68] In some of his earliest writings, namely his unpublished *Essay of the Holy Scriptures*, written from 1649 to 1654, he linked Aristotle and Pompo-nazzi to atheism and, perhaps more pertinent to his times, to anti-Trinitarianism and Socinianism.[69] He associated "divers Modern Socinians" with Aristotle, Galen, Seneca, Epicurus, and Pomponazzi, because he believed they denied the immortality of the soul.[70] He wrote, "Those bigg Names of Aristotle, Galen, Machiavel, Lucretius, Pomponatius or Socinus, doe not much fright me from the Truths they are sayd to oppose," as he accused them of having disputed "with much Disadvantage to Religion."[71] Boyle's concerns with Aristotle's and Pom-ponazzi's influence on those whom he called atheists, libertines, and Socinians continued in his later published works.

Boyle believed that his mechanical philosophy solved the theological prob-lems created by ancient atomism and Aristotelianism. He posited that God cre-ated the corpuscles that comprise the world and devised and sustained principles that govern their movement.[72] According to Boyle, since God designed the world all rationality stems from the divine. Human rational souls and angels remain immaterial in his system, allowing Boyle to distinguish between spiritual and physical realms.[73] In *The Usefulness of Natural Philosophy*, written in the 1650s but not printed until 1663, Boyle outlined these theological considerations, dis-tancing his corpuscular theory from Leucippus's and Epicurus's atomism and from "Adorers of *Aristotle's* Writings, [who] have pretended to be able to explicate the first *Beginning of Things*, and the Worlds *Phaenomena*, without taking in or acknowledging any Divine Author of it."[74] More than a decade later in *Reason and Religion*, 1675, Boyle attacked once again Epicureans and libertines. Boyle believed these "new Libertines" outdid the older "Infidels," who followed Aristo-tle. In spite of this assessment, he painted a negative picture of Aristotelianism. He noted that "*Pomponatius* and *Vaninus*" based their impious teachings on ar-guments borrowed "from *Aristotle*, or the Peripatetick School."[75] Citing Mornay and Vives as furnishing the "proper Weapons" against these impious thinkers, Boyle continued and maintained that "*Aristotle* being himself a deark and dubi-ous Writer" fostered unorthodoxy in religious matters.[76]

In *Free Enquiry into the Vulgarly Received Notion of Nature*, written in the 1660s but not published until 1686, Boyle favorably contrasted the mechanical philosophy to Scholasticism and Aristotle's thought.[77] Tracing the history of idolatry back to the ancient Sabians of Persia, Boyle accused Aristotle of following in their tradition of worshipping nature.[78] Thereby echoing Bacon's and others' view that Aristotle replaced God with nature, he maintained that, "[b]y ascribing the admirable Works of God, to what he calls *Nature*, he [i.e., Aristotle] tacitly denies him [i.e., God] the Government of the World." As a result Aristotle's philosophy is "more unfriendly, not to say pernicious . . . than those of several other Heathen Philosophers." Boyle proudly proclaimed that Aristotle's "Opinion hinders me not at all from acknowledging God to be the Author of the Universe, and the continual Preserver and Upholder of it."[79] Unlike Aristotle, Boyle's system took account of God's providence.

Boyle's late work, *The Christian Virtuoso*, 1691, continued his assault on Aristotle's piety. He contended that the doctrine of some Aristotelians that substantial forms "are educ'd out of the Power or Potentiality of the Matter" leads to the conviction of the mortality of the soul, thereby giving "great advantage to Atheists, and Cavillers." Because "*Aristotle* was a Heathen, and destitute of divine revelation," his philosophy has misled Aristotelians into believing that "philosophy and theology are incompatible." Boyle attributed the success of Aristotle's philosophy among Christian and Jewish philosophers to "luck," just as Campanella had, rather than to any intrinsic value it held for understanding God or nature.[80]

Boyle defended his new philosophy as superior to Scholastic thought. He argued that the mechanical philosophy was not incompatible with the proposition "[t]hat God governs the World he has made." For Boyle, this governing of the world extended not just to human beings but also to "such Small and Abject Ones, as Flies, Ants, Fleas, &c."[81] These views he contrasted to "many (especially *Aristotelian*) Deists," who do not even credit God with creating the universe. His colleagues in the Royal Society agreed. The naturalist John Ray dismissed Aristotle's explanations of monsters as nature's errors by arguing "[a]n Omnipotent Agent would always do its Work infallibly and irresistibly."[82] Similarly, Robert Hooke used observations aided by microscopes to give experiential evidence that "the Wisdom and Providence of the All-wise Creator is not less shewn in these small despicable creatures, Flies and Moths."[83] For Hooke, the microscope offered empirical evidence for God's providence throughout the sublunary world. Experimental and mechanical philosophy offered a pious alternative to Aristotle.

CONTINENTAL CLASHES

In the Netherlands and France, just as in England, proponents of new natural philosophies struck at the universities for their adherence to Aristotelianism. In contrast to England where attacks on universities exaggerated the conservatism of the curriculum, Dutch and French universities in fact banned rivals to Aristotle. In further contrast to England, the Dutch and French controversies largely circled around the acceptability of Cartesian thought. Yet some Dutch thinkers opposed anti-Trinitarianism, just as their English counterparts did, and concerns over natural philosophy partially overlapped with religious disputes. In France, arguments over Aristotelianism divided Jansenists, who favored Augustinian thought and leaned toward Cartesianism, and Jesuits who, by and large, still supported Thomism.[84] In the Netherlands, Remonstrants opposed the Calvinists entrenched in the universities. Dutch professors of theology, however, polemicized against Socinians for their belief that the natural reason of humans could not grasp the truths of faith, a view that resembles the skeptical fideism endorsed by Charron and La Mothe Le Vayer. As a result, the polemics in England and the Continent retained similarities, taking up Aristotle's piety and the *libertas philosophandi* for neoterics.

It was not only Catholic authorities who solidified the curricula of seventeenth-century universities in northern Europe. Even in Calvinist Netherlands, Aristotelianism offered a solution for molding a philosophically coherent theology. In the sixteenth century, Dutch universities, such as at Leiden before 1582, strove to reduce the influence of Scholasticism by not teaching metaphysics. Leaders of universities during the seventeenth century, however, emphasized the possibility of reconciling reason and faith through considerations of Aristotelian natural philosophy and metaphysics.[85] For example, the widely read professor at Leiden, Franco Burgersdijk (1590–1635), believed that considerations of nature and the authority of scripture demonstrated the unity of faith, knowledge, and the sensible world.[86]

By the 1630s, Dutch professors, despite their Calvinism, taught a version of Aristotelianism that largely conformed to the neo-Scholasticism that had taken hold among Jesuits. For example, Gisbertus Voetius, a professor at Utrecht, believed both in the Bible's primacy and that the foundations of theology corresponded to Aristotle's writings. For him, the best understanding of Aristotle and arguments for the compatibility of faith and reason came from Catholic thinkers such as Fonseca, Toletus, Perera, and Suarez.[87] Similar to these Catholic thinkers, Voetius held that the Aristotelian concept of substantial forms furnished the natural reasons for concepts such as body and soul or substance and accident, ideas that theology employed and ultimately derived from interpretations of scripture.[88] As a result, Voetius believed Descartes' philosophy threatened theology, because it explained all natural change, excluding the human soul, by the movement and extension of matter. Voetius and his colleague Martin Schoock argued against

Descartes and his follower Henricus Regius in a series of disputes that resulted in the banning of teaching Cartesian natural philosophy at Utrecht in 1642.[89]

Regius had become an extraordinary professor of medicine and botany at Utrecht in 1638, after returning from studying in Montpellier and Padua. At Padua he studied with the noted professor of medicine Santorio Santorio and the philosopher Cesare Cremonini, who famously defended himself from accusations of impiety by stating that he merely taught Aristotle's thought.[90] Regius, upon his reentry into the Netherlands, was among Descartes' earliest followers and incorporated his rejection of substantial forms into his university lectures. In his defense of Aristotelianism, Voetius not only tried to show that Cartesian novelties undermined Christian theology but also that the skepticism that was the foundation of philosophical discovery for Descartes was dangerous. According to Voetius, the promotion of philosophical liberty could lead to the delusion that something that "is true in theology is false in philosophy," seemingly a reference to the idea Etienne Tempier called "double truth."[91] In a 1636 disputation on the use of reason in matters of faith, Voetius equated Socinianism to skepticism about religious questions.[92] He linked the skeptical aspects of Cartesianism to Remonstrants, who, following Jacobus Arminius, sought toleration from the Calvinist Dutch state in the 1620s and 1630s.[93]

For Voetius, toleration led to atheism, not just as the result of Descartes' philosophy but also resulting from Aristotelians, namely Italians who believed that natural philosophy was, for certain issues, incompatible with the truths of faith. Thus Regius perhaps represented a double danger, since he corresponded with Descartes and his studies in Padua had introduced him to impious versions of Aristotelian thought. Voetius, in a disputation, *On Atheism*, held in 1639, equated atheism with Machiavellianism, skepticism, Epicureanism, libertinism, Socinianism, Remonstrantism, as well as with the rejection of God's providence, the dismissal of the immortality of the human soul, and the denial of bodily resurrection. He associated Italian professors in general with Averroes, "who said, with a loathing for the Christian religion, that he would prefer his soul to be with the philosophers."[94] Although the evidence for Pomponazzi's atheism was not conclusive for Voetius, the degree to which Pomponazzi followed Averroes and Avicenna in *On Incantations* and wavered in theological issues led Voetius to form a negative opinion of the Mantuan philosopher.[95]

Italian philosophers remained a target for Voetius in later writings. Even though they were left unnamed, the positions he rejected are similar to those of Pomponazzi, although Voetius could have had in mind more recent philosophers, such as Antonio Rocco, who favorably discussed Pomponazzi, or Claude

Bérigard, who endorsed Zabarella's ideas on the prime mover.[96] In a 1656 disputation, Voetius accused Italian Aristotelians of siding with atheism against Christianity and natural theology by idly prating about "that which in the study of
nature [in via naturae] is false or uncertain," such as the immortality of the soul,
the existence of demons, providence toward particulars, while conceding these
to be true according to scripture and Church traditions.[97] Voetius, defending
Aristotelianism as ancillary to theology, carefully delineated which versions of
philosophy were acceptable and attributed Pomponazzi's failures to his appropriation of Arabic philosophy.

Elsewhere in the Netherlands, university professors found Descartes to be a
threat to the established Aristotelianism. In 1641 at Leiden, the university banned
lecturing on Descartes, and a new statute demanded that lecturers explain word
for word Aristotle's texts, explicating them aided by the interpretation of the ancient Greeks.[98] Six years later Adriaan Heereboord, a professor who hesitatingly
signed the ban of 1641, defended his right to teach Cartesian physics.[99] The decree's demand that professors teach the ancient Greek commentators followed
humanists' ideals, yet Heereboord perhaps surpassed their evocation of humanism by giving an oration on the libertas philosophandi as the principle that struck
at the university's censoring of Descartes. Tracing this principle to Petrarca, Lorenzo Valla, Gianfrancesco Pico della Mirandola, Juan Luis Vives, and Campanella, his oration applied their rhetoric in defense of the new sciences. Heereboord maintained that the freedom to philosophize was a Christian freedom,
and, echoing Campanella, he argued that "Christians are not obliged to subject
their minds to the authority of Aristotle or any other human."[100] Just as earlier
humanists had done, Heereboord attributed the inflexible obedience to Aristotle
that universities demanded to the restricting influence of Averroes. In a passage
so strongly reminiscent of Gassendi's Paradoxical Exercises as to suggest plagiarism, Heereboord repeated the idea that Aristotelians suppressed the freedom
to philosophize because Averroes contended that Aristotle wrote as if he was an
oracle and believed no one had found any errors in his work.[101]

Heereboord's identification of Averroes as the source of Aristotelian close-
mindedness shows not just the power of humanist rhetoric but the widespread
desire for seventeenth-century scholars to equate their opponents with Averroes.
Voetius, who defended Aristotelianism philosophy as subaltern to theology, criticized Pomponazzi and the "Italian school" for their allegiance to Averroes. In his
counter, Heereboord portrayed the defenders of Aristotelianism, such as Voetius,
in Dutch universities as obedient to Averroes' conception of Aristotle despite the
lack of influence Averroes actually held in the seventeenth century.

Throughout Europe, Averroes' weight diminished during the seventeenth century. Jesuits carefully deemphasized his views, eliminating them from their lectures and commentaries.[102] By the end of the sixteenth century, reliance on Averroes' commentaries weakened even in Italian universities, his traditional stronghold.[103] After having been printed in Italy repeatedly beginning in the 1470s, no one published Averroes' collected works after 1576.[104] In Italian universities after the middle of the sixteenth century, Averroes' position on the intellect was no longer philosophically attractive.[105] Descartes' reading of the Fifth Lateran Council illustrates the diminishing concerns over the unicity thesis. In the preface to the *Meditations*, 1641, Descartes cited the Fifth Lateran Council and its assertions of the primacy of the Catholic Church's teachings of the soul as expressed at the Council of Vienne. However, Descartes did not mention the passage that referred to the condemnation of the view that the soul is "one in all humans" (unica in cunctis hominibus) but only referred to the Council's objections to materialist and mortalist positions regarding the soul.[106] Even if Descartes' concerns with materialist psychology might have been defensive—reflecting the possibility that he was vulnerable to accusations that he proposed a mortal soul but not a single intellect for all humans—they also show that the philosophical attraction of Averroes' views on the soul had failed to register as a possible solution.

By the middle of the seventeenth century, Averroes' reputation would have been unrecognizable to those who partook in the controversies of the fifteenth and sixteenth centuries. The Averroes that appears in Thomas Browne's *Pseudodoxia epidemica*, first printed in 1646, is described as "a man of his owne faith," thus suggesting a degree of independence from the socially mandated norms. His "owne faith" recommended a degree of restraint similar to the Stoics, for example, his admonition that one should only consume wine to produce "a sober incalescence and regulated aestuation from wine"; Averroes' sobriety was in contrast to Avicenna's lushness, who, according to Browne, recommended getting drunk once a month.[107] For Browne, however, the image of Averroes is not just as a sommelier. He also was the source of dubious if not ridiculous theories about the natural world. Echoing Salutati from centuries before, Browne wrote that he had "scarce faith to beleeve" Averroes' assertion that women were capable of conceiving in the bath if there was a "seminall effluxion of a man admitted to bathe in some vicinity unto her."[108] Averroes, in Browne's account, has become unhinged from his notorious position about the unicity of the passive intellect and linked to naive tales of accidental pregnancy. For Browne, the *Averroristas* had become, as Christoph Clavius punned a half-century earlier, *Erroristas*.[109]

Averroes' flagging influence among scholars is further reflected in Théophile

Raynaud's judgment. In his 1653 work *On Good and Bad Books*, a guidebook to censorship, the Jesuit wrote, "Now Averroes, having been tossed from a bridge, has left the schools."[110] In his opinion, Machiavelli, Cardano, Pomponazzi, and proponents of magic had replaced the threat that Averroes once posed. Less than thirty years later, Gottfried Wilhelm Leibniz, in agreement with Raynaud, wrote in the perfect tense about "Averroists" who "once" believed in the theory of double truth, thereby suggesting that they no longer populated lecture halls.[111] Writing in the 1690s, Pierre Bayle supported these assessments. Noting that Ramon Llull had tried without success to have Averroes's commentaries condemned in the fourteenth century, Bayle wrote that there was no longer a need to proscribe his works because "[h]is authority is nonexistent, and no one wastes the time to read him."[112] By the end of the seventeenth century, Averroes might have been irrelevant, but he still remained etched in the collective imagination.

For example, bizarrely, Averroes became a point of reference in debates over the Eucharist, an issue left unexplored in the Muslim's writings. Despite Averroes' silence on this topic, Piotr Skarga (1536–1612), a Jesuit from Vilnius, who wrote against Zwingli's and Calvin's understanding of the Eucharist, likened Calvinists to Averroes. According to Skarga, Averroes, after having travelled the world to learn about various religions, determined that he could find no religion worse than Christianity because its adherents "devour and tear apart with their teeth the God whom they worship."[113] Thus Skarga concluded that Protestants' criticisms of transubstantiation were equivalent to those of this heretical Muslim philosopher. Instead of dismissing Skarga's citation as apocryphal, Calvinists turned the argument on its head. A century after Skarga, the Calvinist Charles Drelincourt cited a quotation attributed to Averroes that he "preferred that his soul would be with the Philosophers, seeing that the Christians eat what they worship," to suggest that the Catholic position on the Eucharist damaged Christianity because it provided material that nonbelievers could easily ridicule.[114] Jean Daillè, also Calvinist, used the witticism Skarga attributed to Averroes to argue that the Catholic position on the Eucharist did not correspond to the views of early Christians. According to Daillè, Averroes' witticism demonstrates that pagans, Jews, and heretics uniformly mock the doctrine of transubstantiation. Since there is no record of ancient pagans jesting about Christians eating God, Daillè concluded, that transubstantiation was not part of the early Church's dogma.[115] That Averroes did not actually write that witticism did not prevent the phrase that "he preferred his soul to be with the philosophers," a play on Balaam's wish in the book of Numbers to have his "soul die with the just," from becoming commonplace.[116]

Just as Averroes found a place in polemics between Calvinists and Catholics over the Eucharist, using Averroes' image of a sycophantic and irreligious Peripatetic became important in Cartesians' demonization of Aristotelianism. The real targets of many seventeenth-century denunciations of Averroes were the neo-Scholastic Aristotelians who held sway in the universities, such as Francisco Suarez and Francisco Toletus in Catholic universities and colleges or Voetius in the Netherlands.[117] Their piety, however, could not be questioned directly. Rather the strategy for anti-Aristotelians included associating the bans of teaching other philosophies with the alleged slavishness and impiety of Averroes. Averroes, a Muslim, long known for unorthodox doctrine and seen by humanists as the epitome of barbarous, poorly translated language, of the improper use of philosophy, and of disregard for Christianity, was ideal for casting doubt on Aristotle's own piety.

Consequently, a number of opponents of traditional university teachings attempted to profit from negative associations of Averroes. Johannes De Raei (1618–1702), a professor at Leiden who combined Cartesian and Aristotelian natural philosophy, prefaced his *Key to Natural Philosophy*, 1654, with a diatribe in which he concluded Averroes possessed "an incredible impiety and most dense ignorance of divine matters" because he allegedly called Moses, Jesus, and Muhammad impostors.[118] De Raei repeated other accusations familiar to humanists, including that Averroes used corrupt texts and distorted Aristotle's thought. Moreover Averroes' faith in Aristotle laid the path for the blind acceptance of "Aristotle's or Averroes' authority" such that no one dared to doubt their views.[119] Averroes had simultaneously sullied Aristotelian thought and removed the freedom of those in universities to dissent from Peripateticism.

Some French thinkers found Averroes to be an emblem of a version of natural philosophy that was more concerned with terminology than nature, just as Vives had. The Oratorian advocate of Cartesian philosophy Bernard Lamy, for example, believed that Aristotle's thought had been contorted so that it was unrecognizable and unintelligible. The blame lay on Averroes. In 1683 he wrote, "Today we understand this Philosopher [Aristotle] in another manner. It is not at all his philosophy that reigns in the schools, it is that of the Arabs."[120] Of these "Arabs," according to Lamy, "Averroes is the most weighty."[121] According to Lamy, "[t]he philosophy of Arabs is merely a form of questioning that applies the prejudices of childhood to the terminology of Aristotle."[122] In contrast to experientially based thought, Aristotle and Averroes merely manipulated words.

French Oratorians, such as Lamy, criticized Aristotle in response to a series of bans of Cartesian philosophy. In 1624 the Sorbonne had condemned De Clave,

Villon, and Bitaud for proposing an alternative matter theory to Aristotle's. By the 1660s attention turned to Descartes' philosophy, for which one of the major concerns was whether his corpuscular matter theory undermined Catholic teachings of the Eucharist as defined in 1551 at the Council of Trent. The Holy Office placed Descartes' works on the index of prohibited books in 1663 "until corrected," which they never were.[123] The index did not prevent Descartes' works from being read in France or in Protestant countries, yet French universities pushed Cartesianism out of their lecture halls. In 1662, the Faculty of Theology issued a censure requiring those in the Faculty of Art to follow Aristotelian philosophy instead of Epicureanism and Cartesianism. The principle problems of Cartesian thought they cited were its ramifications for theories of the Eucharist and animal and human souls.[124]

Affinities between Jansenism and Cartesianism added to worries about the introduction of novelties in philosophy instruction. In the 1660s the Sorbonne reaffirmed its Aristotelianism and in 1671 a royal decree demanded that, in order to protect the teachings of holy mysteries, universities and schools must enforce already existing bans on dangerous opinions, apparently a reference to the 1624 censure of atomism that was now to be applied to Cartesianism.[125] In accordance with the royal decree, the faculties of theology and medicine at the Sorbonne condemned Cartesian thought.[126] Throughout the seventeenth century, French universities censored Descartes. These condemnations often targeted Oratorians, many of whom were Jansenist and Cartesian. The University of Angers banned teaching Descartes in 1675, and Lamy was forced to leave.[127] Four separate times, from the years 1691 to 1705, professors at the Sorbonne were obliged to swear against Cartesian philosophy.[128]

Oratorians, Jansenists, Cartesians, and Gassendists struck back by pointing to the limitations of Aristotelian philosophy and its lack of conformity with Christianity. From the time of his teenage years, the Oratorian philosopher Nicolas Malebranche was dissatisfied with Aristotle's thought because of its perceived impiety.[129] Malebranche followed Descartes in a less than strict manner, developing an occasionalist philosophy. Yet he shared with the Cartesians the same enemy. In his *The Search after Truth*, 1674–75, he noted that many authorities questioned whether Aristotle believed in the immortality of the soul, indicating that Pomponazzi's goal was to show that the human soul is mortal according to Aristotle. As a result, Malebranche expressed disbelief in the great weight that some thinkers placed in Aristotle's views. The efforts made to determine what Aristotle thought about a question of faith was a sign of an "inversion of reason [that] certain men are shocked in philosophy if we speak differently from Aristotle, but not troubled if we speak differently from the Gospel, the Fathers, and the Councils in theology."[130] According to Malebranche, commentators on Aristotle suffered by following Averroes in thinking that Aristotle was infallible.[131]

The Gassendist François Bernier responded to the 1671 ban of Cartesianism and reinforcement of Aristotelianism within French universities by satirizing the royal decree. The mock decree, likely written with the assistance of Molière and Nicolas Boileau Despréaux, declared absurdities such as that all sailors must not circumnavigate the earth on the punishment that they become antipodes, a reference to the discarded Aristotelian view that the antipodes were uninhabited, and that windmills must turn not by the force of wind or water but by a "turnative" faculty, a reference to the multiplicity of qualities and faculties that Aristotelians and Galenists used to explain the powers of substances. The latter witticism anticipated Molière's "dormative virtue" used to ridicule philosophy in the Le Malade imaginaire, which was first performed in 1673.[132] Bernier pointed to critiques of Aristotle's religious views in the satire, mocking Aristotle for positing that the world is eternal and evoking his contemporaries' critiques of Averroes by calling Aristotle "the Genius of nature, the prince of philosophers, and the oracle of the university," which thereby supposedly gave Aristotle authority over and beyond both reason and experience.[133]

Bernier used satire, but it was not the only technique employed to weaken Aristotle's authority. An anonymous short treatise called The Several Reasons for Preventing the Censure or Condemnation of Descartes, which was most likely written by the Cartesian and Jansenist Antoine Arnauld (1612–1694), revisited the history of Aristotelianism in defense of Descartes' philosophy.[134] The treatise follows Launoy's history, citing the various condemnations of Aristotle beginning from the thirteenth century. The author contended that demanding a new philosophy, such as Descartes', to be in accord with the mystery of the Eucharist or other mysteries of faith is unreasonable, especially considering that "the philosophy of Aristotle is not more exempt than the others if one considers it in its purity."[135] A literal reading of Aristotle demonstrates that he deviated from the mysteries of the faith. Arnaud, if indeed he is the author, suggested that those who believe that faith should be submitted to reason were dangerous, giving the example of those Italian philosophers who followed Aristotle and Averroes and produced pestiferous doctrines such as the mortality of the soul and the lack of divine providence over human affairs. Strangely, one such Italian philosopher, Antonio Bernardi, is the example of a philosopher who argued that theology and philosophy should be separate.[136] Thus the treatise vilifies some sixteenth-century Italians for their impious conclusions drawn from Averroes and Aristotle and praises another for his distinguishing philosophical argument from the mysteries of the faith.

CONCLUSION

Throughout Western Europe during the seventeenth century, Aristotle's alleged impiety stimulated the search for alternative natural philosophies. Many of these alternatives looked backward. Some saw models for understanding nature in scripture. Aristotle's irreligion impelled others, such as Bacon, Sébastien Basso, and Bérigard, to base their theories on the Presocratics.[137] The atomism of the Epicureans formed the conceptual foundation for the Gassendists. And Plato served as a guide for More and Cudworth. The promoters of novel natural philosophies, such as Boyle's experimental and mechanical philosophy and Descartes' corpuscular philosophy, noted the advantage they held over Aristotelianism, which, the examples of Pomponazzi and Vanini supposedly had demonstrated, leads to heresy if not outright atheism. Attacking Aristotle as being impious was not merely a stratagem employed by Cartesians and experimentalists to defend themselves from accusations of irreligion. Descartes believed his philosophy could explain the Eucharist better and was surprised by the negative reception it received.[138] Followers of Descartes, such as Malebranche, from early on in their lives, were dissatisfied with Aristotelianism because of its deviations from Christianity. Boyle, as a young man, saw Aristotelianism as fomenting atheism or Socinianism and devised his mechanical conception of the world so that God's providence could play a role in the governing of the world.

The diffusion of historical research into Aristotle's biography aided the spread of the negative views about Aristotle's own person. Despite some histories of philosophy, such as Thomas Stanley's, which portrayed Aristotle as moderate in philosophy and as a good citizen, the cumulative effect of the work of Patrizi, Gassendi, Campanella, and Launoy, as well as the diffusion of new editions of the ancient biographical tradition was that belief in Aristotle's adherence to pagan religion became commonplace.[139] A few dissented. Muzio Pansa, contended that the agreement between Aristotle's and Christian understandings of God stemmed from Aristotle's drawing from Judaism.[140] Fortunio Liceti (1577–1657), a professor at Padua and later Bologna, also contended that correspondences between Aristotle's theology and those found in the Old Testament could be uncovered through a reading of the *De mundo*, a work now generally considered spurious and that by Liceti's time had been rejected as inauthentic by Julius Caesar Scaliger and Daniel Heinsius.[141] In Liceti's view *De mundo* shows that Thomas

was correct in believing that Aristotle held that there is providence in human affairs.[142] Moreover, according to Liceti, Aristotle came to this view because he had
read scripture.[143] Yet by the middle of the sixteenth century most scholars rejected
the evidence of De mundo and followed the Italian Aristotelians of the sixteenth
century, believing that Aristotle rejected divine providence in human affairs and
did not assert the human soul was immortal.

Even a few Jesuits questioned Aristotle's piety by the middle of the seventeenth
century. Niccolò Cabeo, a Jesuit whose 1646 commentary on Aristotle's Meteorology proposed a physics based on corpuscular matter theory derived from experiential evidence, railed against "Peripatetic atheists" (Atheistae peripatetici), who
maintained the eternity of the world.[144] Cabeo extended his critique to Aristotle.
He recounted the most likely apocryphal tale of Aristotle, after having read the
Old Testament, saying, "This author says much and proves nothing."[145] According to Cabeo, by ignoring the historical narration found in the Old Testament,
Aristotle erred in positing that rivers, seas, and mountains change location over
time. Cabeo believed that these geological arguments falsely "introduce the impious eternity of the world and Aristotle's atheism, making God act out of the
necessity of nature," as well as denying God's providence and foreknowledge.[146]
For him, Aristotle's rejection of the Old Testament compounded his confounding
of God with nature, a criticism similar to many seventeenth-century scholars.

The French Jesuit René Rapin published a balanced and thorough historical comparison of Plato and Aristotle in 1671, the same year Cartesianism was
condemned in Paris. His analysis put forward unfavorable characterizations of
Aristotle's life, such as that he quit studying in order to indulge in "libertinage,"
that he began to philosophize because of the oracle, and that he made sacrifices
to honor his wife as though she were a deity.[147] But, just as Launoy and Gassendi
had done, Rapin's history recounted the receptions and condemnations of Aristotle, leading him to conclude that "his fortune was so bizarre in the last centuries,
that it is difficult to understand how in the succession of time it has been possible
to make such different judgments on the same person."[148] Rapin recognized both
the flexibility of Aristotelianism as well as the historical circumstances that led it
to be combined with Christianity.

Rapin lamented the modern philosophers "who believe that the philosophy
of Aristotle is too old for fashionable minds and that it is possible to be knowledgeable without Plato and Aristotle."[149] Yet, shortly thereafter, members of his
order joined the fashionable minds of the day. The conversion of Parisian Jesuits
to Cartesianism in the first decades of the eighteenth century signaled the end of
the Aristotelian curriculum in French colleges.[150] The view that Aristotle was a

heathen—that his philosophy led to impiety or atheism—was not the only cause of his downfall. The motivations, however, of reformers of natural philosophy were in part religious and the increased difficulty in placing Aristotle among the pious added to the desire to dispense with him as an authority for understanding nature and the mysteries of the faith.

Conclusion

As early as Petrus Ramus in the sixteenth century, Aristotle's detractors labeled his followers as "atheists." Atheism in early modern Europe is notoriously difficult to define as well as to detect. Since atheism was a crime punishable by death, extremely few publicly declared their lack of belief in the divine. The supposed discoveries of atheists by polemicists, inquisitors, and civil prosecutors cannot be simply trusted given that they are likely to hide or twist the views of the prosecuted. The records of investigations and trials present the voice of inquisitors rather than that of the accused. Prudence, Nicodemism, and imposture undoubtedly led many to hide their views. Reading between the lines to uncover their supposedly true belief, however, creates the risk of introducing anachronism or the historian's biases into blank spaces of the textual record.[1]

The association of Aristotelianism with atheism followed the contours of polemics around heresy. Ramus linked the atheism of Aristotelianism with a denial of providence and absolute divine omnipotence. Mersenne grouped impious Aristotelians together with deists, skeptics, and Machiavellians; Boyle with Socinians and anti-Trinitarians. When Benedictus Spinoza produced sophisticated arguments that undermined deism, scholars linked Aristotle to the incipient forms of atheism that converge with the modern conception of atheism as the denial of the existence of the divine. Thus as atheism emerged in its modern sense, prominent thinkers considered Aristotle to be its underpinning.

BAYLE'S ARISTOTLE

Pierre Bayle's scholarly works, in particular his *Dictionnaire historique et critique*, were key to this final early modern transformation of Aristotle. Like his predecessors, Bayle used history and erudition to recast the Aristotelian tradition. He promoted intellectual toleration as one of the primary messages in his several entries on sixteenth-century Italian natural philosophers as well as on Aristotle and Averroes. One of Bayle's goals for his dictionary was to correct Louis Moreri's *Le grand dictionnaire historique*, published some twenty years before Bayle's great work. In a few instances surrounding Aristotelians, Bayle points out errors of fact, for example, Moreri's incorrect dating of Pomponazzi's death.[2] But Moreri's reference work is not rich with details on Aristotelians, and Bayle's inclusion of so many Aristotelians from the previous century reflects his own philosophical interests.

That Bayle included entries on Piccolomini, Pomponazzi, Zabarella, and Nifo was not the result of his work's exhaustiveness. He wrote no entries for medieval and arguably more important Aristotelians such as Thomas Aquinas, Ockham, or Scotus. Rather the inclusion of these Renaissance Italian figures resulted from their position in the polemics of seventeenth-century Europe, particularly in France and the Netherlands, and from Bayle's interest in Spinoza, Cartesian philosophy, and intellectual and religious liberty. Religion looms large in all these entries. Especially striking is the extent to which Bayle considered Aristotle's moral and religious life but discussed his philosophy only in a limited fashion. The portrait of Aristotle and his Italian followers that Bayle paints is one of lack of conformity to Christianity.

Not surprisingly, given Bayle's famed defense of the possibility of an ethical atheist, his skepticism, and perhaps his own unorthodoxy, he did not condemn their lack of conformity to Christianity. Rather, his interpretation of Aristotelianism illustrated the hypocrisy of universities and churches in their intolerance toward Cartesianism. On the one hand, the assimilation of Aristotle's thought with Christianity defies an accurate historical understanding of Greek philosophy. On the other hand, that the leading Aristotelians of the sixteenth century held Aristotelian thought to be incompatible with Christianity, positing views that, according to Bayle, were Spinozist *avant la lettre*, shows the unreasonableness of bans of Cartesian thinking. Whereas the Aristotelians attempted to demonstrate that philosophy could not prove the immortality of the soul or the existence of a providential God, Descartes at least gave what he believed were convincing proofs of these issues. In Bayle's eyes, rejecting Descartes in favor of Aristotle was unreasonable, because the former was a self-proclaimed Christian who believed reason could demonstrate the truth of Christianity and the latter was a pagan whose reconciliation with Christianity stemmed from historical accident rather than from a profound understanding of his life and views.

Bayle's entry on Aristotle highlights the tensions between the entrenched posi-

tion of Aristotelianism in theology and Aristotle's life and philosophical positions
that appear to oppose precepts of Christianity. Bayle noted that the acceptance
of Aristotle was common to both Catholic and Protestant theologians who "run
to help him as if to a fire, and are so strengthened with temporal powers against
the new philosophies, that it does not appear at all that he [Aristotle] will lose his
long-term domination."[3] For Bayle, Aristotle is inseparable from the struggles
between promoters of the new philosophies and orthodox theologies supported
by *parlements* and churches.

The first parts of Bayle's entry on Aristotle correct several common mistakes
about Aristotle's biography. These corrections point to the larger issue of religion
and moral character. Bayle cited those who believed Aristotle showed ingratitude
to Plato, that his marriage involved pagan idolatry, and that he was charged with
crimes of impiety that forced him to flee Athens. Perhaps more pertinent to contemporary debates, Bayle dismissed those who believed Aristotle was Jewish, indicating the origin of that error, and he cast doubt on those who linked Aristotle to
Judaism in general, a view that was still circulating among seventeenth-century
defenders of Aristotle's piety.[4]

From there, Bayle addressed the current status of Aristotle, particularly among
theologians. Bayle noted that a number of theologians admitted that certain ecclesiastical doctrines derive entirely from Aristotle.[5] Bayle was quick to point out
that the assimilation of Aristotle to religion was not unique to Christianity: Muslims also had done so. Bayle believed that Islamic theology had collapsed since
the Middle Ages but in Christian lands parliaments, persuaded of the excellence
of Aristotelianism, banned instruction in other philosophies. According to Bayle,
Aristotle's "most blind supporters" tried to whitewash his thought "by using forced
interpretations of the grandest and most impious absurdities of Aristotle."[6] These
impieties included the eternity of the universe, the denial of divine providence
for sublunary bodies, and the failure to recognize the immortality of the soul.
The abundance of citations in the entry lends additional credence to Aristotle's
impiety without Bayle's having to explicitly endorse the idea. For example, Bayle
noted that "the famed Capuchin" Valeriano Magni wrote on Aristotle's atheism,
Campanella wrote in a book, which was even approved by Vatican censors, that
maintained Aristotle opposed Christian faith, and David Hassel considered Aristotelianism akin to Spinozism.[7]

Despite these assessments, as well as those of the patristic writers, of Aristotle
as impious, according to Bayle, Aristotelianism "finds numerous protectors," the
most zealous of whom make their enemies heretics, thereby making Aristotle's
"well-being inseparable from that of theology, because it has accustomed the soul

to acquiesce without evidence."[8] Pessimistically, Bayle concluded that "the union of interests has given the Peripatetics the assurance of the immortality of their sect and given the new philosophies a reason to lower their hopes."[9] Contrary to Bayle, Aristotelianism was not immortal, and Bayle's historical evaluation of it contributed to its death.

Italian Aristotelians of the sixteenth century strongly colored Bayle's comparison of new philosophies with the traditional. Descartes and Spinoza are the standards to which Pomponazzi, Cesalpino, and Averroes are held. Just as La Mothe Le Vayer had done before him, Bayle defended Pomponazzi's views on the immortality of the soul as being in accordance with Christian doctrine. For Bayle, Pomponazzi was sincere and believed that "Scripture was . . . an unshakeable foundation for the belief that our soul is immortal."[10] Pomponazzi strongly believed in the immortality of the soul, but "he only sustained that natural reasons given for it are not solid and convincing."[11] In this manner Pomponazzi became a heroic ally to Bayle's skepticism and toleration because Pomponazzi demanded that "[f]or those things for which there are not philosophical proofs, everyone must enjoy the freedom to submit them to dispute, to examine them, and to say what seems true."[12] Unlike Campanella, Bayle saw in sixteenth-century Aristotelianism freedom of thought instead of a monster that tried to chase away all competitors.

Although Bayle noted that some had accused Pomponazzi of atheism, he found accusations of impiety unfounded. Considering that his contemporary Cartesians also did not accept that Aristotelian philosophy adequately proves the immortality of the soul, Bayle rhetorically asked, "Would it not be an extravagance to say that a Cartesian who said that is impious or atheist?"[13] Bayle ironically concluded that Pomponazzi therefore was only "guilty of not having invented a hypothesis according to which all that is thought to be incorporeal is spiritual."[14] Pomponazzi's treatise on the immortality of the soul established foundations neither for impiety nor for atheism, according to Bayle. Yet he found a parallel in Pomponazzi's psychology to his own defense of the possibility of a virtuous atheist. Bayle cited Pomponazzi's consideration of whether belief in the mortality of the soul will lead to criminality, relaying that Pomponazzi not only wrote that numerous wicked men believed in the immortality of the soul but also believed that knowing the practice of virtue leads to happiness and the practice of vice leads to misery "is sufficient to make a honest man."[15]

Pomponazzi was for Bayle an exemplar for affirming the limits to human reason and the possibility of intellectual freedom, as well as being a source of arguments that separate morality from the consequentialism intertwined with the possibility of eternal life. Pomponazzi's biography also provided an example of unjust prosecution. For Bayle, "there is nothing so ridiculous than to pretend that one cannot form such a judgment on the doctrine of Aristotle without being impious."[16] Understanding Aristotle as a pagan was historically accurate rather than blasphemous.

Bayle believed other sixteenth-century Italians showed the failures of Aristotelianism and the existence of an earlier tradition of freedom of thought. Their public philosophical discourse was also evidence that the Catholic Church once practiced greater toleration than it did in Bayle's time. Bayle noted that, paradoxically, Cartesians were harassed because of Descartes' proof of the existence at God, yet Zabarella was left untouched by the Inquisition, because there was greater liberty to criticize particular proofs of God's existence during the sixteenth century.[17] Despite the potential danger of his position, Zabarella exercised philosophical liberty. According to Bayle, Zabarella's failure to accept that the immateriality of the prime mover can be proven through natural philosophy demonstrates a weak point of Aristotelianism that contrasts with Descartes. Cartesian philosophy demands that God has immediate powers throughout the universe. The limitation placed on God conceived of as Aristotelian prime mover, however, gives rise to the tenet that there are other "movers." This belief "can lead insensibly to the most dangerous Atheism" of Chinese philosophy, in which the attribution of natural powers to the visible bodies has caused them to forsake the immaterial God and make the visible heavens "their great God."[18]

Bayle associated Chinese philosophy not just with Aristotelians but Spinoza, believing that Spinozism was a world view that emerged throughout history in diverse contexts including in China and in the thought of Greek thinkers such as Xenophanes, Diogenes the Cynic, and Aristotelians.[19] Bayle maintained that Cesalpino's reduction of the intellects of the celestial spheres to one substance meant that his "principles hardly differ from those of Spinoza."[20] Regardless of the idiosyncrasy of Bayle's understanding of Spinoza, he equated Cesalpino and Zabarella with Spinoza and also hinted that Averroes' unicity argument had affinities to pantheism.[21] Bayle considered the doctrine that there is only one rational soul to be both impious and absurd; it thus provided an example of pure reason leading astray the credulous away from beliefs that can only be known through faith.[22]

Bayle's transformation of Averroes into a Spinozist was just one of many transformations of the Muslim thinker during these years in which few read his commentaries. The historical details about Averroes' life were better known by the end of the seventeenth century than they had before, as evidenced not just by

Bayle's attempt to correct Moreri but by reference works that took greater care in writing histories of the Muslim world.[23] For example Johannes Hottinger's work of universal history included Leo Africanus's account of Averroes' life that was based exclusively on Arabic sources. Leo portrayed Averroes as "prudent, patient, liberal, and pious."[24] Such accounts as well as the fact that few read Averroes (even Bayle did not cite any primary texts in his entry on him) meant that for some he ceased to be the emblem of impiety that he had been for Renaissance humanists.

Perhaps the most bizarre transformation of Averroes's image is that found in the so-called *Averroeana*, a collection of spurious letters printed in London in English in 1695. All sense of Averroes as a figure of controversy is absent from one of the most curious depictions of him. Thomas Tryon, an autodidact who preached mysticism, peace, and vegetarianism, wrote a collection of letters, allegedly written by Monsieur Grinau, from Porte Royal.[25] The preface purports that a Monsieur Petit found these letters during an earlier stay in Andalusia. The letters themselves are supposed to be from a correspondence between Averroes and a Greek, presumably invented, named Metrodorus, who was in Cordova in 1148 in order to seek out Averroes' opinions about whether the "menstruum of the stomach" is composed of acid and whether "fixt salts" play a significant role in the physiology of plants and animals.

Although why specifically Averroes was chosen to be the figurehead for this fictional epistolary exchange remains unclear, the representation of him is unconventional. Tryon presented him perhaps most importantly as a physician whose reputation is enhanced by his "being an *Arabian*, and having a nearer Communication with *Phoebus*, whatsoever there is of Secret in this Business, that has created so much trouble to our Modern Physicians, could not be unknown to him."[26] Averroes' "Arabic" heritage increased his authority, in part because of its links with Greece. In the letters Averroes reveals some of the inadequacies of recent chemical innovations in medicine, which were in fact unknown to the real Averroes. Here he appears as a learned and wise ancient physician who, having been able to set aside the prejudices of his the physicians of his age, can also undermine the novelties of the present.

The choice of Averroes for the letters illustrates the extent to which Tryon in this work either disregarded or was ignorant of the controversial positions that followers of Averroes had taken over the previous half millennium. Still, shadows of some of these views linger in these letters. Here Averroes is made to put forth a view of religion that differs sharply from his actual one, as he claims that the mysteries of his religion are "unalterably to be believed upon the sole Credit and Authority of their Tradition, withal being of such a Nature as no humane Inquisition can arrive at any further knowledge of." Just decades earlier Gassendi, Heere-boord, and others had deemed Averroes emblematic of recalcitrantly dogmatic Aristotelians. Here, Averroes is a proponent of philosophical liberty, whereby "we are obliged to the Authority of Person, but at Liberty to pick and chuse, change and resettle our Opinion as oft as our Reason directs." He even presented himself, anachronistically, as an opponent of Scholasticism, as he stated that he was "resolved now not to be bound to the Custom of the Schools."[27] His liberty from such customs allow him in the final letters to get past the advice for moderate consumption of food and drink and to relate the Pythagorean practices of the "Indian Brachmans."

Anyone who had been caught up in earlier debates about the intellective soul would be greatly puzzled to see Averroes seemingly side with Pythagoreans or metempsychosis. The "Brachmans," whose dietary practices had stirred considerably interest in Europe as the result of travelers' reports, are described in these letters as ascetics who, like the Pythagoreans, affirm "that God hath made all things by Number, Weight and Measure."[28] These final letters favorably recount the vegetarian practices of Indians and their allegedly peaceful habits, keeping the reader reminded of their belief, both the Pythagoreans' and the Indians', in the transmigration of souls. Tryon, perhaps unwittingly, has transformed Averroes into, if not a proponent, then at least a communicator of the position that immortal individual souls move from one earthly being to another. Why he chose Averroes, remains a matter of question, but it seems that the name evoked, on the one hand, foreign wisdom and, on the other hand, a willingness to question conformity.

Although Averroes had been transformed into a symbol of Oriental free thinking, Bayle's version of Aristotle colored his reception during the eighteenth century. By the time that French universities had implemented Cartesian thought in their curriculum and the success of Newtonian and mechanical philosophies transformed English natural philosophy, few read Aristotle in the same manner that university professors of the previous century had. Aristotle was a subject for the history of philosophy more than he was for philosophy itself. Thus whether his thought could be reconciled with Christianity concerned few and the desire to understand his philosophy in a specific historical context led to a broader acceptance that he was impious. The success of Bayle's account of Aristotle is perhaps most evident in Diderot's and d'Alembert's *Encyclopédie*. As atheism solidified in meaning the denial of divinity by the eighteenth century, this notion became applied to Aristotle. Claude Yvon believed Aristotle's "ideas on divinity indirectly tend to undermine it and destroy it."[29] Such a judgment meshed with the Enlightenment's negative assessment of medieval thought, suggesting that Scholasticism was built on foundations of a misunderstanding of ancient thought. Nevertheless, unlike during Pomponazzi's time, because Aristotelianism no longer dominated universities, the musings on Aristotle of even a devout materialist such as d'Alembert provoked little if any controversy.

During the nineteenth century, various thinkers and historians, such as Ernest Renan, Francesco Fiorentino, and John Addington Symonds, cited Averroes or Pomponazzi as emblems or forerunners of philosophical liberty, naturalism, or positivism.[30] Meanwhile the question of whether Aristotle's conception of the soul or God can be reconciled with Christianity has largely fallen from contemporary scholarship on Aristotle. In the 1920s, W. D. Ross dismissed Franz Brentano's theistic interpretation of Aristotle and concluded that "Aristotle has no theory either of divine creation or of divine providence."[31] Despite a few exceptions, most scholars of today follow Ross and believe that Aristotle stands out from his contemporaries as even rejecting traditional Greek view of the gods.[32] Thus Aristotle's credentials as a pagan worshipper are even in doubt. Although Aristotle's theory of the soul remains of great interest, whether he held the rational human soul to be immortal is no longer apparently a *quaestio vexata*. A highly influential volume of essays on the *De anima* published in the late twentieth century shows philosophers and historians of philosophy more devoted to explaining hylomorphism and to debating whether Aristotle was a functionalist than pondering whether his hylomorphism could be compatible with individual immortality.[33]

These contemporary debates are incommensurable to those of the Renaissance. Entirely different motivations spur inquiry into Aristotle's text. Whether or not Aristotle believed in immortality is irrelevant to contemporary discussions of mind and thus ignored. Or perhaps the reconciliation of Aristotle with Christianity is so distant from the goal of modern philosophy because the work of sixteenth-century Aristotelians, early modern historians of philosophy, and promoters of the new natural philosophy of the scientific revolution convincingly demonstrated the difficulties in combining pagan and Christian thought into a coherent theology that conformed to the experiments and experiences recorded by naturalists. The problem for moderns was, and perhaps still is, reconciling the results of science, rather than ancient philosophy, with faith.

RENAISSANCE LEGACY

The term "Renaissance" entered modern historiography with the work of the nineteenth-century Swiss scholar Jacob Burkhardt. His view of the Italian Renaissance was colored by ideological battles of the nineteenth century, when polemics raged between secular thinkers, who saw themselves as the fruit of modernity, and conservatives, who wished that the traditional authorities of monarchs and church held sway. Emerging from this context, he characterized the Renaissance, and its intellectual tradition, as the first stage of modernity and concomitantly the starting point of secularization.[34] Now, few endorse the position that modern secularism has irrevocably defeated organized religion or that the Italian Renaissance and its intellectual traditions led to modern thought. Yet the intellectual activities of Renaissance Italy influenced subsequent generations of philosophers.

Renaissance ideals fostered the emergence of modern science by laying the foundations for religious critiques of traditional natural philosophy. The plying of the tools of historian and philologist transformed knowledge about antiquity and the Middle Ages. Greater historical sensibilities and the new wealth of ancient texts gave polemicists sufficient evidence that Aristotle behaved no differently, or even worse, than his fellow pagans. The hermeneutical ideals of sixteenth-century commentators of Aristotle of discovering his true thought regardless of its truth made it evident for many that alternatives to Aristotelianism that were in greater conformity to Christianity must be sought.

From the nineteenth century until the present, scholars have seen Pomponazzi, Porzio, and Paduan Aristotelians in general as forerunners of modern science because of their materialism and secularism.[35] In a sense Pomponazzi's and Porzio's views fostered the emergence of the natural philosophies of the scientific revolution but not in the way that positivists have envisioned. Pomponazzi's arguments for the probability of the mortality of the human soul did not convince future generations to give up the study of metaphysics or adopt a thorough materialist understanding of the world. Rather, Pomponazzi and Porzio convinced that Aristotle could not be reconciled with Christianity and that, if understood accurately, Aristotle subverted religious dogma. The motivations of seventeenth-century innovators in natural philosophy, whether Protestant or Catholic, were deeply religious. Their abandonment of Peripatetic philosophy arose, at least in part, from the conviction that the best historical studies of the day demonstrated that Aristotle deviated from Christianity giving permission to seek more pious alternatives.

Acknowledgments

Several fellowships supported the research and writing of this book. A Rockefeller Foundation Fellowship in the Humanities funded my initial research at the History of Science Department at the University of Oklahoma. A Hanna Kiel Fellowship at Villa I Tatti gave me an additional year to investigate the Renaissance reception of Averroes. The Gladys Krieble Delmas Foundation supported a summer's research in Venice and Padua. Research awards and fellowships from Oakland University aided my travels to libraries during summers. Most of the book was written while at the American Academy in Rome, where the Andrew W. Mellon Foundation supported me. I thank these institutions and foundations for their organizational and financial support.

Many were indispensable aides for my scholarly development and assisted me in innumerable ways. At Oklahoma, Jamil Ragep, Steve Livesey, and Sonja Brentjes were excellent mentors. At I Tatti, Katy Park helped improve my writing and thinking. There, Ron Witt, the other visiting professors, and fellows provided me with many insights into the history of the Renaissance, as well fostering a enjoyable environment. Chris Celenza was the ideal director at the American Academy in Rome, and I thank him for both moral support and sharing his vast knowledge. At all three institutions I gave papers on my research and I thank the attendees, the organizers, and all who posed questions and commented on my preliminary ideas.

I gave papers related to this book at the Warburg Institute, at the Department of Philosophy and Moral Sciences at Ghent University, at the Center for the History of Philosophy and Science at Radboud University, and at the annual meetings of the Renaissance Society of America and of the History of Science Society. While there were too many present to name, I thank Guido Giglioni, Anna Akasoy, Hiro Hirai, Christoph Lüthy, Paul Bakker, Cees Leijenhorst, Adam Takahashi, and Maarten van Dyck for their invitations and their many suggestions at these conferences.

I thank the staffs of the History of Science Collections at the University of Oklahoma, the Biblioteca Berenson, Kresge Library at Oakland University, the Huntington Library, the Biblioteca Ambrosiana, the Biblioteca Marciana, the Biblioteca Nazionale di Roma, the Biblioteca Nazionale di Firenze, the Biblioteca Apostolica Vaticana, the Biblioteca Universitaria di Padova, the Biblioteca Ariostea, the Archivio di Stato di Bologna, and the Arthur and Janet C. Ross Library at the American Academy in Rome for helping me gain access to the materials needed for writing this book. The image used for the cover was provided courtesy of the History of Science Collections, University of Oklahoma Libraries.

A number of scholars, besides those already mentioned, aided my research, pointing out gaps in argument and bibliography, while suggesting new approaches and directions. William Newman, John Monfasani, Pamela Long, Eva Del Soldato, Gideon Manning, Brad Bouley, Nick Popper, Nick Dew, Fabio Forner, Matthew Klemm, Monica Azzolini, Evan Ragland, Sandro La Barbera, and Dag Nikolaus Hasse all helped me with my research. Michele T. Callaghan improved the text and saved me from many errors. I am grateful to A. I. Sabra for introducing me to Averroes' writings while I was a graduate student, and John Murdoch's lessons in the history of science, medieval thought, and paleography have had a profound effect on my historical work. Don Smith has consistently been a source of inspiration for my scholarly activities and constant as a dear friend.

I thank my parents and my brother, Russell, for their care and affection. Chiara Bariviera proofread the Latin in footnotes, for which I thank her, but most of all I am grateful for her love that made the time I spent writing this book the most pleasant period of my life. I dedicate this book to her.

Notes

INTRODUCTION

Epigraph. Denis Diderot and Jean Le Rond d'Alembert, eds. *Encyclopédie, ou dictionnaire raisonnée des sciences, des artes et des métiers* (Paris, 1751–65), 1:659–60.

1. William B. Ashworth Jr., "Natural History and the Emblematic World View," in *Reappraisals of the Scientific Revolution*, ed. David C. Lindberg and Robert S. Westman (Cambridge: Cambridge University Press, 1990), 324; Eric Jorink, "In the Twilight Zone: Isaac Vossius and the Scientific Communities in France, England and the Dutch Republic," in *Isaac Vossius (1618–89) between Science and Scholarship*, ed. Eric Jorink and Dirk van Miert (Leiden: Brill, 2012), 155–56.

2. Roger Ariew, *Descartes and the Last Scholastics* (Ithaca, NY: Cornell University Press, 1999); Cees Leijenhorst, *The Mechanisation of Aristotelianism: The Late Aristotelian Setting of Thomas Hobbes's Natural Philosophy* (Leiden: Brill, 2002); Michael Edwards, "Aristotelianism, Descartes, and Hobbes," *Historical Journal* 50 (2007): 449–64.

3. Alfred North Whitehead, *Science and the Modern World* (Cambridge: Cambridge University Press, 1926), 10–14; A. R. Hall, *The Scientific Revolution, 1500–1800: The Formation of the Modern Scientific Attitude* (London: Longmans, 1954), 1–33.

4. William R. Newman, *Atoms and Alchemy* (Chicago: University of Chicago Press, 2006), 85–98.

5. Andrew Dickson White, *A History of the Warfare of Science with Theology in Christendom* (New York: Appleton, 1901).

6. Stephen Menn, "The Intellectual Setting," in *The Cambridge History of Seventeenth-Century Philosophy*, ed. Daniel Garber and Michael Ayers (Cambridge: Cambridge University Press, 1988), 1:41. Ann Blair writes that "most" were; see "Mosaic Physics and the Search for a Pious Natural Philosophy in the Late Renaissance," *Isis* 91 (2000): 33.

7. Fortunio Liceti, *De pietate Aristotelis erga Deum & homines* (Udine, 1645), 88.

8. Amos Funkenstein, *Theology and the Scientific Imagination from the Middle Ages to the Seventeenth Century* (Princeton, NJ: Princeton University Press, 1986), 3–5.

9. Richard H. Popkin, *The History of Scepticism: From Savonarola to Bayle*, 2nd ed. (Oxford: Oxford University Press, 2003).

10. William R. Shea and Mariano Artigas, *Galileo in Rome: The Rise and Fall of a Troublesome Genius* (Oxford: Oxford University Press, 2003), 158–200; Mario Biagioli, *Galileo, Courtier: The Practice of Science in the Culture of Absolutism* (Chicago: University of Chicago Press, 1993), 313–52.

11. Margaret J. Osler, "When Did Pierre Gassendi Become a Libertine?" in *Heterodoxy in Early Modern Science and Religion*, ed. John Brooke and Ian Maclean (Oxford: Oxford University Press, 2005), 169–92.

12. Stephen Gaukroger, *Francis Bacon and the Transformation of Early-Modern Philosophy* (Cambridge: Cambridge University Press, 2001), 74–83; Steven Matthews, *Science and Theology in the Thought of Francis Bacon* (Aldershot, UK: Ashgate, 2008), 51–74.

13. Peter Harrison, *The Bible, Protestantism, and the Rise of Natural Science* (Cambridge: Cambridge University Press, 1998), and *The Fall of Man and the Foundation of Science* (Cambridge: Cambridge University Press, 2007).

14. Roger Ariew, "Descartes and the Jesuits: Doubt, Novelty, and the Eucharist," in *Jesuit Science and the Republic of Letters*, ed. by Mordechai Feingold (Cambridge, MA: MIT Press, 2003), 157–94; Pietro Redondi, *Galileo eretico* (Turin: Einaudi, 1983), 289–344.

15. Christoph Lüthy and Cees Leijenhorst, "The Erosion of Aristotelianism: Confessional Physics in Early Modern Germany and the Dutch Republic," in *The Dynamics of Aristotelian Natural Philosophy from Antiquity to the Seventeenth Century*, ed. Cees Leijenhorst et al. (Leiden: Brill, 2002), 375–411; Sachiko Kusukawa, *The Transformation of Natural Philosophy: The Case of Philip Melanchthon* (Cambridge: Cambridge University Press, 1995).

16. David C. Lindberg, "Medieval Science and Its Religious Context," *Osiris*, 2nd ser., 10 (1995): 77.

17. Pierre Bayle, *Dictionnaire historique et critique*, 5th ed. (Amsterdam, 1740), article on "Aristote," 1:327. This and all subsequent translations are mine unless noted.

18. Silvia Berti, "At the Roots of Unbelief," *Journal of the History of Ideas* 56 (1995): 555–75; Paul Oskar Kristeller, "The Myth of Renaissance Atheism and the French Tradition of Free Thought," *Journal of the History of Philosophy* 6 (1968): 233–43; Lucien Febvre, *Le problème de l'incroyance au XVIe siècle, la religion de Rabelais* (Paris: Michel, 1947).

19. Charles B. Schmitt, *Aristotle and the Renaissance* (Cambridge, MA: Harvard University Press, 1983), 10–33.

20. Edward Grant, "Ways to Interpret the Terms 'Aristotelian' and 'Aristotelianism' in Medieval and Renaissance Natural Philosophy," *History of Science* 25 (1987): 335–58.

21. J. M. M. Hans Thijssen, "Some Reflections on Continuity and Transformation of Aristotelianism in Medieval (and Renaissance) Natural Philosophy," *Documenti e studi sulla tradizione filosofica medievale* 2 (1991): 503–28.

22. Craig Martin, "With Aristotelians Like These, Who Needs Anti-Aristotelians? Chymical Corpuscular Matter Theory in Niccolò Cabeo's Meteorology," *Early Science and Medicine* 11 (2006): 135–61.

23. Craig Martin, "Francisco Vallés and the Renaissance Reinterpretation of Aristotle's *Meteorologica* IV as a Medical Text," *Early Science and Medicine* 7 (2002): 1–30.

24. Newman, *Atoms and Alchemy*, 23–44.

25. Richard Lemay, *Abu Ma'shar and Latin Aristotelianism in the Twelfth Century: The Recovery of Aristotle's Natural Philosophy through Arabic Astrology* (Beirut: American University of Beirut Press, 1962); H. Darrel Rutkin, "Astrology," in *The Cambridge*

History of Science, vol. 3, *Early Modern Science*, ed. Katharine Park and Lorraine Daston (Cambridge: Cambridge University Press, 2006), 541–61.

26. Luca Bianchi, *Studi sull'aristotelismo del rinascimento* (Padua: Il Poligrafo, 2003), 133–72.

27. Ibid., 101–32.

28. Thijssen, "Some Reflections on Continuity," 515.

29. *Statuta almae universitatis dd. philosophorum, et medicorum cognomento artistarum patavini gymnasii* (Padua, 1607), 161.

30. *Statuta antiqua universitatis oxoniensis*, ed. Strickland Gibson (Oxford: Clarendon Press, 1931), 439; Francis A. Yates, "Giordano Bruno's Conflict with Oxford," *Journal of the Warburg and Courtauld Institutes* 2 (1939): 230.

31. Ernest Renan, *Averroès et l'averroïsme: Essai historique*, 2nd ed. (Paris: Lévy, 1861), 479.

32. Thomas Hobbes, *Leviathan* (English version), in vol. 3 of *Collected Works*, ed. William Molesworth (London: Routledge, 1994), 670.

33. Ann Blair and Anthony Grafton, "Reassessing Humanism and Science," *Journal of the History of Ideas* 53 (1992): 535–40.

34. Edith Dudley Sylla, "Walter Burley's Practice as a Commentator," *Medioevo* 27 (2002): 300–71.

35. Bianchi, *Studi sull'aristotelismo*, 180; Charles B. Schmitt, "Aristotelian Textual Studies at Padua: The Case of Francesco Cavalli," in *The Aristotelian Tradition and Renaissance Universities* (London: Variorum, 1984), article VIII, 288–89.

36. For Vimercati's emendation, see Aristotle, *Aristotle's Physics*, ed. W. D. Ross (Oxford: Clarendon Press, 1936), 692. The Renaissance commentaries of Zabarella, Lodovico Boccadiferro, Francesco de' Vieri, and Jacques Charpentier are cited in F. H. Fobes, ed., *Meteorologicorum libri quattuor* (Cambridge, MA: Harvard University Press, 1919). Zabarella informed the interpretation of Aristotle's theory of mixture in Harold H. Joachim, "Aristotle's Conception of Chemical Combination," *Journal of Philology* 29 (1904): 72–86.

37. Jill Kraye, "Aristotle's God and the Authenticity of *De mundo*: An Early Modern Controversy," *Journal of the History of Philosophy* 28 (1990): 339–58.

38. Edward P. Mahoney, "Neoplatonism, the Greek Commentators, and Renaissance Aristotelianism," in *Neoplatonism and Christian Thought*, ed. Dominic J. O'Meara (Albany: State University of New York Press, 1982), 169–77.

39. Giovanni Domenico Mansi, *Sacrorum conciliorum nova et amplissima collectio* (Paris: Welter, 1901–27. Reprint, Graz: Akademische Druck- u. Verlagsanstalt, 1960–62), 11:501–2.

40. Giovanni Santinello, ed., *Models of the History of Philosophy: From Its Origins in the Renaissance to the "Historia philosophica"* (Dordrecht: Kluwer, 1993), 14–65.

41. Lorenzo Valla, *Repastinatio dialectice et philosophie*, ed. Gianni Zippel (Padua: Antenore, 1982), 1:3–8.

42. Jean de Launoy, *De varia Aristotelis in academia parisiensi fortuna* (Paris, 1653).

43. Marin Mersenne, *Quaestiones celeberrimae in Genesim* (Paris, 1623), cols. 8–9.

44. Liceti, *De pietate Aristotelis*, 88.

CHAPTER 1: Scholasticism, Appropriation, and Censure

1. Marie-Thérèse d'Alverny, "Translations and Translators," in *Renaissance and Renewal in the Twelfth Century*, ed. Robert L. Benson et al. (Cambridge, MA: Harvard University Press, 1982), 421–62.

2. Aleksander Birkenmajer, "Le rôle joué par les médecins et les naturalistes dans la réception d'Aristote au XIe et XIIIe siècles," *Etudes d'histoire des sciences et de la philosophie du môyen age, Studia Copernicana* 1 (1970 [1930]): 73–87; Paul Oskar Kristeller, "The School of Salerno: Its Development and Importance for the History of Learning," *Bulletin for the History of Medicine* 17 (1945): 138–94; Piero Morpurgo, "The Salernitan School between Hippocrates, Aristotle and Magic," *Quaderni catanesi* 6 (1984): 197–212; Mark D. Jordan, "Medicine as Science in the Early Commentaries on 'Johannitius,'" *Traditio* 43 (1987): 121–45.

3. Nancy G. Siraisi, *Avicenna in Renaissance Italy* (Princeton, NJ: Princeton University Press, 1987).

4. For Iamblichus's view, see Simplicius, *In Aristotelis categorias commentarium*, ed. Karl Kalbfleisch, vol. 8 of *Commentaria in Aristotelem graeca* (Berlin: Reimer, 1907), 2, 9–25; John Marenbon, *Medieval Philosophy: An Introduction* (London: Routledge, 2007), 23.

5. Pierre Hadot, "The Harmony of Plotinus and Aristotle according to Porphyry," in *Aristotle Transformed: The Ancient Commentators and their Influence*, ed. Richard Sorabji (Ithaca, NY: Cornell University Press, 1990), 125–40. A debate remains over the possibility of distinguishing the approach to Aristotle and Plato in the schools at Athens and Alexandria. See Ilsetraut Hadot, "La vie et l'oeuvre de Simplicius d'après des sources grecques et arabes," in *Simplicius: Sa vie, son oeuvre, sa survie*, ed. Ilsetraut Hadot (Berlin: De Gruyter, 1987), 274–303; R. M. van den Berg, "Smoothing over the Differences: Proclus and Ammonius on Plato's *Cratylus* and Aristotle's *De interpretation*," in *Philosophy, Science and Exegesis in Greek, Arabic and Latin Commentaries*, ed. Peter Adamson et al. (London: Institute of Classical Studies, 2004), 1:191–201. For Avicenna's relation to the Greek commentators, see Robert Wisnovsky, "Avicenna and the Avicennian Tradition," in *The Cambridge Companion to Arabic Philosophy*, ed. Peter Adamson and Richard C. Taylor (Cambridge: Cambridge University Press, 2004), 97–98. For the inauthenticity of the text Iamblichus and Simplicius used, see Carl A. Huffman, *Archytas of Tarentum: Pythagorean, Philosopher, and Mathematician King* (Cambridge: Cambridge University Press, 2005), 596.

6. A. I. Sabra, "The Appropriation and Subsequent Naturalization of Greek Science in Medieval Islam," *History of Science* 25 (1987): 223–43.

7. Wisnovsky, "Avicenna and the Avicennian Tradition," 92.

8. For his response to al-Ghazālī, see Averroes, *Tahafut al-tahafut*, trans. Simon van den Bergh, 2 vols. (Oxford: Oxford University Press, 1954).

9. Averroes, *De anima*, suppl. 2 of *Aristotelis opera cum Averrois commentariis* (Venice, 1562–74. Reprint, Frankfurt: Minerva, 1962), 159v, and *Commentarium magnum in Aristotelis de anima libros*, ed. F. Stuart Crawford (Cambridge, MA: Medieval Academy of America, 1953), 433; translation from David Knowles, *Evolution of Medieval Thought* (Baltimore: Helicon, 1962), 200.

10. Averroes, *Prooemium in libros physicorum Aristotelis*, vol. 4 of *Aristotelis opera cum Averrois commentariis* (Venice, 1562–74. Reprint, Frankfurt: Minerva, 1962), 5r.

11. "Et per hanc virtutem divinam inventam in ipso fuit ipse inventor scientiae, & complens, seu perficiens eam. & hoc raro invenitur in artibus, quaecunque ars fuerit, & maxime in hac arte magna." Averroes, *Meteorologica*, vol. 5 of *Aristotelis opera cum Averrois commentariis* (Venice, 1562–74. Reprint, Frankfurt: Minerva, 1962), 452r.

12. Josep Puig Montada, "El Proyecto vital de Averroes: Explicar e interpretar a Aristóteles," *al-Qanṭara* 32 (2002): 11–52; Steven Harvey, "Averroes' Use of Examples in His *Middle Commentary on the Prior Analytics*, and Some Remarks on his Role as Commentator," *Arabic Sciences and Philosophy* 7 (1997): 91–113; Henri Hugonnard-Roche, "Méthodes d'argumentation et philosophie naturelle chez Averroès," in *Orientalische Kultur und europäisches Mittelalter*, ed. Albert Zimmerman and Ingrid Craemer-Ruegenberg (Berlin: De Gruyter, 1985), 240–53.

13. A. I. Sabra, "The Andalusian Revolt against Ptolemaic Astronomy: Averroes and al-Bitrūjī," in *Transformation and Tradition in the Sciences: Essays in Honor of I. Bernard Cohen*, ed. Everett Mendelsohn (Cambridge: Cambridge University Press, 1984), 143.

14. "In philosophia vero transcendit humanam mensuram, nihil diminute de ea tractat, sed et multa ei adjiciens ex sua solertia totam direxit philosophiam." L. Robbe, ed., *Vita Aristotelis ex codice marciano* (Leiden: Van Leeuwen, 1861), 17. For the text similar to the *Vita latina*, see Ingemar Düring, *Aristotle in the Biographical Tradition* (Gothenburg: Acta universitatis gothoburgnesis, 1957), 156–57. For the Greek of the *Vita marciana*, see Düring, *Aristotle in the Biographical Tradition*, 104. Düring concluded that the work was not Ammonius's but rather "a collective product of several generations of Neoplatonic students" (119). In a text unavailable to Averroes, Cicero referred to Aristotle as having "seen the force and nature of all things," at *De oratore*, 2.38, 160.

15. Y. Tzvi Langermann, "Another Andalusian Revolt? Ibn Rushd's Critique of Al-Kindi's *Pharmacological Computus*," in *The Enterprise of Science in Islam: New Perspectives*, ed. Jan P. Hogendijk and A. I. Sabra (Cambridge, MA: MIT Press, 2003), 366.

16. Averroes, *Physica*, vol. 4 of *Aristotelis opera cum Averrois commentariis* (Venice, 1562–74. Reprint, Frankfurt: Minerva, 1962), 57r.

17. E.g., Averroes, *Metaphysica*, vol. 8 of *Aristotelis opera cum Averrois commentariis* (Venice, 1562–74. Reprint, Frankfurt: Minerva, 1962), 34r, 55r, 315r; Averroes, *Physica*, 4:340v; Averroes, *Tahafut al-tahafut*, 1:279, 1:359–60.

18. Averroes, *Metaphysica*, 8:304r.

19. "Imaginatio ergo super creationes formarum induxit homines dicere formas esse & datorem esse formarum: & induxit Loquentes trium legum, quae hodie quidem sunt, dicere aliquid fieri ex nihilo." Averroes, *Metaphysica*, 8:305r.

20. Ibid., 8:322v.

21. Averroes, *De anima*, supp. 2:171r; Averroes, *Commentarium*, 470.

22. Averroes, *De anima*, supp. 2:171r; Averroes, *Commentarium*, 470.

23. Averroes, *Metaphysica*, 8:181v.

24. Averroes, *De coelo*, vol. 5 of *Aristotelis opera cum Averrois commentariis* (Venice, 1562–74. Reprint, Frankfurt: Minerva, 1962), 227r.

25. Averroes, *Tahafut al-tahafut*, 1–112.

26. Averroes, *Commentary on Plato's Republic*, trans. Erwin I. J. Rosenthal (Cambridge: Cambridge University Press, 1966), 112.

27. Charles Burnett, "Arabic into Latin: The Reception of Arabic Philosophy into Western Europe," in *The Cambridge Companion to Arabic Philosophy*, ed. Peter Adamson and Richard C. Taylor (Cambridge: Cambridge University Press, 2004), 372–76.

28. Bonaventure, *In Hexaemeron*, vol. 5 of *Opera omnia* (Quaracchi: Collegium S. Bonaventurae, 1882–1902), cols. 6, 360–61.

29. Luca Bianchi, "Loquens ut naturalis," in *Le verità dissonanti: Aristotele all fine del Medioevo*, ed. Luca Bianchi and Eugenio Randi (Rome: Laterza, 1990), 37–41.

30. "Quaereamus igitur, quae causa est, quod generatio sit semper, et illa quae est substantiae universaliter, et illa quae est secundum partem ut elementi; numquam enim secundum naturam cessavit nec cessabit generatio." Albertus Magnus, *De generatione et corruptione*, ed. Paul Hossfeld, vol. 5, part 2, of *Opera omnia*, ed. Bernhard Geyer and Wilhelm Kübel (Münster: Aschendorf, 1980), 129.

31. "Si autem quis dicat, quod cessabit voluntate dei aliquando generatio, sicut aliquando non fuerit et post hoc coepit, dico, quod nihil ad me de dei miraculis, cum ego de naturalibus disseram." Albertus Magnus, *De generatione et corruptione*, 5,2:129. Ian Maclean, "Heterodoxy in Natural Philosophy and Medicine: Pietro Pomponazzi, Guglielmo Gratarolo, Girolamo Cardano," in *Heterodoxy in Early Modern Science and Religion*, ed. John Brooke and Ian Maclean (Oxford: Oxford University Press, 2005), 7; Bruno Nardi, "La posizione di Alberto Magno di fronte all'averroismo," in *Studi di filosofia medievale* (Rome: Storia e letteratura, 1960), 119–20.

32. "Sunt autem quidem qui omnia haec divinae dispositioni tantum attribuunt et aiunt non debere nos de huiusmodi rebus quaerere aliam causam nisi voluntatem dei." Albertus Magnus, *De natura loci*, ed. Paul Hossfeld, vol. 5, part 2, of *Opera omnia*, ed. Bernhard Geyer and Wilhelm Kübel (Münster: Aschendorf, 1980), 76.

33. "Sed tamen dicimus haec deum facere propter causam naturalem, cuius primus motor est ipse qui cuncta dat moveri. Causas autem suae voluntatis non quaerimus nos, sed quaerimus causas naturales, quae sunt sicut instrumenta quaedam per quae sua voluntas in talibus producitur ad effectum." Albertus Magnus, *De natura loci*, 5,2:76.

34. Ibid., 5,2:76–79.

35. Albertus Magnus, *De somno et vigilia*, vol. 9 of *Opera omnia*, ed. August Borgnet (Paris: Vives, 1890), 122–23; Nardi, "La posizione di Alberto Magno," 120–21.

36. "eo quod hoc ex physicis rationibus nullo modo potest cognosci: physica enim tantum suscepimus dicenda, plus secundum Peripateticorum sententiam persequentes ea quae intendimus . . . si quid enim forte propriae opinionis haberemus, in theologicis magis quam in physicis, Deo volente a nobis proferetur." Albertus Magnus, *De somno*, 195.

37. "declarabimus opiniones Peripateticorum de istis substantiis, reliquentes aliis iudicium, quid verum vel falsum sit de his quae dicunt." Albertus Magnus, *Metaphysica*, ed. Bernhard Geyer, vol. 16, part 2, of *Opera omnia*, ed. Bernhard Geyer and Wilhelm Kübel (Münster: Aschendorff, 1964), 482.

38. "Non suscepimus in hoc negotio explanare nisi viam Peripateticorum. . . . Theologica autem non conveniunt cum philosophicis in principiis, quia fundantur super revelationem et inspirationem et non super rationem, et ideo de illis in philosophia non possumus disputare." Albertus Magnus, *Metaphysica*, 16,2:542. Nardi, "La posizione di Alberto Magno," 121–22.

39. Thomas Aquinas, *Contra gentiles*, lib. 1, cap. 2, 12.

40. For the proof of the existence of God, see Thomas Aquinas, *Summa theologiae*, Ia22; for his theories of the Eucharist, see Marilyn McCord Adams, *Some Later Medieval Theories of the Eucharist: Thomas Aquinas, Giles of Rome, Duns Scotus, and William Ockham* (Oxford: Oxford University Press, 2010), 85–110.

41. Thomas Aquinas, *Tractatus de unitate intellectus contra Averroistas*, ed. Bruno Nardi and Paolo Mazzantini (Spoleto: Centro italiano di studi sull'Alto Medioevo, 1988), 174.

42. Ibid., 173.

43. Roger Bacon, *Communia naturalium*, vol. 4 of *Opera hactenus inedita*, ed. Robert Steele (Oxford: Clarendon Press, 1913), 3.

44. Thomas Aquinas, *Tractatus de unitate intellectus*, 174.

45. Bruno Nardi, "Introduzione," in Thomas Aquinas, *Tractatus de unitate intellectus*, 65–67.

46. Thomas Aquinas, *Tractatus de unitate intellectus*, 174.

47. Translation, with slight modification from Averroes, *Ibn Rushd's Metaphysics*, 155. The Latin text is found at Averroes, *Metaphysica*, 8:320v.

48. "Sed haec opinio expresse tollit judicium Dei de operibus hominum." Thomas Aquinas, *Scriptum super sententiis*, I. d. 39. q. 2. a. 2.

49. Thomas Aquinas, *De aeternitate mundi contra murmurantes*, in *Opuscula philosophica*, ed. Raimondo M. Spiazzi (Rome: Marietti, 1954), 105.

50. Thomas Aquinas, *De aeternitate mundi*, 108.

51. Thomas Aquinas, *Summa contra gentiles*, lib. 2, c. 31.

52. John Duns Scotus, *De spiritualitate et immortalitate*, in *Philosophical Writings: A Selection*, trans. Allan Wolter (Indianapolis: Hackett, 1990), 158.

53. William of Ockham, *Quodlibeta septem*, 1, q. 1, 7, q. 17–23.

54. "Sed nihil ad nos nunc de Dei miraculis, cum de naturalibus naturaliter disseramus." Siger of Brabant, *Quaestiones in tertium de anima. De anima intellectiva. De aeternitate mundi*, ed. Bernardo Bazán (Louvain: Publications universitaires, 1972), 84. Nardi, "La posizione di Alberto Magno," 119.

55. "secundum viam Philosophi procedendo." Siger, *Quaestiones in tertium de anima*, 113.

56. Pietro d'Abano, *Conciliator* (Venice, 1565. Reprint, Padua: Antenore, 1985), 7v–8r.

57. Pietro d'Abano, *Trattati di astronomia*, ed. Graziella Federici Vescovini (Padua: Programma, 1992), 214. Marie-Thérèse d'Alverny, "Pietro d'Abano et les 'Naturalistes,'" in *Dante e la cultura veneta*, ed. Vittore Branca and Giorgio Padoan (Florence: Olschki, 1966), 207–19.

58. "Propter tertium vero sciendum, quod cum de naturalibus disseratur secundum Albertum in primo de generatione nihil ad nos de miraculis divinis, nec etiam legum persuasionibus magis obsequentes, quod multoties apparet quam quae causarum causa voluntate quadam produxit antecedente, nil ad praesens de miraculis ipsius et voluntate scrutandum consequenti." Pietro d'Abano, *Conciliator*, 14v.

59. "Cum leges vim solam obtineant persuadendi. Polit. 2. Inquiratur igitur veritas quaesiti ex dictis astrologorum, philosophorum et medicorum deinceps." Pietro d'Abano, *Conciliator*, 14v; Nardi, "La posizione di Alberto Magno," 120.

60. Pietro d'Abano, *Conciliator*, 15r. For Albertus on the universal flood, see *De causis proprietatum elementis*, ed. Paul Hossfeld, vol. 5, part 2 of *Opera omnia*, ed. Bernhard Geyer and Wilhelm Kübel (Münster: Aschendorf, 1980), 78; Craig Martin, *Renaissance*

Meteorology: Pomponazzi to Descartes (Baltimore: Johns Hopkins University Press, 2011), 71–72.

61. "Et propterea Ioannis illud. II. Cum Lazarus fuerit morte quatriduanus, ac etiam foetens, quae nequeunt in apopletico reperiri, miraculo magis ascribenudm quam naturae." Pietro d'Abano, *Conciliator*, 239v.

62. "Et ideo apparet hic erroneus intellectus Iacobitarum me persequentium, tamquam posuerim animam intellectivam de potentia educi materiae, differenti 9 cum aliis mihi 54 ascriptis erroribus, a quorum annibus gratia dei & apostolica mediante laudabiliter evasi." Pietro d'Abano, *Conciliator*, 71v; Matthew Klemm, "Medicine and Moral Virtue in the *Expositio problematum Aristotelis* of Peter of Abano," *Early Science and Medicine* 11 (2006): 304.

63. Joseph Polzer, "The 'Triumph of Thomas' Panel in Santa Caterina, Pisa: Meaning and Date," *Mitteilungen des Kunsthistorischen Institutes in Florenz* 37 (1993): 29–70.

64. Bruno Nardi, "Le dottrine filosofiche di Pietro d'Abano," in *Saggi sull'aristotelismo padovano dal secolo XIV al XVI* (Florence: Sansoni, 1958), 21–23.

65. Paolo Marangon, "Per una revisione dell'interpretazione di Pietro d'Abano," in *Il pensiero ereticale nella marca trevigiana e a Venezia dal 1200 al 1350* (Abano Terme: Francisci, 1984), 82.

66. David of Dinant, *I testi di David di Dinant: Filosofia della natura e metafisica a confronto col pensiero antico*, ed. Elena Cadei (Spoleto: Centro italiano di studi sull'alto medioevo, 2008), 60–120.

67. "nec libri Aristotelis de naturali philosophia nec commenta legantur Parisius publice vel secreto, et hoc sub penae xcommunicationis inhibemus," *Chartularium universitatis parisiensis* (Paris: Delalain, 1889–97), 1:70.

68. "Non legantur libri Arisotelis de methafisica et de naturali philosophia, nec summe de eisdem, aut de doctrina magistri David de Dinant, aut Amalrici heretici, aut Mauricii hyspani." *Chartularium*, 1:11.

69. Renan, *Averroès et l'averroïsme*, 222.

70. *Chartularium*, 1:115.

71. "et libris illis naturalibus, qui in Concilio provinciali ex certa causa prohibiti fuere, Parisius non utantur, quousque examinati fuerint et ab omni errorum suspitione purgati." *Chartularium*, 1:138.

72. Roger Bacon, *Opus majus*, trans. Robert Belle Burke (Philadelphia: University of Pennsylvania Press, 1928), 22.

73. Richard Lemay, "Roger Bacon's Attitude toward the Latin Translations and Translators of the Twelfth and Thirteenth Centuries," in *Roger Bacon and the Sciences*, ed. Jeremiah Hackett (Leiden: Brill, 1997), 36.

74. *Chartularium*, 1:278.

75. Giles of Rome, *Errores philosophorum*, ed. Josef Koch (Milwaukee: Marquette University Press, 1944), 11, 25.

76. Pierre Duhem, *Etudes sur Léonard de Vinci* (Paris: Hermann, 1906–13), and *Le système du monde: Histoire des doctrines cosmologiques de Platon a Copernic* (Paris: Hermann, 1913–59), 8:7–121; John E. Murdoch, "Pierre Duhem and the History of Late Medieval Science and Philosophy in the Latin West," in *Gli studi di filosofia medievale fra Otto e Novecento*, ed. Ruedi Imbach and Alfonso Maierù (Rome: Storia e letteratura, 1991), 253–302.

77. Roland Hissette, *Enquête sur les 219 articles condamnés à Paris le 7 mars 1277* (Louvain: Publications universitaires, 1977).

78. J. M. M. Hans Thijssen, *Censure and Heresy at the University of Paris, 1200–1400* (Philadelphia: University of Pennsylvania Press, 1998), 51.

79. "Item quod non est inventum ab Aristotele, quod intellectiva manet post separacionem," *Chartularium*, 1:559.

80. "Quod contraria simul possunt esse vera in aliqua materia," *Chartularium*, 1:558. For the interpretation that the condemnations of 1277 were pointed primarily at Siger and his approach to natural philosophy, see Calvin Normore, "Who Was Condemned in 1277?," *The Modern Schoolman* 72 (1995): 273–81.

81. Thijssen, *Censure and Heresy*, 41. For the historiography of the term "double truth," see Luca Bianchi, *Pour une histoire de la "double vérité"* (Paris: Vrin, 2008), 7–22.

82. All quotations in this paragraph, Ramon Llull, *Opera latina*, ed. Josep Enric Rubio Albarracin, vol. 32 of *Corpus christianorum: continuatio mediaevalis* (Turnholt: Brepols, 1975), 170–71.

83. All quotations in this paragraph, Llull, *Opera latina*, 32:246.

84. *Conciliorum oecumenicorum decreta*, ed. Giuseppe Alberigo et al. (Bologna, Italy: Istituto per le scienze religiose, 1973), 360–61. For Thomas, see "Respondeo dicendum quod necesse est dicere quod intellectus, qui est intellectualis operationis principium, sit humani corporis forma." *Summa theologiae*, I, 1 q. 76 a. 1.

85. David Burr, "The Persecution of Peter Olivi," *Transactions of the American Philosophical Society*, n.s., 66 (1976): 73–80.

86. "quod nulla philosophia legeretur contra theologiam sed legeretur philosophia naturalis, que concordaret cum theologia"; "antiqui philosophi dixerunt multa contra fidem," Helene Wieruszowski, "Ramon Lull et l'idée de la Cité de Dieu: Quelques nouveaux écrits sur la croisade," in *Politics and Culture in Medieval Spain and Italy* (Rome: Storia e letteratura, 1971), 168.

87. James P. Etzwiler, "John Baconthorpe, 'Prince of the Averroists,'" *Franciscan Studies* 36 (1976): 148–76.

88. Thijssen, *Censure and Heresy*, 55.

89. Ibid., 57–89.

90. "Item, jurabitis quod statuta facta per facultatem artium contra scientiam Okamicam observabitis, neque dictam scientiam et consimiles sustinebitis quomodo, sed scientiam Aristotelis et sui Commentatoris Averrois et aliorum commentatorum antiquorum et expositorum dicti Aristotelis, nisi in casibus qui sunt contra fidem." *Chartularium*, 2:680; Thijssen, *Censure and Heresy*, 61.

CHAPTER 2: Humanists' Invective and Aristotle's Impiety

1. *Conciliorum oecumenicorum decreta*, ed. Giuseppe Alberigo et al. (Bologna: Istituto per le scienze religiose, 1973), 206.

2. Francesco Petrarca, *Invectives*, trans. David Marsh (Cambridge, MA: Harvard University Press, 2003), 29, 67.

3. Francesco Petrarca, *Rerum senilium libri*, ed. Ugo Dotti (Turin: Nino Aragno, 2010) (15,6) 3:1958.

4. Petrarca, *Invectives*, 69.

5. Ibid.

6. Petrarca, *Invectives*, 323.

7. Francesco Bausi, "Medicina e filosofia nelle *Invective contra medicum* (Petrarca, l'averroismo, l'eternità del mondo)," in *Petrarca e la medicina*, ed. Monica Berté et al. (Messina: Centro interdipartimentale di studi umanistici, 2006), 19–52.

8. Dante, *Inferno*, canto 4, lines 33–42, 130–33, and 143–44.

9. "Veramente Aristotile, che Stagirite ebbe sopranome, e Zenocrate Calcedonio suo compagnone,[e per lo studio loro], e per lo 'ngegno [singulare] e quasi divino che la natura in Aristotile messo avea." Dante, *Convivio*, book 4, chapter 6, 15.

10. Dante, *Paradiso*, canto 10, lines 98–99, 136–38.

11. William H. Donahue, *The Dissolution of the Celestial Spheres, 1595–1650* (New York: Arno, 1981), 25–28.

12. Dante, *Monarchia*, book 1, chapter 3,4–9; Bruno Nardi, *Saggi di filosofia dantesca* (Florence: La Nuova Italia, 1967), 231–40.

13. [Ottimo]. *L'ultima forma dell'Ottimo commento: Chiose sopra* La Comedia *di Dante Alleghieri fiorentino tracte da diversi ghiosatori*, ed. Claudia Di Fonzo (Ravenna: Longo, 2008), 87.

14. Giovanni Boccaccio, *Boccaccio's Expositions on Dante's Comedy*, trans. Michael Papio. Toronto: University of Toronto Press, 2009, 237, 248.

15. For Benvenuto's relation to Boccaccio and Petrarca, see Franco Quartieri, *Benvenuto da Imola: Un moderno antico commentatore di Dante* (Ravenna: Longo, 2001), 25–27.

16. Dante, *Inferno*, canto 4, lines 103–5.

17. "Aliqui magni philosophi et magistri artium vadunt ad aliquem excellentem et famosum doctorem theologiae, et privatim in camera sua vel in studio suo conferunt et tractant de rebus naturalibus cum rationibus et demonstrationibus naturalibus, sicut de origine animae, de productione mundi, de felicitate humana, de eternitate motus, et multis talibus." Benvenuto da Imola, *Commentum super Dantis Aldigherii comoediam*, ed. James Philip Lacaita (Florence: Barbera, 1887), 1:157.

18. Ibid.

19. Ibid., 1:169.

20. "Damnavit etiam omnem sectam fidei." Benvenuto da Imola, *Commentum*, 1:181–82.

21. "Sed hic statim obicitur: quomodo autor posuit istum sine pena, qui tam impudenter et impie blasfemat Christum dicens, quod tres fuerunt baratores mundi, scilicet Christus, Moyses, et Macomettus, quorum Christus, quia juvenis et ignarus, crucifixus fuit?" Benvenuto da Imola, *Commentum*, 1:181–82.

22. Averroes, *Metaphysica*, 8:305r.

23. Matthew of Paris, *Chronica majora*, ed. Henry Richard Luard (London: Longman, 1872–83), 3:520–21; Alberic of Trois-Fontaines, *Chronicon*, ed. P. Scheffer-Boichorst, vol. 23 of *Monumenta Germaniae Historica*, ed. Georg Heinrich Pertz (Hanover: Hahn, 1874), 944. For the accusations made against Simon of Tournai, see George Minois, *The Atheist's Bible: The Most Dangerous Book That Never Existed*, trans. Lys Ann Weiss (Chicago: University of Chicago Press, 2012), 30–31.

24. Ernest Renan, *Averroès et l'averroïsme: Essai historique*, 2nd ed. (Paris: Lévy, 1861), 297; Charles Burnett, "The 'Sons of Averroes with the Emperor Frederick' and

the Transmission of the Philosophical Works by Ibn Rushd," in *Averroes and the Aristotelian Tradition*, ed. Gerhard Endress and Jan A. Aertsen (Leiden: Brill, 1999), 259–99.

25. "Irreligiosissimum tamen Averroym non sine motu cachinnationis admiror, qui cum de Deo et anime eternitate pessime senserit." Coluccio Salutati, *Epistolario*, ed. Francesco Novati (Rome: Forzani, 1893), 3:191.

26. "illius muliercule crediderit iuramento, que se iactavit ex emisso contra naturam semine in livelli balneo concepisse." Salutati, *Epistolario*, 3:191. For Averroes' position on the possibility of conceiving in a bath, see Averroes, *Collectanea*, suppl. 1 of *Aristotelis opera cum Averrois commentariis* (Venice, 1562–74. Reprint, Frankfurt: Minerva, 1962), 187r.

27. Coluccio Salutati, *De fato et fortuna*, ed. Concetta Bianca (Florence: Olschki, 1985), 172, 178.

28. For Salutati's knowledge of scholasticism, see Charles Trinkaus, "Coluccio Salutati's Critique of Astrology in the Context of His Natural Philosophy," *Speculum* 64 (1989): 46–68; Ronald G. Witt, *Hercules at the Crossroads: The Life, Works, and Thought of Coluccio Salutati* (Durham, NC: Duke University Press, 1983), 295–96.

29. "Sequatur turba philosophantium Aristotelem vel Platonem, sequatur venenosum Averroim et si quem habent . . . meliorem. michi vero solus placeat Iesus Christus." Salutati, *Epistolario*, 4b:215. Berthold L. Ullman, *The Humanism of Coluccio Salutati* (Padua: Antenore, 1963), 90; Witt, *Hercules at the Crossroads*, 412–13.

30. "Incipiam a grammatica, . . . queve ostiaria est omnium liberalium artium omnisque doctrine sive divina dixerimus sive humana." Salutati, *Epistolario*, 4b:215.

31. Coluccio Salutati, *De laboribus Herculis*, ed. Berthold L. Ullman (Zurich: Thesaurus mundus, 1951), 3; Virgil, *Eclogues*, 1,67.

32. Salutati, *De laboribus Herculis*, 3–4.

33. Both quotations Leonardo Bruni, *Dialoghi ad Petrum Histrum*, in *Prosatori Latini del Quattrocento*, ed. Eugenio Garin. (Milan: Ricciardi, 1952), 56; translation from Leonardo Bruni,, *The Humanism of Leonardo Bruni: Selected Texts*, trans. Gordon Griffiths, James Hankins, and David Thompson (Binghamton, NY: Medieval & Renaissance Texts & Studies, 1987), 67–68.

34. Ibid., 58; translation from Bruni, *Humanism of Leonardo Bruni*, 69.

35. Ibid., 54; translation from Bruni, *Humanism of Leonardo Bruni*, 67.

36. Ingemar Düring, *Aristotle in the Biographical Tradition* (Gothenburg: Acta universitatis gothoburgnesis, 1957), 142–63.

37. Ibid., 94–95.

38. John of Wales, *Florilegium de vita et dictis illustrium philosophorum et breviloquium de sapientia sanctorum* (Rome, 1655), 198–205.

39. Ibid., 214, 224.

40. Mario Grignaschi, "Lo pseudo Walter Burley e il 'Liber de vita et moribus philosophorum.'" *Medioevo* 16 (1990): 131–69.

41. "In philosophia vero transcendit humanam mensuram." Pseudo-Burley, *Liber de vita et moribus philosophurm*, ed. Hermann Knust (Tübingen: Litterarischer Verein, 1886), 244; Düring, *Aristotle in the Biographical Tradition*, 156.

42. "Demonstrans quod divinum neque corpus est neque passibile." Düring, *Aristotle in the Biographical Tradition*, 156.

43. Pseudo-Burley, Liber de *vita*, 242–44.

44. "Fuit enim solicitus omnium sapiencium scripturas inquirere." Ibid., 244.

45. Diogenes Laertius, *Lives of Eminent Philosophers*, 5,3; 5,6; 5,10.

46. Ibid., 5,31.

47. Marcello Gigante, "Ambrogio Traversari interprete di Diogene Laerzio," in *Ambrogio Traversari nel sesto centenario della nascita*, ed. Gian Carlo Garfagnini (Florence: Olschki, 1986), 367–459.

48. Gary Ianziti, "Leonardo Bruni and Biography: The 'Vita Aristotelis,'" *Renaissance Quarterly* 55 (2002): 805–32.

49. Bruni, *Humanism of Leonardo Bruni*, 284–85, 288–89.

50. Bruni, *Humanism of Leonardo Bruni*, 288–89; Diogenes Laertius, *Lives of Eminent Philosophers*, 5,2; Plato, *The Republic*, 457c–d.

51. Bruni, *Humanism of Leonardo Bruni*, 290–91.

52. Ianziti, "Leonardo Bruni and Biography," 828.

53. Lodi Nauta, *In Defense of Common Sense: Lorenzo Valla's Humanist Critique of Scholastic Philosophy* (Cambridge, MA: Harvard University Press, 2009), 91–99, 194–95, 225–27.

54. Lorenzo Valla, *Repastinatio dialectice et philosophie*, ed. Gianni Zippel (Padua: Antenore, 1982), 1:17.

55. Ibid., 1:2.

56. Valla, *Repastinatio dialectice*, 1:2. For Valla's use of the term *libertas dicendi*, see Christopher Celenza, *The Lost Italian Renaissance: Humanists, Historians, and Latin's Legacy* (Baltimore: Johns Hopkins University Press, 2004), 88.

57. Valla, *Repastinatio dialectice*, 1:3–5.

58. Lodi Nauta, "Lorenzo Valla's Critique of Aristotelian Psychology," *Vivarium* 41 (2003): 120–43; John Monfasani, "Disputationes vallianae," in *Greeks and Latins in Renaaisance Italy: Studies on Humanism and Philosophy in the 15th Century* (Aldershot, UK: Ashgate, 2004), article VII, 232–35.

59. Valla, *Repastinatio dialectice*, 1:16; Nauta, *In Defense of Common Sense*, 136–39.

60. Valla, *Repastinatio dialectice*, 1:16, 1:42.

61. Valla, *Repastinatio dialectice*, 1:54; Aristotle, *Metaphysics*, 11.7.1072b28–29.

62. For accounts of the controversy, see John Monfasani, *George of Trebizond: A Biography and a Study of His Rhetoric and Logic* (Leiden: Brill, 1976), 201–29; James Hankins, *Plato in the Italian Renaissance* (Leiden: Brill, 1990, 1:161–263); Peter Schulz, "George Gemisthos Plethon (ca. 1360–1454), George of Trebizond (1396–1472), and Cardinal Bessarion (1403–1454): The Controversy between Platonists and Aristotelians in the Fifteenth Century," in *Philosophers of the Renaissance*, ed. Paul Richard Blum (Washington, DC: Catholic University of America Press, 2010), 23–32.

63. B. Lagarde, "Le 'De differentiis' d'après l'autographe de la marcienne," *Byzantion* 43 (1973): 321; translation from C . M. Woodhouse, *Georgios Gemistos Plethon: The Last of the Hellenes* (Oxford: Clarendon, 1986), 192.

64. Lagarde, "De differentiis," 321–22; translation from Woodhouse, *Georgios Gemistos Plethon*, 192–93; Aristotle, *Metaphysics* 12.7.1072b10.

65. Lagarde, "De differentiis," 322; translation from Woodhouse, *Georgios Gemistos Plethon*, 193; Aristotle, *Metaphysics*, 12.7.1072a22–23.

66. Lagarde, "De differentiis," 324–25; translation from Woodhouse, *Georgios Gemistos Plethon*, 195–96.

67. Lagarde, "De differentiis," 327; translation from Woodhouse, *Georgios Gemistos Plethon*, 198; Aristotle, *De anima*, 1.4.408b19, and *Metaphysics*, 12.3.1070a26–27.

68. "aperte perspicitur, alienissimum a veritate Platonem, convenientissimum Aristotelem, cuius ex scriptis plurimum ecclesiae dogmata iuvantur, quae a platonicis aperte oppugnantur." George Trapezuntius, *Comparationes phylosophorum Aristotelis et Platonis* (Venice, 1523), D4r.

69. "inimicum naturae, bonorum morum eversorem, in puerorum clunibus haerentem, phylosophorum principem appellant, admirantur, colunt, ad astra laudibus." Trapezuntius, *Comparationes phylosophorum*, N6v.

70. Trapezuntius, *Comparationes phylosophorum*, T5r–V5r.

71. Monfasani, *George of Trebizond*, 219–20.

72. "Quoniam mihi quoque non syllogismi, neque probabilitates et demonstrationes, sed ipsa nuda dicta sanctorum persuaserunt." Basilios Bessarion, *De processu spritus sancti*, in *vol. 161 of Patrologiae cursus completus, series graeca posterior*, ed. J. P. Migne (Paris: Garnier, 1844–66), 359–60B.

73. Basilios Bessarion, *In calumniatorem Platonis*, vol. 2 of *Kardinal Bessarion als Theologe, Humanist und Staatsman: Funde und Forschungen*, ed. Ludwig Mohler (Paderborn: Schöningh, 1927) (2.5,3) 95.

74. Ibid.

75. John Monfasani, "Cardinal Bessarion's Greek and Latin Sources in the Plato-Aristotle Controversy of the 15th Century and Nicholas of Cusa's Relation to the Controversy," in *Knotenpunkt Byzanz: Wissensformen und kulturelle Wechselbeziehungen*, ed. Andreas Speer and Philipp Steinkrüger (Berlin: De Gruyter, 2012), 473–75.

76. Bessarion, *In calumniatorem Platonis* (3.29,1) 2:414–15.

77. Ibid. (2.8,9) 2:147.

78. Eckhard Kessler, "Alexander of Aphrodisias and his Doctrine of the Soul: 1400 Years of Lasting Significance," *Early Science and Medicine* 16 (2011): 10–18.

79. Monfasani, *George of Trebizond*, 228–29.

80. Hankins, *Plato in the Renaissance*, 1:300–18.

81. Dag Nikolaus Hasse, "Arabic Philosophy and Averroism," in *The Cambridge Companion to Renaissance Philosophy*, ed. James Hankins (Cambridge, MA: Harvard University Press, 2007), 307–31.

82. John Monfasani, "The Averroism of John Argyropolous and His *Quaestio utrum intellectus humanus sit perpetuus*," in *Greeks and Latins in Renaissance Italy: Studies on Humanism and Philosophy in the 15th Century* (Aldershot, UK: Ashgate, 2004), article II, 157–208.

83. Armando Verde, *Lo studio fiorentino 1473–1503: Ricerche e documenti* (Florence: Olschki, 1973–2010), 4,1:88–90. For Dominicus's rejection of this theory of the soul, see Dominicus of Flanders, *Acutissimae quaestiones super tres libros de anima & Sancti Thomae commentaria* (Venice, 1560), 359–80.

84. "Opinio est Aristotelis & Commentatoris, scilicet, quod unicus est intellectus in omnibus hominibus." Niccolò Tignosi, *In libros Aristotelis de anima commentarii* (Florence, 1551), 406; "sed fides Christianorum alium morem postulat, qui secundum

conscientiam verissimus est." Tignosi, *In libros Aristotelis de anima*, 409; Verde, *Lo studio fiorentino*, 91; Dag Nikolaus Hasse, "Aufstieg und Niedergang des Averroismus in der Renaissance: Niccolò Tignosi, Agostino Nifo, Francesco Vimercato," in *Herbst des Mittelalters? Fragen zur Berwertung des 14. und 15. Jahrhunderts*, ed. Jan A. Aertsen and Martin Pickavé (Berlin: De Gruyter, 2004), 451–54.

85. Verde, *Lo studio fiorentino*, 105–19.

86. "Nunc vero quotidie in Plotini libris similiter laboramus, huic operi nos, sicut & illi suo, divinitus destinati, ut hac Theologia in lucem prodeunte & poetae desinant gesta mysteriaque pietatis impie fabulis suis annumerare, & Peripatetici, id est, Philosophi omnes admoneantur non esse de religione tanquam de anilibus fabulis sentiendum." Marsilio Ficino, *Opera* (Basel, 1576), 1:872.

87. "Totus enim terrarum orbis a Peripateticis occupatus in duas plurimum divisus est sectas, Alexandrinam & Averroicam." Ficino, *Opera*, 1:872.

88. "Illi quidem intellectum nostrum esse mortalem existitmant, hi vero unicum esse contendunt. Utrique religionem omnem funditus aeque tollunt." Ficino, *Opera*, 1:872.

89. Ficino, *Opera*, 1:803–4; Nella Giannetto, *Bernardo Bembo: Umanista e politico veneziano* (Florence: Olshki, 1985), 94, 136.

90. Marsilio Ficino, *Platonic Theology*, trans. Michael J. B. Allen and ed. James Hankins (Cambridge, MA: Harvard University Press, 2001–6) (14.10.5) 4:307.

91. Ibid. (15.1.2) 5:9.

92. Ibid. (15.14.3) 5:409.

93. Ficino, *Platonic Theology* (2.97) 1:155. Aristotle, *Metaphysics*, 12.10.1075a10.

94. "Clearchus Peripatheticus scribit Aristotel. fuisse Iudaeum." Ficino, *De christiana religione* in *Opera*, 1:30. For Clearchus's fragment, see Fritz Wehrli, ed. *Die Schule des Aristoteles* (Basel: Schwabe, 1944) (fr. 5) 1:10; Eusebius. *De evangelica praeparatione*, trans. George Tapezuntius (Venice, 1497), 9,5. For Trapezuntius's translation, see Eusebius, *De evangelica praeparatione*, I2r: "Ille igitur subiunxit Aristoteles iudaeus erat." Bayle noted that the lack of commas around "subiunxit Aristoteles" makes the phrase ambiguous, although he reads Trapezuntius's translation as "inquit Aristoteles" and has slightly different punctuation than in the 1497 edition. Pierre Bayle, *Dictionnaire historique et critique*, 5th ed. (Amsterdam, 1740), article on "Aristote," com. C, 1:324. The annotations made on the copy of Eusebius in Biblioteca Apostolica Vaticana (shelf mark: Inc. II. 513) place parentheses around the phrase.

95. Brian Copenhaver, "Ten Arguments in Search of a Philosopher: Averroes and Aquinas in Ficino's *Platonic Theology*," *Vivarium* 47 (2009): 444–79.

96. James Hankins, "Marsilio Ficino on *Reminiscentia* and the Transmigration of Souls," *Rinascimento*, 2nd ser., 45 (2005): 3–17.

97. Giovanni Pico della Mirandola, *Syncretism in the West: Pico's 900 Theses (1486)*, ed. S. A. Farmer (Tempe, AZ: Medieval & Renaissance Texts & Studies, 1998).

98. Bohdan Kieszkowski, "Les rapports entre Elie del Medigo et Pico de la Mirandole," *Rinascimento*, 2nd ser., 4 (1964): 41–91; F. Edward Cranz, "Editions of the Latin Aristotle Accompanied by the Commentaries of Averroes," in *Philosophy and Humanism: Renaissance Essays in Honor of Paul Oskar Kristeller*, ed. Edward P. Mahoney (Leiden: Brill, 1976), 118–20; Elia del Medigo, *Parafrasi della "Repubblica" nella traduzione latina di Elia del Medigo*, ed. Annalisa Coviello and Paolo Edoardo Fornaciari

(Florence: Olschki, 1992), vi; Josep Puig Montada, "Elia del Medigo and his Physical *Quaestiones*," in *Was ist Philosophie im Mittelalter?*, ed. Jan A. Aertsen and Andreas Speer (Berlin: De Gruyter, 1998), 929–36.

99. Albano Biondi, "La doppia inchiesta sulle *Conclusiones* e le traversie romane di Pico nel 1487," in *Giovanni Pico della Mirandola: Convegno internazionale di studi nel cinquecentesimo anniversario della morte (1494–1994)*, ed. Gian Carlo Garfagnini (Florence: Olschki, 1997), 197–212.

100. For Savonarola's skepticism, see Richard H. Popkin, *The History of Scepticism: From Savonarola to Bayle*, 2nd ed. (Oxford: Oxford University Press, 2003), 17–23.

101. "Plato enim ad animi insolentiam, Aristoteles vero ad impietatem instruit." Pietro Crinito, *De honesta disciplina* (Paris, 1508), 10v.

102. Gianfrancesco Pico della Mirandola, *Vita Hieronymi Savonarolae*, ed. Elisabetta Schisto (Florence: Olschki, 1999), 116.

103. Girolamo Savonarola, *Prediche sopra l'Esodo* (Rome: Belardetti, 1955–56), 2:291.

104. Girolamo Savonarola, *Triumphus crucis*, ed. Mario Ferrara (Rome: Belardetti, 1961), 490.

105. Ibid., 493.

106. Charles B. Schmitt, *Gianfrancesco Pico della Mirandola (1469–1533) and His Critique of Aristotle* (The Hague: Nijhoff, 1967).

107. Gianfrancesco Pico della Mirandola, *Examen vanitatis doctrinae gentium, & veritatis Christianae disciplinae* (Mirandola, 1520), B2r.

108. Ibid., 115r.

109. Ibid., B2r, 195v>N>97v.

110. For Ficino's and Pico's influence on Aristotelians, see Edward P. Mahoney, "Giovanni Pico della Mirandola and Elia del Medigo, Nicoletto Vernia and Agostino Nifo," in *Giovanni Pico della Mirandola*, ed. Gian Carlo Garfagnini (Florence: Olschki, 1997), 127–56; D. P. Walker, *Spiritual and Demonic Magic: From Ficino to Campanella* (University Park: Pennsylvania State University Press, 2000), 107–11; Cristiana Innocenti, "Una fonte neoplatonica nel 'De incantationibus' di Pietro Pomponazzi: Marsilio Ficino," *Interpres* 15 (1996): 439–71; Francesca Lazzarin, "Vate e filosofo: Riflessi ficiniani nel 'De incantationibus,'" in *Pietro Pomponazzi: Tradizione e dissenso*, ed. Marco Sgarbi (Florence: Olschki, 2010), 93–103.

111. Marc van der Poel, *Cornelius Agrippa, the Humanist Theologian and His Declamations* (Leiden: Brill, 1997), 119–20.

112. Cornelius Agrippa, *De incertitudine et vanitate scientiarum et artium* (Paris, 1531), 13r–13v; 62v.

113. "ipsis Daemonibus dignum factus sacrificium, qui docuerunt illum scire." Agrippa, *De incertitudine*, 67v.

114. "ex impio Aristotele et perfido Averroe sapientiam?" Agrippa, *De incertitudine*, 158v.

115. Juan Luis Vives, *Über die Gründe des Verfalls der Künste / De causis corruptarum artium*, trans. and ed. Wilhelm Sendner (Munich: Fink Verlag, 1990), 492.

116. Ibid.

117. Ibid.

118. Ambrogio Leone, *Castigationes adversus Averroem* (Venice, 1532), 1.

119. Ibid., 11.

120. Leone, *Castigationes adversus Averroem*, +1r, 2, 9,173, 211. Desiderius Erasmus, *Opus epistolarum*, ed. P. S. Allen (Oxford: Clarendon, 1906–58), letter 868, 3:402.

121. Erasmus, *Opus*, letter 1581, 6:89.

122. Desiderius Erasmus, *Opera omnia* (Amsterdam: North-Holland, 1969–), ordo 5 vol. 2:248–49; Erasmus, *Opus*,letter 2771, 10:163; István Bejczy, *Erasmus and the Middle Ages: The Historical Consciousness of a Christian Humanist* (Leiden: Brill, 2001), 68–69.

123. Erasmus, *Opus*, letter 2513, 9:294–95.

124. "ha dottrina tutta barbara e confusa ed è semplice Averroista." Pietro Bembo, *Lettere* (Milan: Classici Italiani, 1809), 2:102.

CHAPTER 3: **Renaissance Aristotle, Renaissance Averroes**

1. Arnaldo Segarizzi, "Lauro Quirini, umanista veneziano del secolo XV," *Memorie della reale accademia delle scienze di Torino*, 2nd ser., 54 (1904): 17.

2. Ibid.

3. "Non tamen rationibus aliquibus contra philosophos disputamus." Segarizzi, "Lauro Quirini," 17.

4. "Sed ipso Aristotele referente contenti fuimus, potius auctoritati eius fidem tribuentes." Segarizzi, "Lauro Quirini," 17.

5. Segarazzi, "Lauro Quirini," 18.

6. On the rediscovery of the Greek commentators, see Charles H. Lohr, "Renaissance Translations of the Greek Commentaries on Aristotle," in *Humanism and Early Modern Philosophy*, ed. Jill Kraye and M. W. F. Stone (London: Routledge, 2000), 24–40.; F. Edward Cranz, "Alexander of Aphrodisias," in vol. 1 of *Catalogus translationum et commentariorum*, 77–135 (Washington, DC: Catholic University of America Press, 1960), 77–135; Charles B. Schmitt, "Olympiodorus," in vol. 2 of *Catalogus translationum et commentariorum* (Washington, DC: Catholic University of America Press, 1971), 199–204; "Il commento di Simplicio al *De anima* nelle controversie della fine del secolo XV e del secolo XVI," in *Saggi sull'aristotelismo padovano dal secolo XIV al XVI* (Florence: Sansoni, 1958), 365–442.

7. Bruno Nardi, "La fine dell'averroismo," in *Saggi sull'aristotelismo padovano dal secolo XIV al XVI* (Florence: Sansoni, 1958), 452; Antonino Poppi, *Introduzione all'aristotelismo padovano*, 2nd ed. (Padua: Antenore, 1970), 35–38.

8. For Averroism and humanism, see Dag Nikolaus Hasse, "Arabic Philosophy and Averroism," in *The Cambridge Companion to Renaissance Philosophy*, ed. James Hankins (Cambridge, MA: Harvard University Press, 2007), 129–30; Craig Martin, "Humanism and the Assessment of Averroes in the Renaissance," in *Renaissance Averroism and Its Aftermath: Arabic Philosophy in Early Modern Europe*, ed. Anna Akasoy and Guido Giglioni (Dordrecht: Springer, 2013), 65–79.

9. Thérèse-Anne Druart, "Averroes: The Commentator and the Commentators," in *Aristotle in Late Antiquity*, ed. Lawrence P. Schrenk (Washington, DC: Catholic University of America Press, 1994), 184–200. For attacks on *kalām*, see Averroes, *Metaphysica*, vol. 8 of *Aristotelis opera cum Averrois commentariis* (Venice, 1562–74. Reprint, Frankfurt: Minerva, 1962), 34v–35r, 181v, 305r.

10. Agostino Nifo, *Expositio super octo Aristotelis Stagiritae libros de physico auditu, Averrois etiam Cordubensis in eosdem libros prooemium, ac commentaria* (Venice, 1552), ***ii verso–***iii recto.

11. Averroes, *Prooemium in libros physicorum Aristotelis*, vol. 4 of *Aristotelis opera cum Averrois commentariis* (Venice, 1562–74. Reprint, Frankfurt: Minerva, 1962), 1r. For the fourteenth-century Hebrew translation and an English translation of this work, see Steven Harvey, "Arabic into Hebrew: The Hebrew Translation Movement and the Influence of Averroes upon Medieval Jewish Thought," in *The Cambridge Companion to Medieval Jewish Philosophy*, ed. Daniel H. Frank and Oliver Leaman (Cambridge: Cambridge University Press, 2003), 55–84.

12. Cranz, "Alexander of Aphrodisiensis," 77–135.

13. Francesco Patrizi, *Discussiones peripateticae* (Venice, 1571), 163.

14. Eugene F. Rice Jr., "Humanist Aristotelianism in France: Jacques Lefèvre and His Circle," in *Humanism in France at the End of the Middle Ages and in the Early Renaissance*, ed. Anthony H. T. Levi (Manchester: Manchester University Press, 1970), 132–49; Eckhard Kessler, "Introducing Aristotle to the Sixteenth Century: The Lefèvre Enterprise," in *Philosophy in the Sixteenth and Seventeenth Centuries: Conversations with Aristotle*, ed. Constance Blackwell and Sachiko Kusukawa (Aldershot, UK: Ashgate, 1999), 1–21. For Barbaro and the Greek commentators, see Jill Kraye, "Philologists and Philosophers," in *The Cambridge Companion to Renaissance Humanism*, ed. Jill Kraye (Cambridge: Cambridge University Press, 1996), 144–47.

15. "Et quando graecos nos latini non habebamus. innixi sumus in hoc homine: propter fragmenta graecorum: quem collegit." Agostino Nifo, *Commentationes in librum de substantia orbis* (Venice, 1508), 2r.

16. "potissimum a themistio: quem Averoes in toto sectatus est: vero enim Averoes themistius expansus." Nifo, *Commentationes in librum de substantia orbis*, 2r.

17. "Ille enim paraphrastice verba Aristotelis enodavit: Averoes autem commentando atque ampliando." Nifo, *Commentationes in librum de substantia orbis*, 2r.

18. "Licet ipse barbarus Aristotelis sensus passim (ut puto) non acceperit: multa Alexandri Themistii et caeterorum fragmenta colligens quaedam (et si non ad verba) satis ad aures Aristotelis aducit." Agostino Nifo, *In duodecimum metaphysices Aristotelis [et] Auerrois volumen commentarii* (Venice, 1518), 1v.

19. "nostro tempore famosus est: ita ut nullus videatur peripatheticus nisi Averroicus." Nifo, *In duodecimum metaphysics*, 2r.

20. Alexander of Aphrodisias, *Enarratio de anima*, trans. Girolamo Donato (Brescia, 1495), A2r; F. Edward Cranz, "The Prefaces to the Greek Editions and Latin Translations of Alexander of Aphrodisias, 1450–1575," *Proceedings of the American Philosophical Society* 102 (1958): 517–20.

21. "magis ex religione quam ex aristotelis doctrina acutissime philosophati sunt." Alexander of Aphrodisias, *Enarratio*, A3r.

22. Alexander of Aphrodisias, *Enarratio*, A3r–A3v.

23. "[Averroes] dixit quod anima intellectiva est simpliciter aeterna et una numero in omnibus homnibus . . . non coniuncta sibi velut forma substantialis sed tanquam nauta navi et intelligentia orbi: quam opinionem puto esse Aristotelis." Nicoletto Vernia, *Quaestio utrum anima intellectiva humano corpori coniuncta tanquam unita forma substantialis*, Venice, Biblioteca Marciana, MS Lat. VI, 105 (=2656) 156r, col. 2; Dag Nikolaus Hasse, "The Attraction of Averroism: Vernia, Achillini, Prassicio," in vol. 2 of *Philosophy, Science and Exegesis in Greek, Arabic and Latin Commentaries*, ed. Peter Adamson et al. (London: Institute of Classical Studies, 2004), 135; Edward P. Mahoney,

"Nicoletto Vernia on the Soul and Immortality," in *Philosophy and Humanism: Renaissance Essays in Honor of Paul Oskar Kristeller*, ed. Edward P. Mahoney (Leiden: Brill, 1976), 146.

24. Vernia, *Quaestio utrum anima intellectiva*, 158r–58v; Mahoney, "Nicoletto Vernia on the Soul," 147.

25. Vernia, *Quaestio utrum anima intellectiva*, 159v, col. 1; Giulio F. Pagallo, "Nicoletto Vernia e un'anonima quaestio sull'anima," in *La filosofia della natura nel medioevo* (Milan: Vita e pensiero, 1966), 670–82.

26. Vernia, *Quaestio utrum anima intellectiva*, 160r, col. 1; Mahoney, "Nicoletto Vernia on the Soul," 148–49.

27. "Mandamus ut nullus vestrum, sub pena excommunicationis late sententie quam si contrafeceritis ipso facto incurratis audeat vel presumat de Unitate intellectus quovis quesito colore publice disputare." Pietro Ragnisco, "Documenti inediti e rari intorno alla vita ed agli scritti di Nicoletto Vernia e Elia del Medigo," *Atti e memorie della R. Accademia di Scienze, Lettere ed Arti in Padova*, n.s., 7 (1891): 279. Bruno Nardi, "La miscredenza e il carattere morale di Nicoletto Vernia," in *Saggi sull'aristotelismo padovano dal secolo XIV al XVI* (Florence: Sansoni, 1958), 100.

28. "sublatis ita tum premiis virtutum, tum vero supliciis vitiorum existimant se liberius maxima queque flagitia posse committere." Ragnisco, "Documenti inediti," 279.

29. "Videte ne quis vos decipiat per philosophiam et inanem falaciam secundum traditionem hominum, secundum elementa mundi et non secundum Christum. Et scientes sic inter disputandum solere animos perturbari. . . . Volentesque ut et hi qui Philosophiam discunt sic discant ut christianam philosophiam." Ragnisco, "Documenti inediti," 278. Compare with Colossians 2:8.

30. Antonio Trombetta, *Tractatus singularis contra Averroystas de humanarum animarum plurificatione* (Venice, 1498). For Trombetta's regard for Averroes' theory of intelligible species, see Leen Spruit, *Species intelligibilis: From Perception to Knowledge* (Leiden: Brill, 1995), 2:75–76.

31. "In hoc Averroym malificae opinionis perfidum et vanum auctore certissimis argumentis refutavi. Nec mihi unquam persuadet potui esse quemquam tam fatui tanquam stolidi ingenii ut eius mentem tam vesana sententia." Nicoletto Vernia, *Contra perversam Averroys opinionem de unitate intellectus et de anime felicitate* (Venice, 1505), 2r.

32. "in hoc conveniunt tria lumina religionis christiane." Vernia, *Contra perversam Averroys opinionem*, 10r.

33. "Dico secundum sacrosanctam Romanam ecclesiam et veritatem. Quod intellectiva anima est forma substantialis corporis humani." Vernia, *Contra perversam Averroys opinionem*, 9v.

34. "Dans ei esse formaliter et intrinsece a sublimi deo creata et in humano corpore infusa: multiplicataque in ipsis secundum multitudinem eorum et in ipsis individuata." Vernia, *Contra perversam Averroys opinionem*, 9v.

35. "et non tantum credo haec omnia dicta ex fide: sed physice dico omnia possunt probari." Vernia, *Contra perversam Averroys opinionem*, 10r.

36. For the influence of Greek commentators, see Mahoney, "Nicoletto Vernia on the Soul," 162–63. For Barozzi as cause, see John Monfasani, "Aristotelians, Platonists, and the Missing Ockhamists: Philosophical Liberty in Pre-Reformation Italy," *Renais-*

sance Quarterly 46 (1993): 250. For Vernia's supposed dishonesty, see Nardi, "La miscredenza di Nicoletto Vernia," 95–114.

37. Hasse, "Attraction of Averroism," 136–37.

38. "Qui cum prius et disputando et docendo: unum esse in omnibus intellectum: sic explanaveris ut totam pene italiam errare feceris." Vernia, *Contra perversam Averroys opinionem*, 2v.

39. "Nunc opusuculum composuisti quo sentire te contrarium non solum dicis: verum etiam probas." Vernia, *Contra perversam Averroys opinionem*, 2v.

40. Agostino Nifo, *In Averrois de animae beatitudine* (Venice, 1524), 2r.

41. "fortasse loquitur more catholico." Nifo, *In Averrois de animae beatitudine*, 2r.

42. "Verumtamen animadverte hunc non esse spiritum sanctum: qui est tertia persona in divinis: quem catholici describunt." Nifo, *In Averrois de animae beatitudine*, 22v.

43. "& quia fundamenta Aver. sunt convenientia principiis Aristo. ideo consuetus sum dicere meis scholaribus quod Aver. est Aristo. transpositus. quando enim homo considerat fundamenta Averro. & collegaverit ea perfecte cum verbis Aristote. non inveniet discrepantia nisi phantastice." Agostino Nifo, *In librum destructio destructionum Averrois commentarii* (Lyon, 1542), 250r. Edward P. Mahoney, "Agostino Nifo's Early Views on Immortality," *Journal of the History of Philosophy* 8 (1970): 453–54. For Nifo's commentary on the *Destructio*, see Heinrich C. Kuhn, "Die Verwandlung der Zerstörung der Zerstörung: Bemerkungen zu Augustinus Niphus' Kommentar zur 'Destructio destructionum' des Averroes," in *Averroes im Mittelalter und in der Renaissance*, ed. Friedrich Niewöhner and Loris Sturlese (Zurich: Spur, 1994), 291–308.

44. "Et debes scire quod istud est secundum Aristotelem & Averroem aliter autem secundum nos christianos." Nifo, *In librum destructio destructionum*, 166v; "Et debes scire quod haec est opinio Arist. & Aver. & est error purus, secundum nos christianos." Nifo, *In librum destructio destructionum*, 57v.

45. "Sed videtur quod Aver. falsum imponat nobis christiani: nos enim non ponimus quod deus sit unum realiter, & tres secundum rationem. . . . Forte Averroes ignorat opinionem nostram cum non sit apud alios valde famosa, nisi apud christianos: & perfecte legis eius, protulit sibi illo modo." Nifo, *In librum destructio destructionum*, 184v.

46. "dixi enim quod philosophi totam doctrinam a sensibus accipiunt: & quia dicta eorum sunt ex sensibus. ideo potest dici quod ratione naturali sunt dicta." Nifo, *In librum destructio destructionum*, 58v.

47. "omnia ergo principia hic dicta sunt falsa simplicitier. . . . Omnes rationes Averrois sunt fatuae." Nifo, *In librum destructio destructionum*, 58v.

48. "Illa autem quae ponit lex nostra: cum sint supra sensus: & valde veriora: & per testimonium prophetiae: quod est multo firmius quam per sensus sunt tenenda." Nifo, *In librum destructio destructionum*, 58v; Mahoney, "Agostino Nifo's Early Views," 454.

49. "Sensus enim in multis contradicit rationi: ut apparet in moribus. Sensus enim vult fornicari. ratio vero vult esse continens." Nifo, *In librum destructio destructionum*, 58v.

50. Mahoney, "Agostino Nifo's Early Views," 451–60.

51. "Quia secundum rationem naturalem, aut intellectus est unus, ut intellexit Averrois Aristotelem." Alessandro Achillini, *De elementis*, in *Opera omnia* (Venice, 1568).

52. "Haec est circumstantia, propter quam dixi oportere relinquere Philosophum, & inter duas illas opiniones falsas probabiliorem eligendo: illa est opinio Averrois, quam-

vis bene dixerit." Achillini, *De elementis*, 223; Dag Nikolaus Hasse, "The Attraction of Averroism: Vernia, Achillini, Prassicio," in vol. 2 of *Philosophy, Science and Exegesis in Greek, Arabic and Latin Commentaries*, ed. Peter Adamson et al. (London: Institute of Classical Studies, 2004), 138.

53. Paolo Giovio, *Gli elogi degli uomini illustri*, vol. 8 of *Opera*, ed. Renzo Meregazzi (Rome: Istituto poligrafico dello stato, 1972), 84; Francesco Vimercati, *De anima rationali peripatetica disceptatio* (Venice, 1566), 44; Francesco Venier, *I discorsi sopra i tre libri dell'anima d'Aristotele* (Venice, 1555), 116v; Rinaldo Odoni, *Discorso per via peripatetica ove si dimostra se l'anima, secondo Aristotele, è mortale, o immortale* (Venice, 1557), 31r; Jacopo Mazzoni, *De triplici hominum vita, activa nempe, contemplativa, & religiosa methodi tres* (Cesena, 1576), 318; Giacomo Zabarella, *De rebus naturalibus* (Frankfurt, 1597), 955.

54. Achillini, *De elementis*, 205–8.

55. Alessandro Achillini, *De orbibus*, in *Opera omnia* (Venice, 1568), 48–60; Peter Barker, "The Reality of Peurbach's Orbs: Cosmological Continuity in Fifteenth and Sixteenth Century Astronomy," in *Change and Continuity in Early Modern Cosmology*, ed. Patrick Boner (Dordrecht: Springer, 2011), 16–17.

56. A. I. Sabra, "The Andalusian Revolt against Ptolemaic Astronomy: Averroes and al-Bitrūjī," in *Transformation and Tradition in the Sciences: Essays in Honor of I. Bernard Cohen*, ed. Everett Mendelsohn (Cambridge: Cambridge University Press, 1984), 133–53.

57. Alessandro Achillini, *De intelligentiis*, in *Opera omnia* (Venice, 1568), 36.

58. "sic error Aristotelis a veritate fidei parum distat. sed in philosophia naturali non apparet error." Achillini, *De intelligentiis*, 312.

59. Nelson Minnich, "Concepts of Reform Proposed at the Fifth Lateran Council," *Archivum historiae pontificiae* 7 (1969): 168–73.

60. John W. O'Malley, *Giles of Viterbo on Church and Reform* (Leiden: Brill, 1968), 40–49.

61. John Monfasani, "Giles of Viterbo as *Alter Orpheus*," in *Forme del neoplatonismo: Dall'eredità ficiniana ai platonici di Cambridge*, ed. Luisa Simonutti (Florence: Olschki: 2007), 97–115, and "The Augustinian Platonists," in *Marsilio Ficino: Fonti, testi, fortuna,* ed. Sebastiano Gentile and Stéphane Toussaint (Rome: Instituto nazionale di studi sul rinascimento, 2006), 319–37.

62. John Monfasani, *Nicolaus Scutellius, O.S.A., as Pseudo-Pletho: The Sixteenth-Century Treatise Pletho in Aristotelem and the Scribe Michael Martinus Stella* (Florence: Olschki, 2005).

63. Charles B. Schmitt, "Gianfrancesco Pico della Mirandola and the Fifth Lateran Council," *Archiv für Reformationsgeschichte* 2 (1970): 161–78.

64. "Cum itaque diebus nostris (quod dolenter ferimus) zizaniae seminator, antiquus humani generis hostis, nonnullos perniciosissimos errores a fidelibus semper explosos in agro Domini superseminare et augere sit ausus, de natura praesertim animae rationalis, quod videlicet mortalis, sit aut unica in cunctis hominibus; et nonnulli temere philosophantes, secundum saltem philosophiam verum id esse asseverant." *Conciliorum oecumenicorum decreta*, ed. Giuseppe Alberigo et al. (Bologna: Istituto per le scienze religiose, 1973), 581.

65. "hoc sacro approbante concilio damnamus et reprobamus omnes asserentes ani-

mam intellectivam mortalem esse, aut unicam, in cunctis hominibus et haec in dubium vertentes." *Conciliorum oecumenicorum decreta*, 581.

66. Felix Gilbert, "Cristianesimo, umanesimo e la bolla 'Apostolici regiminis' del 1513," *Rivista storica italiana* 79 (1967): 976–90. Cf. Eric A. Constant, "A Reinterpretation of the Fifth Lateran Council Decree Apostolici regiminis (1513)," *The Sixteenth Century Journal* 33 (2002): 353–79.

67. "Insuper omnibus et singulis philosophis in universitatibus studiorum generalium, et alibi publice legentibus, districte praecipiendo mandamus, ut cum philosophorum principia aut conclusiones, in quibus a recta fide deviare noscuntur, auditoribus suis legerint, seu explanaverint, quale hoc de animae mortalitate aut unitate, et mundi aeternitate, ac alia huiusmodi, teneantur eisdem veritatem religionis christianae omni conatu manifestam facere, et persuadendo pro posse docere, ac omni studio huiusmodi philosophorum argumenta cum omnia solubilia existant, pro viribus excludere atque resolvere." *Conciliorum oecumenicorum decreta*, 582. For interpretations of this passage see Bianchi, *Pour une histoire de la "double vérité"*, 120–22.

68. *Conciliorum oecumenicorum decreta*, 582.

69. Bruno Nardi, "Letteratura e cultura veneziana del Quattrocento," in *Saggi sulla cultura veneta del quattro e cinquecento* (Padua: Antenore, 1971), 37.

70. On De Vio's presence at the Fifth Lateran Council, see Nelson Minnich, "The Last Two Councils of the Catholic Reformation: The Influence of Lateran V on Trent," in *Early Modern Catholicism: Essays in Honour of John W. O'Malley, S.J.*, ed. Kathleen M. Comerford and Hilmar M. Pabel (Toronto: University of Toronto Press, 2001), 4.

71. "Et reverendus pater dominus Thomas generalis ordinis Praedicatorum dixit, quod non placet secunda pars bullae, praecipiens philosophiis, ut publice persuadendo doceant veritatem fidei." Giovanni Domenico Mansi, *Sacrorum conciliorum nova et amplissima collectio* (Paris: Welter, 1901–27. Reprint, Graz: Akademische Druck- u. Verlagsanstalt, 1960–62), 32: col. 843.

72. "Neque mihi tribuant ut sectator sim opinionum quas exponendo textum affirmavero: sed quod hec sint de mente Aristotelis. quicquid sit quo ad veritatem." Tommaso De Vio, *Commentaria in libros Aristotelis de anima* (Venice, 1514), 42v; Jill Kraye, "Pietro Pomponazzi (1462–1525): Secular Aristotelianism in the Renaissance," in *Philosophers of the Renaissance*, ed. Paul Richard Blum (Washington, DC: Catholic University of America Press, 2010), 97.

73. Cesare Oliva, "Note sull'insegnamento di Pietro Pomponazzi," *Giornale critico della filosofia italiana* 7 (1926): 83–103, 179–90, 254–75.

74. Bartolomeo Podestà, "Di alcuni documenti inediti resguardanti Pietro Pomponazzi lettore nello studio bognese cavati dall'antico archivio del reggimento in oggi della prefettura," *Atti e memorie della regia deputazione di storia patria per le provincie di Romagna* 6 ([1867] 1868): 174. For these negotiations, see also Oliva, "Note," 255–57.

75. For Pomponazzi's doubting, see Stefano Perfetti, "*Docebo vos dubitare*. Il commento di Pietro Pomponazzi al *De partibus animalium* (Bologna 1521–24)," *Documenti e studi sulla tradizione filosofica medievale* 10 (1999): 439–66.

76. Craig Martin, *Renaissance Meteorology: Pomponazzi to Descartes* (Baltimore: Johns Hopkins University Press, 2011), 30–33.

77. Podestà, "Di alcuni documenti inediti," 140–41.

78. "ut temporis angustia non petit omnia dicta illorum dicere; tantum enim se-

cundum principia naturalia opinionem istam pertractabo." Pietro Pomponazzi, *Corsi inediti dell'insegnamento padovano*, ed. Antonino Poppi (Padua: Antenore, 1970), 2:2.

79. "dico in via peripatetica." Pomponazzi, *Corsi inediti*, 2:10; "Scotus . . . dicat quod in via peripatetica animam rationalem esse immortalem est unum problema, non credatis quod hoc dicat secundum fidem, quia fuit vir magni ingenii etc., sed hoc dixit in via peripatetica." Pomponazzi, *Corsi inediti*, 2:9–10.

80. "Cum hoc dico etiam, ut credo, quod opinio Commentatoris sit opinio Aristotelis infallanter." Pomponazzi, *Corsi inediti*, 2:10.

81. "Et, domini, ut solvam hoc argumentum faciam sicut faciebat Mahumetus, qui modo istorum dicta, modo aliorum dicta accipiebat ac aggregabat legesque condebat; sic ego faciam." Pomponazzi, *Corsi inediti*, 2:15.

82. "accipiam dicta aliqua christianorum et aliqua dicta Alexandri, et rationem solvam." Pomponazzi, *Corsi inediti*, 2:15.

83. "elevat se supra materiam . . . et ut sic intellectus est immaterialis et abstractus." Pomponazzi, *Corsi inediti*, 2:17.

84. "Et sic patent responsiones ad rationes Philosophi: falsae tamen sunt responsiones et impossibiles, nec rationes Philosophi possunt solvi, ideo dimittantur et accipiantur pro constanti animam rationalem esse immortalem etc." Pomponazzi, *Corsi inediti*, 2:25.

85. Pomponazzi, *Corsi inediti*, 2:42: "Oh dixisti quod est opinio Aristotelis! Dico quod verum est, sed dico quod ipse Aristoteles fuit homo et potuit errare et, . . . opinio Aristotelis de anima intellectiva est multum chimerica et bestialis. . . . Dico quod opinio Commentatoris est in extremo fatuitatis." Pomponazzi, *Corsi inediti*, 2:42; "Dixi quod volui paleare, quia non est vera, sed falsa et erronea; tamen unum dico, quod credo firmiter et indubitanter quod opinio Commentatoris, Themistii et Theophrasti sit opinio Aristotelis infallanter." Pomponazzi, *Corsi inediti*, 2:46; "Quare infallanter credo quod opinio Themistii, Theophrasti et Commentatoris sit opinio Aristotelis, falsa tamen et bestialis." Pomponazzi, *Corsi inediti*, 2:47. For the trope of Aristotle's erring, see Luca Bianchi, *Studi sull'aristotelismo del rinascimento* (Padua: Il Poligrafo, 2003), 101–32.

86. "Oh, Philosophus non concedit resurrectionem! Dico quod verum dicis, nec ista opinio est Aristotelis, ut infinities dixi, sed resurrectio ex solo lumine fidei est nobis manifesta." Pomponazzi, *Corsi inediti*, 2:54.

87. "quod stando in puris naturalibus videtur quod consimiliter possemus dicere, substentando opinionem Alexandri, quod anima intellectiva sit materialis. . . . Dico tamen hoc stando in puris naturalibus, licet in rei veritate sit opinio falsa." Pomponazzi, *Corsi inediti*, 2:50; "Dico tamen quod opinio ista Alexandri est falsissima." Pomponazzi, *Corsi inediti*, 2:60.

88. "Nec possum credere quod divus Thomas non vidit quia ista opinio non est Aristotelis, sed voluit fortassis zelo fidei paleare istam opinionem secundum Philosophum." Pomponazzi, *Corsi inediti*, 2:56.

89. "Ecce igitur quod Thomas fatetur tandem veritatem coactus, quod hec non fuerit opinio Aristotelis." Paul Oskar Kristeller, "Two Unpublished Questions on the Soul of Pietro Pomponazzi," in vol. 3 of *Studies in Renaissance Thought and Letters* (Rome: Storia e letteratura, 1993), 391.

90. "Conatus est autem ipsam substentare de mente Aristotelis. Conatus est autem

ipsam substentare de mente Aristotelis motus zelo fidei." Kristeller, "Two Unpublished Questions on the Soul," 391.

91. "Boni tamen Cristiani non movebuntur auctoritate Aristotelis." Kristeller, "Two Unpublished Questions on the Soul," 391.

92. Giacomo Filippo Tomasini, *Bibliothecae patavinae manuscriptae publicae & privatae* (Udine, 1639), 97; Bruno Nardi, "Filosofia e religione," in *Studi su Pomponazzi* (Florence: Le Monnier, 1965), 137.

93. "mihi tamen videtur quod nedum in se sit falsissima, verum [in]intelligibilis & monstruosa, & ab Aristotele prorsus aliena." Pietro Pomponazzi, *Tractatus de immortalitate animae* (Bologna, 1516), 1v.

94. "Gregorius nicenus [sic] sicut Divus Thomas refert de eo cum vidit Aristotelem dicere animam esse actu corporis. dixit Aristotelem credere animam humanam esse corruptibilem. . . . aliqui dicunt hoc etiam sensisse Gregorium nazarenum [sic] de Aristotele." Pomponazzi, *Tractatus de immortalitate animae*, 7v. Gregory of Nazianus, *Oratio*, 27:10, in vol 36 of *Patrologiae cursus completus, series graeca posterior*, ed. J. P. Migne (Paris: Garnier, 1844–66), cols. 23–24; Gregory of Nyssa, *De anima* in vol. 45 of *Patrologiae cursus completus, series graeca posterior*, cols. 187–88.

95. "quare haec sola via firmissima in concussa & stabilis est. caetere vero sunt fluctuante." Pomponazzi, *Tractatus de immortalitate animae*, c5v.

96. "sed animam esse immortalem est articulus fidei . . . ergo probari debet per propria fidei. Medium autem cui inititur fides est revelatio. Et scriptura canonica ergo tantum vere & proprie per haec habet probari: caetere vero rationes sunt extranee." Pomponazzi, *Tractatus de immortalitate animae*, c5v. For the prohibition of *metabasis*, see Amos Funkenstein, *Theology and the Scientific Imagination from the Middle Ages to the Seventeenth Century* (Princeton, NJ: Princeton University Press, 1986), 35–37.

97. "Semper tamen me & in hoc & in aliis subiciendo sedi apostolicae." Pomponazzi, *Tractatus de immortalitate animae*, c5v.

98. Charles B. Schmitt, "Alberto Pio and the Aristotelian Studies of his Time," in *The Aristotelian Tradition and Renaissance Universities* (London: Variorum, 1984), article VI, 47–50.

99. Agostino Nifo, *De immortalitate animae libellus* (Venice, 1518), 9r–9v.

100. "De articulis fidei non neutro modo opinamur, immo ipsos firmiter credimus. Ergo si de immortalitate esset problema neutrum eam non firmiter crederemus, quare non est articulus fidei." Nifo, *De immortalitate animae*, 23r.

101. "Bessarion vir utraque lingua dissertissimus." Nifo, *De immortalitate animae*, 23r.

102. Pietro Pomponazzi, *Apologiae libri tres* (Bologna, 1518), F4v.

103. "Petrus de Mantua asseruit quod anima rationalis secundum propria [sic] philosophiae et mentem Aristotelis sit seu videatur mortalis, contra determinationem concilii Lateranensis. papa mandat ut dictus Petrus revocet: alias contra ipsum procedatur. 13 Junii 1518." Leopold von Ranke, *Die römischen Päpste, ihre Kirche und ihr Staat im sechszehnten und siebzehnten Jahrhudert* (Berlin: Bunder und Humblot, 1844), 1:73–74, n1.

104. For Bembo's influence, see Richard Lemay, "The Fly against the Elephant: Flandinus against Pomponazzi on Fate," in *Philosophy and Humanism: Renaissance Essays in Honor of Paul Oskar Kristeller*, ed. Edward P. Mahoney (Leiden: Brill, 1976), 74;

Giovanni Di Napoli, *L'immortalità dell'anima nel Rinascimento* (Turin: Società editrice internazionale, 1963), 267–68; Kraye, "Pietro Pomponazzi," 102.

105. Podestà, "Di alcuni documenti inediti," 178. His salary set in 1515 was four hundred ducats; see Podestà, "Di alcuni documenti inediti," 171–73.

106. "liceat ac licitum sit ipsi D. M. Petro totum Aristotelem legere et interpretari sive eius librorum partes magis utiles et necessarias, seu easdem quas alii Doctores ad philsophiam ordinariam rotulati legere solent." Podestà, "Di alcuni documenti inediti," 176.

107. For the typical sequence of books taught at Bologna, see *Statuti delle università e dei collegi dello studio bolognese*, ed. Carlo Malagola (Bologna: Zanichelli, 1888), 274–75.

108. "il pereto [i.e., Pomponazzi] avea fato stampare certa opera che son contro la fede, e che volean intendere con qualle autorità el pereto la havess fata stampare, perche li e prohibitione che non se po stampare simile cossa senza licentia del superiore." Podestà, "Di alcuni documenti inediti," 148.

109. Podestà, "Di alcuni documenti inediti," 153; Bruno Nardi, "Corsi manoscritti e ritratto di P. Pomponazzi," in *Studi su Pomponazzi* (Florence: Le Monnier, 1965), 25.

110. "Item statuimus, quod nullus Religiosus, cuiuscunque ordinis fuerit, et sit, non possit, necque valeat legere aliquam lecturam ordinariam in studio Bonon., nisi lecturam methaphysicae, et theologiae, ac moralis philosophiae, si illas legi contingat in praefato studio: aliae lecturae studii secularibus intelligantur esse destinatae, et ita volumus, etiam mandamus, eo hoc, ut seculares non desistant a studio Artium, et iam Medicinae, a quo desisterent, et in quo pauci, aut nulli studerent, si sibi per multitudinem Religiosorum loca eorum repleri viderent, et intelligerent." *Statuta Collegium Artium*, cap. 18, Bologna, Archivio di Stato, Studio, busta 216..

111. Etienne Gilson, "Autour de Pomponazzi: Problématique de l'immortalité de l'ame en Italie au début du XVIe siècle," in *Humanisme et renaissance* (Paris: Vrin, 1983), 196–206, 254–58.

112. "quod scilicet animorum immortalitas repugnet principiis naturalibus: immo hec est heretica." Bartolomeo Spina, *Opuscula* (Venice, 1519), K5r; see also K6r.

113. "Ex mandato enim Leonis decimi: et Senatus Bononiensis teneor legere, interpretari et scilicet iudicium meum sententiare: quod senserit Aristoteles quod propter principia naturalia haberi potest: et de hoc quaestione et de aliis astrictus sum ex iuramento fideliter mentem Arist. aperire." Pietro Pomponazzi, *Defensorium autoris* (Venice, 1525), 104r.

114. "Mandata sequor: iuramentum observo. Non est nostri arbitrii dicere Aristoteles sic vel non sit tenuit: scilicet iudicium sumitur ex rationibus et verbis suis." Pomponazzi, *Defensorium*, 104r.

115. "Cum itaque officium nostrum sit Aristotelem interpretari: sicque mihi videtur Aristotelem intellexisse: et quia nullo alio modo intellexit. Mentiar ne: ut aliter dicam quam sententiam. . . . Quare obsecro hos viros qui in tabernis et in conventibus virorum vulgarium me de heresi damnant iam suis latratibus finem imponere: longe enim maiorem heresim isti profitentur. Cum innocentem accusent." Pomponazzi, *Defensorium*, 104r.

116. Francesco Paolo Raimondi, "Pomponazzi's Criticism of Swineshead and the Decline of the Calculatory Tradition in Italy," *Physis*, n.s., 37 (2000): 311–58.

117. "Quoniam autem in superioribus dictum est animam rationalem secundum

quam homo est homo: secundum Aristotelis sententiam esse divisibilem: et extensam: quamquam hoc in se falsum sit: et nostre verissime religionis contrarium." Pietro Pomponazzi, *Tractatus acutissimi, utillimi et mere peripatetici* (Venice, 1525), 130r; "stando in opinionem Aristo. quod humana anima est mortalis." Pomponazzi, *Tractatus acutissimi*, 132r. Rita Ramberti, "Esegesi del testo aristotelico e naturalismo nel *De nutritione et augumentatione*," in *Pietro Pomponazzi: Tradizione e dissenso*, ed. Marco Sgarbi (Florence: Olschki, 2010), 314–45.

118. "Ego quid ut divi Tho. fidelis sectator non existimo sic clarum esse: quod deducis. Quin trahi possit opositum: verum quia nunc intentionis meae non est super Aristo. sensum disputare id ut penitus impertinens pertranseo." Crisostomo Javelli, *Solutiones rationum animi mortalitatem probantium* (Venice, 1525), 108v.

119. "Solvam igitur quascunque rationes formasti mortalitatem probantes: principiis quidem non Aristotelis. pro nunc sed sacrae theologiae et verissimae philosophiae quam arbitramur nostrae catholicae fidei subministrare." Javelli, *Solutiones*, 108v.

120. "fratizare, idest miscere diversa broda." Pietro Pomponazzi, *Commentarii in Aristotelis octo physicorum libros*, Paris, Bibliothèque nationale de France, MS lat. 6533, 567v–68r; Nardi, "Corsi manoscritti," 27, n2.

121. "Domini mei quoniam ego pervenio ad aliquem passum ubi agitur de fide tunc ego me subijcio Fratribus, et quicquid ipsi mihi dicunt, ego credo: subijcio me semper correctioni ecclesiae." Pietro Pomponazzi, *In libros meteorum*, Milan, Biblioteca Ambrosiana, MS R. 96 sup., 71v.

122. "Mihique videtur esse magis consentiendum nostris theologis quam ipsis peripateticis. Et dico non tantum ex fide sed ex puris naturalibus." Pietro Pomponazzi, *Libri quinque de fato*, ed. Richard Lemay (Lugano: Thesaurus mundi, 1957), 174–75.

123. Giancarlo Zanier, *Ricerche sulla diffusione e fortuna del "De incantationibus" di Pomponazzi* (Florence: La Nuova Italia, 1975); Paola Zambelli, "'Aristotelismo eclettico' o polemiche clandestine? Immortalità dell'anima e vicissitudini della storia universale in Pomponazzi, Nifo e Tiberio Russiliano," in *Die Philosophie im 14. und 15. Jahrhundert. In memoriam Konstanty Michalski (1879–1947)*, ed. Olaf Pluta (Amsterdam: Grüner, 1988), 535–72, and "Pietro Pomponazzi's *De immortalitate* and his Clandestine *De incantationibus*: Aristotelianism, eclecticism or libertinism?," *Bochumer philosophisches Jahrbuch für Antike und Mittelatler* 6 (2001): 87–115.

124. Laura Regnicoli, "Produzione e circolazione dei testimoni manoscritti del *De incantationibus*," in *Pietro Pomponazzi: Tradizione e dissenso*, ed. Marco Sgarbi (Florence: Olschki, 2010), 131–80.

125. Pietro Pomponazzi, *De naturalium effectuum admirandorum causis, seu de incantationibus*, in *Opera* (Basel, 1567), 6.

126. "quoniam quanquam aliqua quae referuntur esse facta tam in historia legi Mosis quam legis Christi, superficialiter reduci possunt in causam naturalem, tamen multa sunt quae minime in talem causam reduci possunt . . . de infinitis aliis, quorum nullum potest reduci in causam naturalem." Pomponazzi, *De naturalium effectuum admirandorum causis*, 81.

127. "Talia miracula quae sunt praeter ordinem naturae creatae, & a solo Deo fieri possunt, & fiunt aliquando, veraciter demonstrant insufficientiam doctrinae Aristotelis . . . ipsamque veritatem & firmitatem religionis Christianae aperte declarant." Pomponazzi, *De naturalium effectuum admirandorum causis*, 315.

128. Pomponazzi, *De naturalium effectuum admirandorum causis*, 317–18; Kraye, "Pietro Pomponazzi," 107.

129. Katharine Park, *Secrets of Women: Gender, Generation, and the Origins of Human Dissection* (New York: Zone, 2006), 175–76.

130. Kraye, "Pietro Pomponazzi," 114; Kristeller, "Two Unpublished Questions on the Soul," 368–69.

131. Nardi, "Corsi manoscritti," 25.

132. Stephen Menn, "The Intellectual Setting," in vol. 1 of *The Cambridge History of Seventeenth-Century Philosophy*, ed. Daniel Garber and Michael Ayers (Cambridge: Cambridge University Press, 1988), 51, n42.

133. On the limited enforcement of the Fifth Lateran Council, see Minnich, "Last Two Councils of the Catholic Reformation," 5.

134. Luca Prassicio, *Questio de immortalitate anime* (Naples, 1521), A1r; Hasse, "Attraction," 2:141.

CHAPTER 4: Italian Aristotlelianism after Pomponazzi

1. Dag Nikolaus Hasse, "Aufstieg und Niedergang des Averroismus in der Renaissance: Niccolò Tignosi, Agostino Nifo, Francesco Vimercato," in *Herbst des Mittelalters? Fragen zur Bewertung des 14. und 15. Jahrhunderts*, ed. Jan A. Aertsen and Martin Pickavé (Berlin: De Gruyter), 2004.

2. Massimo Campanini, "Edizioni e traduzioni di Averroè tra XIV e XVI secolo," in *Lexiques et glossaires philosophiques à la Renaissance*, ed. J. Hamesse and M. Fattori (Louvain-la-Neuve: FIDEM, 2003), 21–42; F. Edward Cranz, "Editions of the Latin Aristotle Accompanied by the Commentaries of Averroes," in *Philosophy and Humanism: Renaissance Essays in Honor of Paul Oskar Kristeller*, ed. Edward P. Mahoney (Leiden: Brill, 1976), 116–28.

3. Charles Burnett, "The Second Revelation of Arabic Philosophy and Science: 1492–1562," in *Islam and the Italian Renaissance*, ed. Charles Burnett and Anna Contadini (London: The Warburg Institute, 1999), 185–98; Harry A. Wolfson, "The Twice-Revealed Averroes," *Speculum* 36 (1961): 373–92.

4. Charles Burnett, "Arabic into Latin: The Reception of Arabic Philosophy into Western Europe," in *The Cambridge Companion to Arabic Philosophy*, ed. Peter Adamson and Richard C. Taylor (Cambridge: Cambridge University Press, 2004), 397–400.

5. Noel Swerdlow, "Aristotelian Planetary Theory in the Renaissance: Giovanni Battista Amico's Homocentric Spheres," *Journal for the History of Astronomy* 3 (1972): 36–48; Nicholas Jardine, *The Birth of History and Philosophy of Science: Kepler's A Defence of Tycho against Ursus with Essays on Its Provenance and Significance* (Cambridge: Cambridge University Press, 1984), 231–33; James M. Lattis, *Between Copernicus and Galileo: Christoph Clavius and the Collapse of Ptolemaic Cosmology* (Chicago: University of Chicago Press, 1994), 86–94; Gaspare Contarini, *De homocentricis ad Hieronymum Fracastorium*, in *Opera omnia* (Venice, 1589), 238–52; Christoph Clavius, *In sphaeram Ioannis de sacro bosco commentarius* (Rome, 1585), 432–35, 453–54.

6. Mario Di Bono, *Le sfere omocentriche di Giovan Battista Amico nell'astronomia del cinquecento con il testo del "De motibus corporum coelestium"* (Genoa: CNR–Centro di studio sulla storia della tecnica, 1990), 134–35, 147–49.

7. *Statuti delle università e dei collegi dello studio bolognese*, ed. Carlo Malagola (Bologna: Zanichelli, 1888), 274–75.

8. Pietro Mainardi, *Colliget Averois cum explanationes super V, VI, VII libri*, Ferrara, Biblioteca Ariostea, MS II 84, 2v–287v.

9. Matteo Corti, *Recollectae in septimum colliget Averrois*, Venice, Biblioteca Marciana, MS Lat. VII, 50 (=3570), 1r–65r.

10. Agostino Nifo, *Commentationes in librum de substantia orbis* (Venice, 1508), 2r.

11. Giovanni Battista Confalonieri, *Averrois libellus de substantia orbis expositus* (Venice, 1525); Giovanni Francesco Beati, *Quaesitum in quo Averois ostendit quomodo verificatur corpora coelestia cum finita sint, et possibilia ex se acquirant aeternitatem ab alio* ([Padua], 1542); Mainetto Mainetti, *Commentarii in librum primum Aristotelis de coelo. Necnon librum Averrois de substantia orbis* (Bologna, 1570); Nicolò Vito di Gozze, *In sermonem Averrois de substantia orbis* (Bologna, 1580).

12. Pietro Pomponazzi, *Corsi inediti dell'insegnamento padovano*, ed. Antonino Poppi (Padua: Antenore, 1970), 1:3–5, 1:96.

13. "at noster Aver. sibi ita constat: ut qui eo duce utatur, eundem fere semper doctrinae aristotelicae tenorem persentiat, tanta autem est eius ingenii subtilitas & solertia: ut saepius depravato etiam textu veram tamen aristo. mentem depromat." Confalonieri, *Averrois libellus de substantia orbis*, 2v.

14. "Ponitur vera mens Aver. in proposita difficultate & solvuntur obiecta." Confalonieri, *Averrois libellus de substantia orbis*, 64v.

15. John of Jandun, *In libros Aristotelis de coelo et mundo quaestiones subtilissimae: quibus adiecimus Averrois sermonem de substantia orbis cum commentario ac quaestionibus* (Venice, 1552); *Subtilissime quaestiones in octo libros Aristotelis de physico auditu* (Venice, 1544); and *Quaestiones super Parvis naturalibus ad Aristotelis et Averrois intentionem* (Venice, 1589).

16. Agostino Nifo, *Averroys de mixtione defensio* (Venice, 1505).

17. Vittore Trincavelli, *Quaestio de reactione iuxta Aristotelis sententiam et commentatoris* (Venice, 1520).

18. "Thimestii Averrois videtur eorum sententiam perfectissme habuisse." Beati, *Quaesitum*, D1r.

19. Bruno Nardi, "Il commento di Simplicio al *De anima* nelle controversie della fine del secolo XV e del secolo XVI," in *Saggi sull'aristotelismo padovano dal secolo XIV al XVI* (Florence: Sansoni, 1958), 383–94; Paul J. J. M. Bakker, "Natural Philosophy, Metaphysics, or Something in Between? Agostino Nifo, Pietro Pomponazzi, and Marcantonio Genua on the Nature and Place of Science of the Soul," in *Mind, Cognition and Representation: The Tradition of Commentaries on Aristotle's De anima*, ed. Paul J. J. M. Bakker and Johannes M. M. H. Thijssen (Aldershot, England: Ashgate, 2007), 169–75.

20. "Unde infertur quod nos sequentes Averr. sequimar etiam Graecos. Et monemus, quod in iis sequemur ac imitabimur Aver. ac Graecos, quae erunt conformia ac consonamenti Arist." Girolamo Balduini, *Expositio aurea in libros aliquot physicorum Aristotelis, et Averrois super eiusdem commentationem; et in prologum physicorum eiusdem Averrois* (Venice, 1573), 4.

21. "In commentariis suis in Aristotelem Graecos maxime imitatus est." Konrad Gesner, *Bibliotheca universalis* (Zurich, 1545), 100r.

22. "Averroes volens exponere librum Aristotelis primo proponit se velle imitari grae-cos illos in quibus merito magni commentatoris nomen convenit." Simone Porzio, *Pro-logus Averrois super primum phisicorum Aristotelis*, Milan, Biblioteca Ambrosiana, MS A. 153, 2r.

23. Balduini, *Exposito aurea in libros physicorum*, 1–4.

24. "qui nostra memoria iunctis omnium sententiis merito ab Alexandro mutuatus est, ut NEMO ARISTOTELICUS NISI AVERROISTA." Giovanni Bernardino Longo, *Dilucida expositio in prologum Averrois in posteriora Aristotelis* (Naples, 1551), A1r.

25. Annibale Balsamo, *Dubia aliquot in posteriora circa mentem Averrois*, Milan, Biblioteca Ambrosiana, MS D. 129 inf., 7r–16r.

26. Ermolao Barbaro, *Epistolae, orationes et carmina*, ed. Vittore Branca (Florence: Bibliopolis, 1943), 1:92.

27. "Persevera, mi Nicolae, persevera contemnere, detestari, evertere scelestissimum genus philosophandi." Barbaro, *Epistolae*, 1:45.

28. "eos & sequi & incessere delectatus: velut ex optimis fontibus philosophiam visus est non tam hausisse quam expressisse. Quum solus is fuerit: qui commentatoris nomen adeptus sit." Symphorien Champier, *Cribratio, lima et annotamenta in Galeni, Avicennae et Consiliatoris opera* ([Paris], 1516), 3r.

29. "immo dico, quod quicquid boni habet Averroes sumpsit a graecis, nihil ex se boni dixit, & omnia, quae dicit ex se, vel sumpsit a suis Arabibus. omnia dico, sunt fa-tuitates ambages." Lodovico Boccadiferro, *Explanatio libri primi physicorum Aristotelis* (Venice, 1558), 53v.

30. "In eodem luto haerent fere omnes Graeci expositores, qui Aristotelis doctrinam cum Platonis doctrina permiscent et volunt ipsos fuisse concordes, qui dum vixerunt voluerunt esse discordes." Charles B. Schmitt, "Girolamo Borro's *Multae sunt nostra-rum ignorationum causae* (Ms. Vat. Ross. 1009)," in *Studies in Renaissance Philosophy and Science* (London: Variorum, 1981), article VI, 475.

31. "Ex illis nulla una doctrina nascitur sed permistio quaedam doctrinarum quae non est Accademica nec Peripatetica." Schmitt, "Girolamo Borro," 475.

32. "dignus est enim Averroes, dum digreditur, qui cum Aristotele conferatur." Gi-rolamo Borro, *De motu gravium, et levium* (Florence, 1575), 5.

33. "quia ad gravissimam Aristotelis doctrinam, doctissimamque eiusdem gravitatem, vel omnino, vel saltem quam proxime accedit." Borro, *De motu gravium, et levium*, 5.

34. Burnett, "Arabic into Latin," 397–400.

35. Franceso Storella, *Animadversionum in Averroem, pars prima logicales locos com-prehendens*, Milan, Biblioteca Ambrosiana, MS I. 166 inf. 123r–56r, and *Observationum in Averroem liber secundus locos ad naturalem, medicinam, atque super naturalem philoso-phiam attinensque amplectens*, Milan, Biblioteca Ambrosiana, MS I. 166 inf., 158r–214v.

36. Ruth Glasner, "Levi ben Gershom and the Study of Ibn Rushd in the Four-teenth Century," *Jewish Quarterly Review* 86 (1995): 51–90; Steven Harvey, "Arabic into Hebrew: The Hebrew Translation Movement and the Influence of Averroes upon Medi-eval Jewish Thought," in *The Cambridge Companion to Medieval Jewish Philosophy*, ed. Daniel H. Frank and Oliver Leaman (Cambridge: Cambridge University Press, 2003), 258–80.

37. Benedetto Varchi, *Lezzioni sopra diverse materie, poetiche, e filosofiche* (Flor-ence, 1590), 136.

38. Ibid., 144, 153.

39. Eva Del Soldato, *Simone Porzio: Un aristotelico tra natura e grazia* (Rome: Storia e letteratura, 2010), 30.

40. "multo quippe audies Averroicos, Simplicianos, & Themistianos, qui authoritate magis, nomineque philosophi suam sententiam adstruant, quam eam verborum Arist. fide confirmare conentur. Taceo Latinos, quorum studium perpetuum est, ut contradicant, atque ab aliis semper dissentiant." Simone Porzio, *De humana mente disputatio* (Florence, 1551), 3–4.

41. Ibid., 5–6.

42. "Haec sunt, quae da [sic] humana mente ex ipso Aristotele probanda putamus . . . ut cum Latinis de ea differas; verum ut instructior cum Graecis philosophis congrediaris." Porzio, *De humana mente*, 98.

43. Del Soldato, *Simone Porzio*, 308.

44. Ibid., 102.

45. Neal W. Gilbert, "Francesco Vimercato of Milan: A Bio-Bibliography," *Studies in the Renaissance* 12 (1965): 188–89.

46. "quasi vero non ille multa alia contra fidem nostram dogmata habet, & acerrime defendet." Francesco Vimercati, *De anima rationali peripatetica disceptatio* (Venice, 1566), 292.

47. "Ego certe magis labefactari fidem nostram arbitro, cum testimoniis Aristotelis Platonis, & aliorum exterorum non aptis, non accommodatis, non ad id propositum ab illis conscriptis, eam confirmamus, & tutamur. Quid enim aliud suspicari possunt fidei inimici, quam ea quae credimus vana esse & ridicula, non ab eorum authoribus testimonis non consentanea perquirumus accipimusque?" Vimercati, *De anima rationali*, 292.

48. "Hanc itaque Dei, mentiumque divinarum tractationem liber hic exhibet, non quidem omnem eam, quam fide nos credimus, sed quae naturae lumine a philosopho excellentissimo tradi potuit." Francesco Vimercati, *In eam partem duodecimi libri metaphysices Aristotelis, in qua de deo & caeteris mentibus divinis disseritur, commentarii* (Paris, 1551), 2.

49. "Spontaneous Generation and the Ontology of Forms in Greek, Arabic, and Medieval Latin Sources," in *Classical Arabic Philosophy: Sources and Reception*, ed. Peter Adamson (London: Warburg Institute, 2007), 150–75; Paola Zambelli, *Una reincarnazione di Pico ai tempi di Pomponazzi* (Milan: Il Polifilo, 1994), 79–88.

50. Antonio Bernardi, *Eversiones singularis certaminis* (Basel, 1562), 496.

51. "Dicerem, Aristotelem ex fundamentis naturae loquentem, nunquam posuisse Daemones diversos a Deo." Bernardi, *Eversiones*, 518.

52. "Quare illud negare non possumus, Aristotelem ratione naturali non pervenisse nisi ad formas, quae in corpore aliquo sunt." Bernardi, *Eversiones*, 519.

53. Jill Kraye, "Pietro Pomponazzi (1462–1525): Secular Aristotelianism in the Renaissance," in *Philosophers of the Renaissance*, ed. Paul Richard Blum (Washington, DC: Catholic University of America Press, 2010)," 93; Paul Oskar Kristeller, "Philosophy and Medicine in Medieval and Renaissance Italy," in vol. 3 of *Studies in Renaissance Thought and Letters* (Rome: Storia e letteratura, 1993), 436–38.

54. Jerome L. Bylebyl, "The School of Padua: Humanistic Medicine in the Sixteenth Century," in *Health, Medicine and Mortality in the Sixteenth Century*, ed.

Charles Webster (Cambridge: Cambridge University Press, 1979), 338; David A. Lines, "Natural Philosophy in Renaissance Italy: The University of Bologna and the Beginnings of Specialization," *Early Science and Medicine* 6 (2001): 267–320.

55. Craig Martin, "Francisco Vallés and the Renaissance Reinterpretation of Aristotle's *Meteorologica* IV as a Medical Text," *Early Science and Medicine* 7 (2002): 1–30.

56. "Et quia compertissimum est Aver. nunquam decessisse a magno magistro Aristotele. Ego semper ita existimavi, idcirco eum in hanc sententiam devenisse, quia putavit sic quoque existimasse Arist." Girolamo Mercuriale, *Praelectiones patavinae, de cognoscendis, et curandis humani corporis affectibus* (Venice, 1613), 19.

57. Nancy G. Siraisi, *Avicenna in Renaissance Italy* (Princeton, NJ: Princeton University Press, 1987).

58. "In qua quidem generatione aliter respondent Theologi, aliter Philosophi, aliter Medici . . . dicam ut peripateticus, & ut medicus." Giovanni Battista Da Monte, *In primi libri canonis Avicennae primam fen, profundissima commentaria* (Venice, 1558), 65.

59. "Nihil autem existimo deterius in philosophia posse contingere; quam cum ea Theologiam commiscere. Hinc est, quod nullum unquam vidi recte philosophatum; qui Theologiam cum Philosophia voluerit permiscere. Hinc est quod Avicenna in infinitos pene errores incidit. Hinc etiam; quod omnes, Scotistae, & Thomistae; dum permiscent; perturbant omnia saepissime a vera peripateticorum via aberrantes." Da Monte, *In primi libri canonis*, 65–66.

60. "praeter Averrroem, qui solus cum Graecis sensit; &, qui unus semper in veritate praeponderat; omnibus aliis Arabibus simul collectis. . . . Graeci autem omnes una cum Averroe inter Arabes. (Quorum melior est sententia, & vere peripatetica) voluerunt quidem & ipsi mista ab intelligentiis generari." Da Monte, *In primi libri canonis*, 66.

61. "Est et alia opinio Arist. et sui expositoris fidelissimi Aver." Da Monte, *In primi libri canonis*, 205.

62. "Exerceamus itaque ingenium in Physicis, ut percipiamus Naturae vires, constanter tamen eis credamus, quae prodeunt ex ore Dei, ut supra Naturam elevati, & cum Deo iuncti, ipso bono, vero, & pulchro perennis secula fruamur." Francesco Piccolomini, *Librorum ad scientiam de natura attinentium pars prima* (Venice, 1596), 186v.

63. "Quae Providentia sequitur eminentissimam Dei cognitionem, qua noscens Deus se ut bonum, vult omne, quod est bonum, & sua facultate movendo Coelum, motu totius effundit munera bona in universum orbem, idque facit per motum, non per influxum aliquem ab eo distinctum." Francesco Piccolomini, *Librorum ad scientiam de natura attinentium pars secunda* (Venice, 1600), 41r.

64. "Hanc itaque puto ego de Providentia fuisse opinionem Aristotelis, qui solum per Physica opera progrediens, exactam veritatem inspicere non valuit, quam cum latissime explicent nostri Theologi." Piccolomini, *Librorum ad scientiam pars secunda*, 41v.

65. Branko Mitrović, "Defending Alexander of Aphrodisias in the Age of the Counter-Reformation: Iacopo Zabarella on the Mortality of the Soul according to Aristotle," *Archiv für Geschichte der Philosophie* 91 (2010): 330–54; Eckhard Kessler, "Alexander of Aphrodisias and his Doctrine of the Soul: 1400 Years of Lasting Significance," *Early Science and Medicine* 16 (2011): 49–58.

66. "quod ego monstrum esse arbitror, & Hermaphroditum a doctrina Aristotelica & ab Averroe quoque ipso alienissima." Giacomo Zabarella, *De rebus naturalibus* (Frankfurt, 1597), 260.

67. Zabarella, *De rebus naturalibus*, 269–70; Antonino Poppi, *La dottrina della scienza in Giacomo Zabarella* (Padua: Antenore, 1972), 298–328.

68. Lodovico Boccadiferro, *Lectiones super tres libros de anima* (Venice, 1566), 47v, 69r.

69. Marcantonio Genua, *Disputatio de intellectus humani immortalitate* (Mondovì, 1565.)

70. Guido Giglioni, "Nature and Demons: Girolamo Cardano Interpreter of Pietro d'Abano," in *Continuities and Disruptions between the Middle Ages and the Renaissance*, ed. Charles Burnett et al. (Louvain-la-Neuve: FIDEM, 2008), 89–112.

71. "Haec sunt quae non solum falsa & impia, sed etiam a Peripateticis aliena magna ex parte, ac etiam ridicula invicemque pugnantia scripsit Pomponatius." Girolamo Cardano, *Contradicentium medicorum liber primus [-secundus]*, vol. 6 of *Opera omnia* (Lyon, 1663), 478.

72. "Lex enim nostra vera est . . . imo falsam religionem nostram reddat." Cardano, *Contradicentium medicorum liber*, 6:478. Nancy G. Siraisi, *The Clock and the Mirror: Girolamo Cardano and Renaissance Medicine* (Princeton, NJ: Princeton University Press, 1997), 164.

73. "In via Peripateticorum hoc Pomponatius dicere non potest, cum sensus percipiant intus recipiendo, non extra mittendo, & praecipue oculus." Cardano, *Contradicentium medicorum liber*, 6:479.

74. Sirarisi, *Clock and Mirror*, 164–66; Alfonso Ingegno, *Saggio sulla filosofia di Cardano* (Florence: La Nuova Italia, 1980), 1–78.

75. "Anima quidem quod forma sit corporis omnis mortalis est." Girolamo Cardano, *De immortalitate animorum liber*, vol. 2 of *Opera omnia* (Lyon, 1663), 528.

76. Guido Giglioni, "Libertinismo, censura e scrittura. A proposito di alcune recenti pubblicazioni cardaniane," *Bruniana e campanelliana* 10 (2004): 377–88.

77. Ugo Baldini and Leen Spruit, *Catholic Church and Modern Science: Documents from the Archives of the Roman Congregations of the Holy Office and the Index* (Rome: Libreria Editrice Vaticana, 2009), 1,2:1042–1472; and "Cardano e Aldrovandi nelle lettere del Sant'Uffizio Romano all'inquisitore di Bologna (1571–73)," *Bruniana e campanelliana* 6 (2000): 145–63.

78. Christoph Lüthy, "An Aristotelian Watchdog as Avant-garde Physicist: Julius Caesar Scaliger," *Monist* 84 (2001): 547–48.

79. Ibid., 548.

80. Kuni Sakamoto, "Creation, the Trinity and *Prisca Theologia* in Julius Caesar Scaliger," *The Journal of the Warburg and Courtauld Institutes* 73 (2010): 195–207.

81. "Avenrois autem verba divina sunt. . . . Quid aliud diceret Christianus?" Julius Caesar Scaliger, *Exotericarum exercitationum libri XV* (Paris, 1557), 61,3, 95r. Sakamoto, "Creation in Scaliger," 200.

82. Scaliger, *Exotericarum exercitationum libri*, exer. 3, 4v. Sakamoto, "Creation in Scaliger," 197–98.

83. "Quid enim est aliud Naturae providentia quam Dei perpetua sui ipsius praesentia?" Scaliger, *Exotericarum exercitationum libri*, exer. 77,5, 121r. Sakamoto, "Creation in Scaliger," 198.

84. Paganino Gaudenzio, *De pythagoraea animarum transmigratione* (Pisa, 1641), 199–249; Giovanni Santinello, ed. *Models of the History of Philosophy: From its Origins in the Renaissance to the "Historia philosophica"* (Dordrecht: Kluwer, 1993), 1:120–24.

85. "Respondere possum, Scaligerum in opere de Subtilitate exercuisse stylum contra Cardanum, dedisseque operam, ut de ipso triumpharet; non semper sollicitus de sententia Aristotelis." Gaudenzio, *De pythagoraea animarum transmigratione*, 201.

86. Donato Antonio Altomare, *Omnia opera* (Lyon, 1565), 21.

87. Rinaldo Odoni, *Discorso per via peripatetica ove si dimostra se l'anima, secondo Aristotele, è mortale, o immortale* (Venice, 1557), 35v–36r.

88. "il che gli avviene per un libretto, come egli dice, che già ne scrisse il Peretto con la dottrina secondo lui d'Aristotile. . . . egli il trova abbondevole di cose nuove e degne d'esser considerate." Sperone Speroni, *Opere* (Venice, 1740), 2:221.

89. Ibid., 1:73.

90. Torquato Tasso, "Dialoghi," in *Opere*, ed. Bruno Maier (Milan: Rizzoli, 1965), 416.

91. "Aggiungono a tutte queste ragioni Alessandro, Simplicio e Averroè che da le contemplazioni de le cose naturali e celesti nascono le virtù morali." Tasso, "Dialoghi," 428.

CHAPTER 5: **Religious Reform and the Reassessment of Aristotlenianism**

1. For the use of Aristotelian notions of causation in the definition of justification, see the Sixth Session of the Council of Trent, at *Conciliorum oecumenicorum decreta*, ed. Giuseppe Alberigo et al. (Bologna: Istituto per le scienze religiose, 1973), 649.

2. "se egli on si fosse adoperato, noi mancavammo di molti articoli di fede." Paolo Sarpi, *Istoria del Concilio Tridentino* (Florence: Sansoni, 1966), 1:284.

3. John W. O'Malley, *The First Jesuits* (Cambridge, MA: Harvard University Press, 1993), 252; Ladislaus Lukács, ed., *Monumenta paedagogica Societitatis Iesu*, in *Monumenta historica Societiatis Iesu* (Rome: Institutum historicum Societatis Iesu, 1965–92), 1:151–52.

4. "Ne magistri novas opiniones inducant contra communes, inconsulto superiore, aut etiam pro libitu contra Aristotelem, sine authoritate antiquorum." Lukács, *Monumenta paedagogica*, 2:477.

5. "Item ne laudent nimis, imo ne laudent quidem, Averroin vel alios impios interpretes. sed si qui laudandi sint, potius laudent D. Thomam, Albertum Magnum, vel alios christianos et pios." Lukács, *Monumenta paedagogica*, 2:478.

6. John M. McManamon, *The Texts and Contexts of Ignatius Loyola's Autobiography* (New York: Fordham University Press, 2013), 73–74.

7. Roger Ariew, "Descartes and the Jesuits: Doubt, Novelty, and the Eucharist," in *Jesuit Science and the Republic of Letters*, ed. Mordechai Feingold (Cambridge, MA: MIT Press, 2003), 162.

8. "Item non se ostendant esse averroistas, aut graecorum factionem sectare vel arabum contra latinos aut theologos." Lukács, *Monumenta paedagogica*, 2:478.

9. "Item, prohibeatur ne magistri interpretentur digressiones Averrois vel Simplicii aut alterius; sed simpliciter proponantur opiniones eorum indifferenter." Lukács, *Monumenta paedagogica*, 2:478.

10. "Primus, nimiae libertatis, quae quidem nocet fidei, ut experientia ostendit in accademiis Italiae; secundus, esse adstrictos unius tantum aut alterius authoris doctrinae; nam hoc efficit in Italia odiosos et contemptibiles." Lukács, *Monumenta paedagogica*, 2:478.

11. "Item, sic doceatur philosophia, ut serviat theologiae; et ideo notentur opiniones non tenendae in his quae fidem concernunt, ac eae quae sunt defendendae, ut omnes sic doceant, et totis viribus defendant." Lukács, *Monumenta paedagogica*, 2:478.

12. "et ad id obligentur expresse, etiam secundum Aristotelem, ut de immortalitate animae etc. ac per totam Societatem sic servetur." Lukács, *Monumenta paedagogica*, 2:478.

13. "Quod Deus habeat providentiam istorum inferiorum, etiam rerum singularum ac rerum humanarum." Lukács, *Monumenta paedagogica*, 2:496; "Quod creatio est possibilis secundum veram philosophiam. Item quod ex principiis Aristotelis et ex eius doctrina necessario sequatur creationem esse possibilem." Lukács, *Monumenta paedagogica*, 2:498.

14. "De aeternitate mundi, confutare rationes Aristotelis et modis omnibus veritatem confirmare, scilicet non fuisse ab aeterno; etiam rationibus naturalibus, quoad fieri possit, id ostendere. Item, id asserrere esse secundum veritatem et secundum veram philosophiam. licet non secundum Aristotelem." Lukács, *Monumenta paedagogica*, 2:498.

15. "radicitius extirpetur infoelix lolium Averroycae philosophiae, quae non tam haereticos quam atheos e nostris quosdam fecisse putatur." Lukács, *Monumenta paedagogica*, 3:415.

16. "qui cum Averroystis nimium vixere familiariter, non parum depravati esse credantur." Lukács, *Monumenta paedagogica*, 3:415.

17. "Nunc divinum Averroem nominare quidam e nostris audent." Lukács, *Monumenta paedagogica*, 3:415.

18. "Leggere Averroe è molto utile, si per la sua dottrina, come per la fama che ha in Italia, e per poterlo intendere, leggerà li suoi seguaci, come Janduno, Barleo, Paolo veneto, Zimarra, Nipho." Lukács, *Monumenta paedagogica*, 2:665–66.

19. "Averroes fuit singulare decus, gloria et praesidium Lyceion, cui uni (Alexandrum et Simplicium excipio), ausim dicere, plus debere disciplinam peripateticam quam omnibus aliis explicatoribus." Lukács, *Monumenta paedagogica*, 2:665. Paul Richard Blum, "Benedictus Pererius: Renaissance Culture at the Origins of Jesuit Science," *Science and Education* 15 (2006): 279–304.

20. Lukács, *Monumenta paedagogica*, 4:664–65.

21. "Ego, sicut eos non laudo, qui Averroem fastidiose contemnunt, convitiis lacerant, & veluti pestem ingeniorum, omnibus fugiendum esse clamant." Benito Perera, *De communibus omnium rerum naturalium principiis et affectionibus* (Paris, 1579), A2v.

22. "Sed dicent, Averroem fuisse acri ingenio, iudicio gravi, & singulari studio ac diligentia." Perera, *De communibus principiis*, A3r.

23. "negari tamen non potest, Averoem, interpretando Aristotelem, ob ignorationem linguae graecae, mendososque codices, & bonorum interpretum penuriam, multifariam hallucinatum esse." Perera, *De communibus principiis*, A3r.

24. "R. P. N. Generalis commendatum magnopere habeat ut praeceptores nostri qui interpretantur Aristotelem, non nisi cum magno delectu interpretes eos legant qui contra christiana dogmata impie scripserunt, et maxime ad ea impugnanda quae christianae veritati adversentur, et ita philosophiam interpretentur ut verae theologiae scholasticae quam nobis commendant constitutiones, ancillari et subservire faciant." Lukács, *Monumenta paedagogica*, 4:248.

25. Lukács, *Monumenta paedagogica*, 5:100–101.

26. Ibid., 5:283, 5:397.

27. "Hoc autem loco admonendus es, Lector pie, ne cum in hos, aut alios impios Aristotelis interpretes, sive Graecos, sive Arabes incideres, in iis praesertim, quae ad pietatem attinent, facile illis credas, atque committas. Nam, cum impii fere omnes fuerint, Ethnici, Idololatrae, nonnulli etiam Sarraceni, vel Mahumetani." Francisco Toletus, *Commentaria una cum quaestionibus in octo libros Aristotelis de physica auscultatione* (Venice, 1580), a3r. The author of the preface is unidentified. The passage also appears in Antonio Possevino, *Bibliotheca selecta* (Rome, 1593), 2:106–7.

28. "Nec vero satis mirari possum, sic quosdam in nonnullis Academiis esse Authoribus impiis addictos." Toletus, *Commentaria in libros de physica auscultatione*, a3r.

29. "quod deterius est, viri alioqui catholici, Averroistas se & esse, & dici velint. . . . & ea tanquam oracula sibi edita putent, indignum sane est gravitate & dignitate sapientis. primum enim is, quem dico pluris erroribus fidei pietatique perniciosis plenus est." Toletus, *Commentaria in libros de physica auscultatione*, a3r.

30. Andreas Schott, *Vitae comparatae Aristotelis ac Demosthenis* (Augsburg, 1603), 154–66.

31. "Adde fuisse Mahemetanum, & (quod ipsa scripta facile declarant) conceptum animo adversus Christianam religionem odium semper habuisse." Toletus, *Commentaria in libros de physica auscultatione*, a3v.

32. "Usus praeterea corrupto Aristotelis libro, & pluribus in locis depravato; id quod scripta praese fuerunt. Graecis fere omibus explanatoribus caruit; Latini etiam." Toletus, *Commentaria in libros de physica auscultatione*, a3v.

33. "Deinde plerique, cum audiunt verum quid esse secundum Aristotelem, id omnino intelligunt, quid est, secundum Philosophiam: immo secundum omnium optimam (quam esse putant Aristotelis) Philosophiam: & quod Aristoteles sensit, id credunt esse naturali rationi ac lumini consentaneum." Toletus, *Commentaria in libros de physica auscultatione*, a4r.

34. "atque ideo sic disputantes, quantum in ipsis est idem illud Concilii decretum per imprudentiam evertunt, quod credant, rationi, & lumini naturali fidem adversari." Toletus, *Commentaria in libros de physica auscultatione*, a4r.

35. Alfonso de la Vera Cruz, *Physica speculatio* (Salamaca, 1569), 430.

36. Frans Titelmans, *Compendium philosophiae naturalis* (Lyon, 1558); David A. Lines, "Teaching Physics in Louvain and Bologna," in *Scholarly Knowledge: Textbooks in Early Modern Europe*, ed. Emidio Campi et al. (Geneva: Droz, 2008).

37. "sappiamo che se il mondo fosse eterno, o si darebbe un'infinito in atto, e la revolutione nella Pitagorica, o uno intelletto solo all'averroistica, o la mutabilità dell'anime all Epicurea." Francesco Panigarola, *Prediche quadragesimali* (Venice, 1617), 145.

38. Corrie E. Norman, "The Franciscan Preaching Tradition and Its Sixteenth-Century Legacy: The Case of Cornelio Musso," *The Catholic Historical Review* 85 (1999): 208–32.

39. "questa verità provata da Socrate, da Platone, Pitagora, Homero . . . predicata da Mosè, da tutti i Profeti, dal sommo Filosofo di tutti i Filosofi, Christo Giesù, da San Paolo da tutti gli Apostoli & da Peripatetici ancora da tanti." Cornelio Musso, *Delle prediche quadragesimali* (Venice, 1603), 2:702. Paul F. Grendler, *The Universities of the Italian Renaissance* (Baltimore: Johns Hopkins University Press, 2002), 293.

40. "O sognatori, informa l'anima non assiste. Non est sicut nauta in navi sicut forma in materia. . . . Non quelle che *educuntur de potentia materiae* . . . che non intende senza fantasmati?" Musso, *Prediche quadragesimali*, 2:702.

41. Giacomo Nacchiante, *Opus doctum ac resolutum* (Venice, 1557), 27r.

42. "Convinci possit, apud intellectum recte dispositum, ex principiis scientiae naturalis: & peculiariter secundum mentem Aristotelis, animam, humanam, esse incorruptiblem, & immortalem." Nacchiante, *Opus doctum*, 36r.

43. "Verum haec opinio, non solum est impia, imo maxime impia, sed nec ex lumine naturali vere dependens: ac minus Aristoteli satisfaciens." Nacchiante, *Opus doctum*, 54r.

44. "Sic Saracenus ille deribebat [sic] nostram fidem." Domingo de Soto, *Quaestiones in octo libros physicorum Aristotelis* (Douai, 1613), 412.

45. "Audivimus enim Italos esse quosdam, qui suis & Aristoteli & Averroi tantum temporis dant, quantum sacris litteris." Melchor Cano, *De locis theologicis* (Salamanca, 1563), 312.

46. "Ex quo nata sunt in Italia pestifera illa dogmata de mortalitate animi, & divina circa res humanas improvidentia." Cano, *De locis theologicis*, 312.

47. Cano, *De locis theologicis*, 312–13.

48. "in plerisque tamen veritati, non solum Catholicae fidei, verum etiam rationis naturalis fuisse adversatum." Possevino, *Biblioteca selecta*, 2:78. For objections to Cano, see Possevino, *Biblioteca selecta*, 2:90–93.

49. Pedro de Fonseca, *In libros metaphysicorum Aristotelis Stagiritae, cui praemissi sunt institutionum dialecticarum libri octo* (Lyon, 1597), X4v, 17–18.

50. Charles B. Schmitt, *John Case and Aristotelianism in Renaissance England* (Kingston, Ontario: McGill–Queen's University Press, 1983), 74.

51. Charles H. Lohr, *Renaissance Authors*, vol. 2 of *Latin Aristotle Commentaries* (Florence: Olschki, 1988), 99.

52. Collegium Conimbricense, *In tres libros de anima* (Cologne, 1617), col. 101.

53. Ibid., col. 113–14, 118.

54. "Averroem (quem alioqui raro laudare consuevit) melius hac in re, quam ceteros Peripateticos, Aristotelis mentem assecutum fuisse." Collegium Conimbricense, *In quatuor libros de coelo* (Lyon, 1617), 38.

55. Martin Delrio, *Disquisitionum magicarum libri sex* (Mainz, 1617), 10.

56. Girolamo Dandini, *De corpore animato* (Paris, 1611), col. 1832.

57. Antonio Rubio, *Commentarii in libros Aristotelis de anima* (Brescia, 1626), 488–94.

58. "Quid enim minus convenit, quam, in quo nos Aristotelis interpretes profitemur, in eo Christianos inveniri? Pugnat omnino cum Aristotelica Philosophia nostra persuasio, neque ullo modo conciliari potest." Ottavio Maria Paltrinieri, *Notizie intorno alla vita di Primo del Conte Milanese della congregazione di Somasca teologo al Concilio di Trento* (Rome: Fulgonim, 1805), 98.

59. "Cum Aristotelem interpretandum sumis, fac, ut ipsum Aristotelem loqui, non Christianum audiam. . . . Scimus omnes, Aristotelem Christianum non fuisse." Paltrinieri, *Notizie alla vita di Primo del Conte*, 98; Marcantonio Maioragio, *In quatuor Aristotelis libros de coelo paraphrasis* (Basel, 1554), a4v; Silvana Seidel Menchi, *Erasmo in Italia, 1520–1580* (Turin: Bollati Boringhieri, 1987), 278.

60. Sachiko Kusukawa, *The Transformation of Natural Philosophy: The Case of Philip Melanchthon* (Cambridge: Cambridge University Press, 1995), 33.

61. Philip Melanchthon, *De Aristotele*, vol. 11 of *Opera quae supersunt omnia*, ed. Karl Gottlieb Bretschneider (Halle: Schwetschke, 1834–60), 11:342–49.

62. "Verum ut redeam ad Aristotelem, etsi in his libris, qui extant, nos [sic] satis perspicue dicit, quid sentiat, nec integre tractat hanc causam, tamen & ipsum apparet in hanc sententiam inclinare, cum dicat mentem extrinsecus accedere, & divinum quiddam esse." Philip Melanchthon, *Commentarius de anima* (Wittenberg, 1550), 156v. "Nos" should read "non."

63. Peter Petersen, *Geschichte der aristotelischen Philosophie im protestantischen Deutschland* (Leipzig: Meiner, 1921), 219–58.

64. "Primum enim consyderare cepi, duplex ne posset unius esse rei veritas, ut quod Theologice falsum est, verum possit esse Philosophis." Nicolaus Taurellus, *Philosophiae triumphus, hoc est metaphysica philosophandi methodus* (Arnheim, 1617), *4r.

65. "Demonstrationes autem noveram esse necessario verissimas, ut errasse Philosophos existimarem, cujus culpa non scientiae, vel demonstrationis, sed hominum imperitiae deberet adscribi." Taurellus, *Philosophiae triumphus*, *4v.

66. "Non enim dubium est quin ipsemet Aristoteles, si Christi lumine fuisset edoctus, suos meditando revocasset errores, veritate diligentius inquisita." Taurellus, *Philosophiae triumphus*, E1v.

67. Taurellus, *Philosophiae triumphus*, 2, 74–75.

68. "Negant hoc fieri posse qui Christum & Belilal una mente copulant, homines ineptissimi, quos impostor Sathanas hac futilissima distinctione detinet, ut duplex unius & eiusdem rei veritas sit, alia nimirum Philosophica & alia Theologica." Nicolaus Taurellus, *Medicae praedictionis methodus* (Frankfurt, 1581), 4r. For Taurellus and the question of double truth, see Christoph Lüthy, *David Gorlaeus (1591–1612): An Enigmatic Figure in the History of Philosophy and Science* (Amsterdam: Amsterdam University Press, 2012), 123–24.

69. "Multa namque continentur in Bibliis, quae philosophica sunt: & multa etiam in libris habentur Aristotelicis, quae cum vera non sint, philosophica certe non sunt." Nicolaus Taurellus, *De rerum aeternitate* (Marburg, 1614), 2.

70. "Quaecunque igitur de Christo: de generis humani redemptione: de gratuita peccatorum remissione: de fide: de spe futurae felicitatis, & similibus, quaestiones proponuntur, Theologicae sunt omnes. Caeterae vero, de Deo, de lege, de homine, de mundo, & reliquis rebus a Deo conditius, merito philosophiae tribuuntur." Taurellus, *De rerum aeternitate*, 8.

71. Taurellus, *De rerum aeternitate*, 14.

72. Kristian Jensen, "Description, Division, Definition: Caesalpinus and the Study of Plants as an Independent Discipline," in *Renaissance Readings of the Corpus Aristotelicum*, ed. Marianne Pade (Copenhagen: Tusculanum, 2001), 196–203.

73. Cesalpino, *Quaestionum peripateticorum libri quinque* (Venice, 1593), 21r–24v.

74. Ibid., 31v–32v.

75. "Si fuit olim Averrhois de unitate intellectus in concilio Lateranensi damnata sententia: cur hanc renovare licuit Caesalpino?" Nicolaus Taurellus, *Alpes caesae, hoc est, Andr. Caesalpini Itali monstrosa & superba dogmata discussa & excussa* (Frankfurt, 1597), 24.

76. "Nescio an Christianus fuerit Caesalpinus. Hoc iure tamen affirmare ausim nullam in Ethnicorum scriptis impietatem hac una inveniri maiorem." Taurellus, *Alpes caesae*, 24.

77. "ad percipienda Graecorum sensa non ipsos Graecos, sed barbaros interpretes consulant." Taurellus, *Alpes caesae*, 30.

78. "Cum enim nulli sectae simus addicti: neque nos Averrhois, neque Alexandri, neque ipsius etiam Aristotelis movet auctoritas." Taurellus, *Alpes caesae*, 37.

79. "Tu si Christianus es, necesse est fatearis, saepius errasse Aristotelem." Taurellus, *Alpes caesae*, 38.

80. Nicolaus Taurellus, *Kosmologia. Hoc est physicarum et metaphysicarum discussionum* (Hamburg, 1603), 8–12.

81. Richard A. Muller, "Vera Philosophia cum sacra Theologia nusquam pugnat: Keckermann on Philosophy, Theology, and the Problem of Double Truth," *Sixteenth Century Journal* 15 (1984): 341–65.

82. Otto Casmann, *Marinarum quaestionum tractatio philosophica* (Frankfurt, Germany, 1596), 22; Ann Blair, "Mosaic Physics and the Search for a Pious Natural Philosophy in the Late Renaissance," *Isis* 91 (2000): 45.

83. "Quod si Ethnicus Aristoteles vel alius, oraculis Theosophicis contraria effutivit pronuntiata, ea non sunt philosophica, sed tanquam naevi & excrementa." Casmann, *Marinarum tractatio*, 26.

84. Ugo Baldini and Leen Spruit, *Catholic Church and Modern Science: Documents from the Archives of the Roman Congregations of the Holy Office and the Index* (Rome: Libreria Editrice Vaticana, 2009), 1,1:815–17.

85. Grendler, *Universities of the Italian Renaissance*, 293–94.

86. Baldini and Spruit, *Catholic Church and Modern Science*, 1,3:2279–80.

87. For the myth of Venice, see Edward Muir, *Civic Ritual in Renaissance Venice* (Princeton, NJ: Princeton University Press, 1981), 13–61.

88. Paul F. Grendler, *The Roman Inquisition and the Venetian Press* (Princeton, NJ: Princeton University Press, 1977), 25–62, 78.

89. John Martin, *Venice's Hidden Enemies: Italian Heretics in a Renaissance City* (Berkeley: University of California Press, 1993); Ruth Martin, *Witchcraft and the Inquisition in Venice, 1550–1650* (New York: Blackwell, 1989).

90. Richard Palmer, "Physicians and the Inquisition in Sixteenth-Century Venice: The Case of Donzelli," in *Medicine and the Reformation*, ed. Ole Peter Grell and Andrew Cunningham (London: Routledge, 1993), 118–33.

91. Luigi Firpo, *I processi di Tommaso Campanella*, ed. Eugenio Canone (Rome: Salerno, 1998), 59–87; Saverio Ricci, *Giordano Bruno nell'Europa del Cinquecento* (Rome: Salerno, 2000), 458–510.

CHAPTER 6: Learned Anti-Aristotelianism

1. Charles H. Lohr, *Renaissance Authors*, vol. 2 of *Latin Aristotle Commentaries* (Florence: Olschki, 1988), xiii.

2. Charles B. Schmitt, *Aristotle and the Renaissance* (Cambridge, MA: Harvard University Press, 1983), 36–44; F. Edward Cranz, A *Bibliography of Aristotle Editions, 1501–1600*, 2nd ed. (Baden-Baden: Koerner, 1984), vii–xx; Jill Kraye, "The Printing History of Aristotle in the Fifteenth Century: A Bibliographic Approach to Renaissance Philosophy," *Renaissance Studies* 9 (1995): 189–211.

3. "ad corrumpendam iuventutem dissolvendamque Christianae vitae disciplinam nihil pestilentius induci potuit." Paolo Giovio, *Gli elogi degli uomini illustri*, vol. 8 of *Opera*, ed. Renzo Meregazzi (Rome: Istituto poligrafico dello stato, 1972), 96.

4. Nancy G. Siraisi, *Avicenna in Renaissance Italy* (Princeton, NJ: Princeton University Press, 1987).

5. Vivian Nutton, "The Rise of Medical Humanism: Ferrara, 1464–1555," *Renaissance Studies* 11 (1997): 2–19.

6. "Sumus autem Christiani, non Mahometistae: Galli, non Arabes." Symphorien Champier, *Hortus gallicus, pro Gallis in Gallia scriptus* (Lyon, 1533), 8.

7. "Barbari Barbaris scripsere, & Arabes Arabibus: Galli igitur Gallis." Champier, *Hortus gallicus*, 6.

8. "Mahomethea spurcissima & nephanda secta." Symphorien Champier, *Annotamenta errata et castigationes in Avicenne opera* (Lyon, 1522), 2r.

9. "cum is neget & Mosi praecepta, & Christi, ac suum Mahumetem deserat." Symphorien Champier, *Epistola responsiva pro Graecorum defensione in Arabum errata* (Lyon, 1533), 21v.

10. "Dicuntur et pagani hoc est rusticani: quod non spiritu renati sicut christiani: sed in ea qua geniti sunt rusticitate semper permanserunt: prout fuit ille impius averrhous sine lege infidelis rusticus paganus et gentilis qui unitatem intellectus contra aristotelis mentem cuius se dicit interpretem solus somniavit." Champier, *Annotamenta in Avicenne opera*, 2v.

11. Champier, *Annotamenta in Avicenne opera*, 4r–v.

12. "Avicennae medici aemulus & inimicissimus fuit." Konrad Gesner, *Bibliotheca universalis* (Zurich, 1545), 100r.

13. Dag Nikolaus Hasse, "King Avicenna: The Iconographic Consequences of a Mistranslation," *Journal of the Warburg and Courtauld Institutes* 60 (1997): 230–43.

14. "scilicet vulgares bythinie regem fuisse dicunt: et ab Averoi medico venenatum tradunt: sed ante quem periret ipse illum Averroi interfecisse." Giacomo Filippo Foresti, *Supplementum cronicarum* (Venice, 1486), 224v. The legend is repeated by Franciscus Calphurnius: "Post aliquot dies ne ab officio invidorum extularet potu venefico illum interemit." Avicenna, *Liber canonis* (Lyon, 1522), 1v. He traced this story back to Champier.

15. Paul F. Grendler, *Critics of the Italian World, 1530–1560: Anton Francesco Doni, Nicolò Franco and Ortensio Lando* (Madison: University of Wisconsin Press, 1969), 21–38.

16. Strabo, *Geography*, 13,1,54; Plutarch, *Life of Sulla*, 26,1–3; Jonathan Barnes, "Roman Aristotle," in *Philosophia Togata II: Plato and Aristotle at Rome*, ed. Jonathan Barnes and Miriam Griffin (Oxford: Clarendon, 1997), 1–69.

17. Ortensio Lando, *Paradossi cioe sententie fuori del comun parere* (Venice, 1544), M4r–M6r; Cicero, *Epistulae ad familiares*, 1.9.23; Simplicius, *In physicorum libros quattuor priores*, ed. Hermann Diels, vol. 9 of *Commentaria in Aristotelem graeca* (Berlin, 1882), 8,16–20.

18. "Che Aristotele fusse non solo un'ignorante, ma anche lo piu malvagio huomo di quella eta." Lando, *Paradossi*, M6r.

19. "[O] dotto Simon Portio che col tuo bellissimo ingegno non habbi penetrato mai si avanti, c'habbi conosciuto che questo tuo tanto familiare Aristotele, fusse un bue?" Lando, *Paradossi*, M7r.

20. "nega l'immortalita del'anima, & concede la felicita nel stato presente." Lando, *Paradossi*, M8v.

21. Arrian, *Anabasis of Alexander*, 7,27; Plutarch, *Life of Alexander*, 77.

22. Lando, *Paradossi*, N1r.

23. Ibid., N2v–N3r.

24. "inamorossi d'una sfacciata meretrice detta per nome Hermia." Lando, *Paradossi*, N2r.

25. "la poca barba rappresenta, ch'egli habbi da esser effeminato & impudico, il scuto c'ha sotto e piedi mostra lo dispregio delle cose divine." Lando, *Paradossi*, N1v–N2r.

26. "Albertistas, Thomistas, Scotists, Abenrhoistas, Avicenistas, et reliquam huiusmodi Philosophiae fecem, quos nemo vere doctus, neque docti viri neque Philosophi nomine dignos arbitratur." Mario Nizolio, *De veris principiis et vera ratione philosophandi contra pseudophilosophos*, ed. Quirinus Breen (Rome: Fratelli Bocca, 1956), 1:27–28. For Nizolio in general, see Eugenio Garin, *History of Italian Philosophy*, trans. Giorgio Pinton (Amsterdam: Rodopi, 2008), 1:513–17.

27. "Abenrhois Arabs, qui usque adeo Porphyrii doctrina infatuatus est, ut hunc ipsum generis modum explanans non dubitaverit affirmare, et Adamum esse genus nominum." Nizolio, *De veris principiis*, 1:133.

28. "et contra authoritatem Platonis, et Veterum illorum Graecorum." Nizolio, *De veris principiis*, 2:45.

29. "et Aristotelem, qui Theologiam a Physica, vehementer errasse." Nizolio, *De veris principiis*, 2:45.

30. Sebastian Fox Morcillo, *De naturae philosophia, seu de Platonis et Aristotelis consensione* (Louvain, 1554).

31. Jean Fernel, *On the Hidden Causes of Things*, ed. and trans. John M. Forrester and John Henry (Leiden: Brill, 2005), 1,9:323–331.

32. "Intorno alle cose divine & invisibili. l'anima nostra è come l'occhio pilistrello al lume del sole." Francesco de' Vieri, *Vere conclusioni di Platone conformi alla dottrina Christiana. Et a quella d'Aristotile* (Florence, 1590), 28.

33. "L'universale reale e di due maniere uno essemplare in Dio, l'altro nelle cose particolari." De' Vieri, *Vere conclusioni di Platone*, 33.

34. "Aristotelem naturae genium in Physicis divino Platone tanto superiorem esse, quanto in Metaphysicis, eodem meo quidem iudicio, inferior est." Jacques Charpentier, *Platonis cum Aristotele in universa philosophia comparatio* (Paris, 1573), B1r.

35. "Nisi forte pro Aristotelis Metaphysicis potius habendi sunt libri quatuordecim de secretiore parte divinae sapientiae secundum Aegyptios. Qui cum in huius operibus graecis nusquam, quod sciam extent sub huius nomine ex Arabum lingua ante annos quinquaginta latine descripti & Romae editi, non vulgarem Theologiam continent, nec a Platonica valde alienam." Charpentier, *Platonis cum Aristotele comparatio*, C1r. For the reception of *De mundo*, see Jill Kraye, "Aristotle's God and the Authenticity of *De mundo*: An Early Modern Controversy," *Journal of the History of Philosophy* 28 (1990): 339–58.

36. Marcia L. Colish, *Peter Lombard* (Leiden: Brill, 1993).

37. Charles L. Stinger, *Humanism and the Church Fathers: Ambrogio Traversari (1386–1439) and Christian Antiquity in the Italian Renaissance* (Albany: State University of New York Press, 1977), 88–166.

38. Robert Peters, "Erasmus and the Fathers: Their Practical Value," *Church History* 36 (1967): 254–61; C. A. L. Jarrott, "Erasmus' Biblical Humanism," *Studies in the Renaissance* 17 (1970): 121–24; Eugene F. Rice Jr., "The Humanist Idea of Christian Antiquity: Lefèvre d'Étaples and his Circle," *Studies in the Renaissance* 9 (1962): 126–60; E. P. Meijering, *Melanchthon and Patristic Thought: The Doctrines of Christ and Grace, the Trinity and the Creation* (Leiden: Brill, 1983), 19–108; Elizabeth G. Gleason, *Gaspare Contarini: Venice, Rome, and Reform* (Berkeley: University of California Press, 1993), 161–62.

39. Sachiko Kusukawa, *The Transformation of Natural Philosophy: The Case of Philip Melanchthon* (Cambridge: Cambridge University Press, 1995), 75–123, and "Nature's Regularity in Some Protestant Natural Philosophy Textbooks 1530–1630," in *Natural Law and Laws of Nature in Early Modern Europe: Jurisprudence, Theology, Moral and Natural Philosophy*, ed. Lorraine Daston and Michael Stolleis, 105–22; Charlotte Methuen, "Special Providence and Sixteenth-Century Astronomical Observation: Some Preliminary Observations," *Early Science and Medicine* 4 (1999): 99–113; Craig Martin, "The Ends of Weather: Teleology in Renaissance Meteorology," *Journal of the History of Philosophy* 48 (2010): 259–82.

40. Examples of such works include Jakob Milich, *Liber II, C. Plinii de mundi historia, cum commentariis* (Frankfurt, 1543); Marcus Frytsche, *Meteorum, hoc est impressionum aerearum et mirabilium naturae operum* (Wittenberg, 1598).

41. Ann Blair, *The Theater of Nature: Jean Bodin and Renaissance Science* (Princeton, NJ: Princeton University Press, 1997), 120–26.

42. "Licet ipse sanctus Martyr cuius iam totis mille & quadringentis circiter annis scripta Graeca extant, possit abunde sibi satis esse & in evertenda Aristotelis authoritate, ubi aut divinae authoritati aut rationi contraria est." Guillaume Postel, *Eversio falsorum Aristotelis dogmatum* (Paris, 1552) and *Liber de causis* (Paris, 1552), 75r–75v. For Postel's uneasy relations with both Calvinists and the Inquisition, see William J. Bouwsma, *Concordia Mundi: The Career and Thought of Guillaume Postel (1510–1581)* (Cambridge, MA: Harvard University Press, 1957), 1–29.

43. Postel, *Eversio falsorum*, 4r.

44. "Quum autem vocarit Aben Reis (cuius libentius sententiam meae confirmatoriam tanquam a summo providentiae hoste & recipio & maxime ob Ismaelem, Abrahamicique Sanguinis nomen & jura, quam Aristotelis)." Postel, *Liber de causis*, B1r.

45. "ipsa schola Parisiensi aut doctrina scholastica, tota (quatenus ecclesiae fuit in hanc diem cognitu necessaria & intelligibilis) sacrorum authoritas innumeris rationibus in utramque partem discussa, spoliaret Aristotelem illis veritatibus, quae in ipso liquidae erant, eo quod veritas a quocunque dicatur a spiritu sancto est, & semper divina: ut tandem post resolutiones ipsas sacrorum, ipse Aristoteles omnino neglectus iaceret." Postel, *Eversio falsorum*, 81v.

46. "sono tanto temerarii che . . . negano la Providentia d'Iddio gabbati da Aristotele, da Galeno metafrasta, da Averroi et da Alexandro Aphrodiseo et dagli altri peripatetici." Pirro Ligorio, *Libro di diversi terremoti*, ed. Emanuela Guidoboni (Rome: De Luca, 2005), 5.

47. Jacopo Mazzoni, *In universam Platonis et Aristotelis philosophiam praeludia, sive de comparatione Platonis et Aristotelis*, ed. Sara Matteoli (Naples: M. D'Auria, 2010), 311–25.

48. De' Vieri, *Vere conclusioni di Platone*, 3.

49. "Dio ha providenza di tutte le cose, & in particolare dell'huomo." De' Vieri, *Vere conclusioni di Platone*, 12.

50. De' Vieri, *Vere conclusioni di Platone*, 13–14.

51. For his disputes with Charpentier, see Robert Goulding, *Defending Hypatia: Ramus, Savile and the Renaissance Rediscovery of Mathematical History* (Dordrecht: Springer, 2010), 50–77. For his disputes with Vimercati, see Neal W. Gilbert, "Francesco Vimercato of Milan: A Bio-Bibliography," *Studies in the Renaissance* 12 (1965): 193–94.

52. Howard Hotson, *Commonplace Learning: Ramism and its German Ramifications, 1543–1630* (Oxford: Oxford University Press, 2007), 38–43.

53. Peter Mack, *A History of Renaissance Rhetoric, 1380–1620* (Oxford: Oxford University Press, 2011), 136–38.

54. Petrus Ramus, *Aristotelicae animadversiones* (Paris, 1553), A2v, A4v.

55. Omer Talon, *Dialecticae praelectiones in Porphyrium* (Paris, 1550), 6.

56. Hotson, *Commonplace Learning*, 43.

57. "Item cum saepe ait naturam nihil agere frustra: attamen Physicam & Metaphysicam contra divinam mundi creationem & administrationem composuit." Petrus Ramus, *Commentariorum de religione christiana libri quatuor* (Frankfurt, 1576), 26.

58. "Hac Aristotelis Philosophia imbuti Epicurus, Manichaeus, Caelestinus totam divinae oeconomiae providentiam, praescientiamque calumniantur." Ramus, *De religione christiana*, 26–27.

59. Howard Jones, *The Epicurean Tradition* (London: Routledge, 2001), 95–116.

60. Robert D. Preus, *The Theology of Post-Reformation Lutheranism* (St. Louis, MO: Concordia, 1970), 2:196.

61. "Impiis etiam atque atheis ut Aristoteli, & Epicuro sua pariter contemplatio est, sed humanae vitae finibus conclusa." Ramus, *De religione christiana*, 143.

62. Concetta Bianca, "Per la storia del termine 'atheus' nel cinquecento. Fonti e traduzioni greco-latine," *Studi filosofici* 3 (1980): 71–104; Paul Oskar Kristeller, "The Myth of Renaissance Atheism and the French Tradition of Free Thought," *Journal of the History of Philosophy* 6 (1968): 233–43.

63. "Aristoteles ut mundi creationem, providentiam Dei: sic animi immortalitatem derisit." Ramus, *De religione christiana*, 91.

64. "Sed Aristotelis impietatem, quod homo seipso contentus esset ad beate vivendum." Ramus, *De religione christiana*, 99. Aristotle, *Nicomachean Ethics*, 1.7.1098a13–20.

65. "In quo decima coepit, atque ad nostra tempora perdurat. In qua floruerunt Nicolaus Leonicus, Franciscus Caballus, Alexander Achillinus, M. Antonius Zimara, Augustinus Niphus, Petrus Pomponatius, atque horum discipuli Ludovicus Buccaferreus, Simon Portius, Vincentius Madius, M. Antonius Ianua, Franciscus Vicomercatus, atque alii viri celebres tum munere docendi, tum et labore Aristotelis explanandi." Francesco Patrizi, *Discussiones peripateticae* (Venice, 1571), 145–46.

66. "Aristotelicam Philosophiam Christianae Theologiae iungerent, huiusque praeceptis, decretisque Aristolicas rationes, ac etiam dubitationes quaestionesque permiscerent. Nihilque in Theologia arbitrati sunt posse statui, nisi Aristotelicis fundamentis firmatum esset." Patrizi, *Discussiones peripateticae*, 163.

67. "Hi modi summi huiusce philosophiae, quae postea in varias ac multiplices sectas Nominalium, atque Realium divisa est, quarum sectatorum, qui numerus innumerabilis est." Patrizi, *Discussiones peripateticae*, 163.

68. "si quidem vera est Averrois, omnium interpretum, ut multi putant, Aristotelicis-simi, sententia." Patrizi, *Discussiones peripateticae*, 66; "Averrois autem Alexandrum, huic soli omnia credit, eumque in omnibus fere amplexatur. Huiusmodi rationem com-mentandi Aristotelis aetas haec nostra atque superior tenuit, Averroimque prae caeteris omnibus est admirata, atque secuta." Patrizi, *Discussiones peripateticae*, 149.

69. "uti Aven Rois, & Latinis Graecos interpretes admiscerent." Patrizi, *Discussiones peripateticae*, 163.

70. "Ita sempiternis quaestionibus modo imposito, verborum Aristotelicorum sin-ceriore inventa expositione nonum philosophandi genus ortum, in quo Buccaferreus, Portius, Ianua, Madius, Vicomercatus, atque alii nostrorum temporum philosophiae doctores floruerunt." Patrizi, *Discussiones peripateticae*, 163.

71. "Cur Aristotelis philosophiae, solae eae praeleguntur partes, quae magis & Deo, & Ecclesiae suae sunt hostes?" Francesco Patrizi, *Nova de universis philosophia* (Fer-rara, 1591), A3r.

72. "Quadringentis vero ab hinc circiter annis, Scholasticis Theologi, in contrarium sunt annixi, Aristotelicis impietatibus, pro fidei fundamentis sunt usi." Patrizi, *Nova de universis philosophia*, A3r.

73. "Origenes quoque Aristotelem Epicuro inferiorem dicit esse, quia fuerit in divi-nam providentiam impius." Patrizi, *Nova de universis philosophia*, 49v.

74. "Clemens quoque Alexandrinus Origenis praeceptor, Aristotelem, & Peripatet-icos incusat . . . providentiamque usque ad lunam tantum porrigi." Patrizi, *Nova de universis philosophia*, 49v.

75. "Qui e Parisiana schola . . . nec solum, in eius philosophiam, magnis laboribus incubuerunt, sed vel partem, vel etiam universam Aven Rois imitati, commentariis, quamvis satis infeliciter, explanarunt. Sed etiam ipsum, invitum, & maxime omnium repugnantem, & contradictem, Christianum effecerunt." Patrizi, *Nova de universis philosophia*, 49v.

76. Ugo Baldini and Leen Spruit, *Catholic Church and Modern Science: Documents from the Archives of the Roman Congregations of the Holy Office and the Index* (Rome: Libreria Editrice Vaticana, 2009), 1,3:2234–44.

77. For how Patrizi changed the Nova *de universis philosophia* as a result of the In-quisition, see Francesco Patrizi, *Nova de universis philosophia: Materiali per un'edizione emendata*, ed. Anna Laura Puliafito Bleuel (Florence: Olschki, 1993).

78. Richard J. Blackwell, *Galileo, Bellarmine, and the Bible* (South Bend, IN: Uni-versity of Notre Dame Press, 1991).

79. "dixitque periculum certius a Platone, quam ab Aristotele emanare in ingenia posse si inter Christianos ille cathedram habeat; non quod singula eius Auctoris infecta erroribus putaret; sed quia doctrinae Catholicae magis affinis Plato, quam Aristoteles est." Giacomo Fuligatti, *Vita Roberti Bellarmini politiani* (Liege, 1626), AA2r.

80. "Et fore perinde aiebat, atque experimur plus afferre damni, ex hac similitudine Haereticos Auctores, quam Scriptores Ethnicos. Certe, subdidit, ignorari a nemine, Origenem Platonis dulcedine abreptum, & facundiae eius lenocinio depravatum, libros scatentes erroribus Ecclesiae reliquisse. Quod deinde causae fuit, cur Plato idem in v. Sinodo sit proscriptus." Fulgatti, *Vita Bellarmini*, AA2r.

81. Origen was condemned as a heretic and Theodore was condemned for his com-

parisons of Platonism and Christianity in AD 553; see Heinrich Denzinger and Adolf Schönmetzer, eds. *Enchiridion symbolorum*, 23rd ed. (Barcelona: Herder, 1965), 149–51.

82. Pedro de Fonseca, *In libros metaphysicorum Aristotelis Stagiritae, cui praemissi sunt institutionum dialecticarum libri octo* (Lyon, 1597), 11–13.

83. "Aristoteles igitur, quanquam Plato illis in hac parte a multis praefertur, tamen libro 12. de sapientia multa accurate de deo, & reliquis mentibus disputat, atque etiam ratione & via apertiori quam Plato, non fabulis, symbolisve Pythagoricis illa involvens." Pedro Juan Nuñez, *Vita Aristotelis per Ammonium, seu Philoponum* (Leiden, 1621), 132.

84. For Pontano's connection to Bellarmino, see Pietro Pampilio Rodotà, *Dell'origine progresso, e stato presente del rito greco in Italia* (Rome, 1758–63), 3:164.

85. Girolamo Pontano, *De immortalitate animae ex sententia Aristotelis* (Rome, 1597), 3.

86. Giulio Cesare La Galla, *De immortalitate animorum ex Aristotelis sententia* (Rome, 1621), 312–13.

87. "et rerum humanarum administrationem cognitionemque Deo demit omnem." Bernardino Telesio, *De rerum natura iuxta propria principia*, ed. Luigi de Franco (Cosenza, Italy: Casa del libro, 1965), 1:20.

88. Eva Del Soldato, *Simone Porzio: Un aristotelico tra natura e grazia* (Rome: Storia e letteratura, 2010), 23–24

89. "Hic est ille princeps Aristoteles naturae miraculum, . . . eccelsi creatoris contemplator." Giacomo Antonio Marta, *Pugnaculum Aristotelis adversus principia Bernardini Telesii* (Rome, 1587), 3.

90. Ibid., 45.

91. Baldini and Spruit, *Catholic Church and Modern Science*, 1,3:2419.

92. Ibid.

93. Francesco Patrizi, *Apologia contra calumnias Theodori Angelutii* (Ferrara, 1584), A2r.

94. "Sic est etiam de Aristotele, quem non dico simpliciter veritatem attigisse, quippe qui erravit quandocumque contra fidem sensit." Cesare Cremonini, *Apologia dictorum Aristotelis de quinta caeli substantia adversus Xenarcum, Ioannem Grammaticum, & alios* (Venice, 1616), 4–5.

95. Heinrich C. Kuhn, "Cesare Cremonini: Volti e maschere di un filosofo scomodo per tre secoli e mezzo," in *Cesare Cremonini: Aspetti del pensiero e scritti*, ed. Ezio Riondato and Antonino Poppi (Padua: La Garangola, 2000), 153–68.

96. "Quinimo magnopere in errorem lapsus, ut constat ex fide, & ex decretis sacrae Theologiae, quibus omnia, & singula Dei providentia administrari recte & vere statuitur. & Averroes: ad bonum sensum nempe Aristotelicum reducendum proponimus, secundum veritatem minime, & secundum eam, quam profitetur, interpretis Aristotelici doctrinam." Cremonini, *Apologia de caeli substantia*, 78.

97. "Aristoteles quidem, votum solvendum iudicavit, putans actionem studiosam sibi ipsi esse praemium, quippe. qui ut ignoravit verae religionis cultum, ita nescivit operibus ex vera religione esse aliud praemium aeternum." Cremonini, *Apologia de caeli substantia*, 79.

98. Gregorio Piaia, "Aristotelismo, 'Heresia' e giurisdizionalismo nella polemica del Antonio Possevino contro lo Studio di Padova," *Quaderni per la storia dell'Università di Padova* 6 (1973): 125–45.

99. Piaia, "Aristotelismo," 61, 64.

100. Antonino Poppi, *Cremonini, Galilei e gli inquisitori del Santo a Padova* (Padua: Centro Studi antoniani, 1993), 10; Giorgio Spini, *Ricerca dei libertini: La teoria dell'impostura delle religioni nel seicento italiano*, 2nd ed. (Rome: La Nuova Italia, 1983), 155.

101. Poppi, *Cremonini, Galilei e gli inquisitori*, 13–15, 19–20.

102. Leonard A. Kennedy, "Cesare Cremonini and the Immortality of the Human Soul," *Vivarium* 18 (1980): 143–58; Heinrich C. Kuhn, *Venetischer Aristotelismus im Ende der aristotelischen Welt: Aspekte der Welt und des Denkens des Cesare Cremonini (1550–1631)* (Frankfurt: Peter Lang, 1996), 166–85, 555–66.

103. Cesare Cremonini, *Quaestio de animi moribus et facultatibus*, Padua, Biblioteca universitaria, MS 2075, 178r–95r; Kuhn, *Venetischer Aristotelismus*, 244–63.

104. Kennedy, "Cremonini and Immmortality of the Human Soul," 143–58.

105. "Quod Doctores omnes sub poena privationis lecturae, quam legunt, teneantur, ac debeant legere, et clare exponere, ac declarare tex. Authorum, quos legere tenentur de verbo ad verbum, neque amplius aliquem expositorem." *Statuta almae universitatis dd. philosophorum, et medicorum cognomento artistarum patavini gymnasii* (Padua, 1607), 161.

106. Maurizio Sangalli, "Cesare Cremonini, la compagnia di Gesù e la repubblica di Venezia: Eterodossia e protezione politica," in *Cesare Cremonini: Aspetti del pensiero e scritti*, ed. Ezio Riondato and Antonino Poppi (Padua: Accademia Galileiana, 2000), 207–18.

107. Giorgio Raguseo, *Peripateticae disputationes* (Venice, 1613), 252; Marko Josipović, *Il pensiero filosofico di Giorgio Raguseo nell'ambito del tardo aristotelismo padovano* (Milan: Massimo, 1985).

108. Antonis Fyrigos, "Joannes Cottunios di Verria e il neoaristotelismo padovano," in *Renaissance Readings of the Corpus Aristotelicum*, ed. Marianne Pade (Copenhagen: Tusculanum, 2001), 225–40; Giovanni Cottunio, *De triplici statu animae rationalis* (Bologna, 1628).

109. Robert S. Westman, "The Copernicans and the Churches," in *God and Nature: Historical Essays on the Encounter Between Christianity and Science*, ed. David C. Lindberg and Ronald L. Numbers (Berkeley: University of California Press, 1986), 76–113.

CHAPTER 7: **History, Erudition, and Aristotle's Past**

1. Pedro Juan Nuñez, *Vita Aristotelis per Ammonium, seu Philoponum* (Leiden, 1621), 64.

2. Charles B. Schmitt, *Aristotle and the Renaissance* (Cambridge, MA: Harvard University Press, 1983), 43–47, 83–85.

3. "Mirare, Lecto, hominis Ethnici Theologiam." Guillaume Du Val, *Synopses analyticae universae doctrinae peripateticae*, in *Aristotelis operum*, ed. Isaac Casaubon and Giulio Pace (Paris, 1619), 2:122.

4. "Hinc enim manifeste colligeretur, Deum non videre actiones nostras; . . . At haec asserere impium est, & divinam providentiam iustitiamque tollere." Du Val, *Synopses*, 2:122–23.

5. Francesco Arias, *Dell'immitatione di Christo*, trans. Tiberio Putignano (Rome, 1609–15), 1:288.

6. Ibid., 1:307.

7. Emanuel do Valle de Moura, *De incantationibus seu ensalmis* (Evora, 1620), 96.

8. Ibid., 97, 153.

9. Francesco Patrizi, *Discussiones peripateticae* (Venice, 1571), 145–46.

10. René Rapin, *La comparaison de Platon et d'Aristote* (Paris, 1671), 239.

11. *Statuta antiqua universitatis oxoniensis*, ed. Strickland Gibson (Oxford: Clarendon Press, 1931), 437; *Statuta almae universitatis dd. philosophorum, et medicorum cognomento artistarum patavini gymnasii* (Padua, 1607), 161; Francis A. Yates, "Giordano Bruno's Conflict with Oxford," *Journal of the Warburg and Courtauld Institutes* 2 (1939): 230.

12. Didier Kahn, "Entre atomisme, alchimie et théologie: La réception des thèses D'Antoine de Villon et Étienne De Clave contre Aristote, Paracelse et les 'Cabalistes' (24–25 août 1624)," *Annals of Science* 58 (2001): 241–86.

13. Pierre Charron, *Le trois veritez* (Paris, 1595), 2.

14. Donald R. Kelley, *Foundations of Modern Historical Scholarship: Language, Law, and History in the French Renaissance* (New York: Columbia University Press, 1970), 151–214.

15. Johann Wier, *De praestigiis daemonum et incantationibus ac venificiis* (Basel, 1566), 716; Marino Sanudo, *I diarii*, ed. Rinaldo Fulin (Venice: Visentini, 1879–1903), 38: cols. 387–88; Martin L. Pine, *Pietro Pomponazzi: Radical Philosopher of the Renaissance* (Padua: Antenore, 1986), 51.

16. Nicholas S. Davidson, "'Le plus beau et le plus meschant esprit que ie aye cogneu': Science and Religion in the Writings of Giulio Cesare Vanini, 1585–1619," in *Heterodoxy in Early Modern Science and Religion*, ed. John Brooke and Ian Maclean (Oxford: Oxford University Press, 2005), 59–79.

17. Nancy G. Siraisi, *The Clock and the Mirror: Girolamo Cardano and Renaissance Medicine* (Princeton, NJ: Princeton University Press, 1997), 197.

18. For a defense of Vanini, see Dagmar von Wille, "Apologie häretischen Denkens: Johann Jakob Zimmermanns Rehabilitierung der 'Atheisten' Pomponazzi und Vanini," in *Atheismus im Mittelalter und in der Renaissance*, ed. Friedrich Niewöhner and Olaf Pluta (Wiesbaden: Harrassowitz, 1999), 129–44.

19. "cum sola Religione mihi persuasum sit adesse daemones." Giulio Cesare Vanini, *De admirandis naturae* (Paris, 1616), 472.

20. William L. Hine, "Mersenne and Vanini," *Renaissance Quarterly* 29 (1976): 53–54.

21. "Petrus Pomponatius Philosophus acutissimus, in cuius corpus animam Averrois commigrasse Pythagoras iudicasset." Giulio Cesare Vanini, *Amphiteatrum aeternae providentiae divino-magium* (Lyon, 1615), 36. For Pomponazzi and oracles, see Anthony Ossa-Richardson, "Pietro Pomponazzi and the Rôle of Nature in Oracular Divination," *Intellectual History Review* 20 (2010): 435–55.

22. Vanini, *Amphiteatrum*, 50–78.

23. Vittoria Perrone Compagni, "Introduzione, la fondazione 'scientifica' della magia nel *De incantationibus*," in *De incantationibus*, ed. Vittoria Perrone Compagni (Florence: Olschki, 2011), lxxii–lxxi.

24. Cesare Vasoli, "Vanini e il suo processo per ateismo," in *Atheismus im Mittelalter und in der Renaissance*, ed. Friedrich Niewöhner and Olaf Pluta (Wiesbaden: Harrassowitz, 1999), 129–44.

25. Giles of Rome, *Errores philosophorum*, ed. Josef Koch (Milwaukee: Marquette University Press, 1944), 25.

26. Gregorio Piaia, "'Averroisme politique': Anatomie d'un mythe historiographe," in *Orientalische Kultur und europäisches Mittelalter*, ed. Albert Zimmerman and Ingrid Craemer-Ruegenberg (Berlin: De Gruyter, 1985), 289–300.

27. David Wootton, *Paolo Sarpi: Betweeen Renaissance and Enlightenment* (Cambridge: Cambridge University Press, 1983), 41–42; *Theophrastus redivivus*, ed. Guido Canziani and Gianni Paganini (Florence: La Nuova Italia, 1981), 333, 409, 676–713.

28. Tullio Gregory, "Pierre Charron's 'Scandalous Book,'" in *Atheism from the Reformation to the Enlightenment*, ed. Michael Hunter and David Wootton (Oxford: Clarendon Press, 1992), 87–109; Popkin, *History*, 57–63. For the view that skeptical fideism was not responsible for causing the emergence of modern atheism , see Silvia Berti, "At the Roots of Unbelief," *Journal of the History of Ideas* 56 (1995): 555–75.

29. René Pintard, *Le libertinage érudit dans la première moitié du XVIIe siècle* (Paris: Boivin, 1943); Jacques Denis, *Sceptiques ou libertins de la première moitié du XVIIe siècle: Gassendi, Gabriel Naudé, Gui-Patin, Lamothe-Levayer, Cyrano de Bergerac* (Caen: Le Blanc-Hardel, 1884). For libertines, see also Richard H. Popkin, *The History of Scepticism: From Savonarola to Bayle*, 2nd ed. (Oxford: Oxford University Press, 2003), 80–98; Jacques Charbonnel, *La pensée italienne au XVIe siècle et le courant libertin* (Paris: Champion, 1919), 220–388.

30. Henri Busson, *Le rationalisme dans la littérature française de la Renaissance (1533–1601)*, 2nd ed. (Paris: Vrin, 1957), 44–69, 191–233; Giorgio Spini, *Ricerca dei libertini: La teoria dell'impostura delle religioni nel seicento italiano*, 2nd ed. (Rome: La Nuova Italia, 1983), 15–25.

31. Perez Zagorin, *Ways of Lying: Dissimulation, Persecution, and Conformity in Early Modern Europe* (Cambridge, MA: Harvard University Press, 1990), 288–330; David Wootton, "Lucien Febvre and the Problem of Unbelief in the Early Modern Period," *The Journal of Modern History* 60 (1988): 695–730.

32. *Naudaeana et Patiniana*, 2nd ed. (Amsterdam, 1703), 7–8, 31–32, 53–57.

33. Paul Oskar Kristeller, "The Myth of Renaissance Atheism and the French Tradition of Free Thought," *Journal of the History of Philosophy* 6 (1968): 233–43; Margaret J. Osler, "When Did Pierre Gassendi Become a Libertine?" in *Heterodoxy in Early Modern Science and Religion*, ed. John Brooke and Ian Maclean (Oxford: Oxford University Press, 2005), 169–92.

34. Frédéric Lachèvre, *Le procès du poète Théophile de Viau: (11 juillet 1623–1er septembre 1625)* (Paris: Champion, 1909), 132.

35. François Garasse, *La doctrine curieuse des beaux esprits de ce temps, ou prendus tel* (Paris, 1623. Reprint, Westmead : Gregg, 1971), 2:650–51.

36. Ibid., 2:651.

37. Popkin, *History of Scepticism*, 101–3.

38. "The Confessionalization of Physics: Heresies, Facts, and the Travails of the Republic of Letters," in *Heterodoxy in Early Modern Science and Religion*, ed. John Brooke and Ian Maclean (Oxford: Oxford University Press, 2005), 81–114.

39. Martin Delrio, *Disquisitionum magicarum libri sex* (Mainz, 1617), 9, 17, 24, 29, 50.

40. Hine, "Mersenne," 52–69; Robert Lenoble, *Mersenne ou la naissance du mécanisme* (Paris: Vrin, 1943), 110–33.

41. "ut ostenderem Campanellae, Bruni, Telesii, Kepleri, Galilaei, Gilberti, & aliorum recentiorum discipulis, falsum esse, quod aiunt, Doctores videlicet Catholi-

cos, & Theologos solum Aristotelem sequi, & in eius verba iurare, licet experientiae, atque phaenomena contrarium evincant." Marin Mersenne, *Quaestiones celeberrimae in Genesim* (Paris, 1623), Er.

42. "enimvero Theologi nunquam ulli authori assentiuntur, si ratione careat." Mersenne, *Quaestiones in Genesim*, Er.

43. "Athei metaphysicam ignorant." Mersenne, *Quaestiones in Genesim*, col. 38.

44. "Atheismi Politicorum, Cardanistarum, Pomponistarum." Mersenne, *Quaestiones in Genesim*, cols. 8–9.

45. Mersenne, *Quaestiones in Genesim*, col. 395.

46. Ibid., cols. 497–98, 537–42, 562–63, 640–42.

47. Ibid., cols. 400–403. For Vanini and monsters, see Tristan Dagron, "Nature and Its Monsters during the Renaissance: Montaigne and Vanini," in *Monsters and Philosophy*, ed. Charles T. Wolfe (London: College Publications, 2005), 37–59.

48. Marin Mersenne, *L'impiété des déistes, athées et libertins de ce temps* (Paris, 1624. Reprint, Stuttgart-Bad Cannstatt: Frommann, 1975).

49. Marin Mersenne, *La vérité des sciences. Contre les sceptiques ou pyrrhoniens* (Paris, 1625, Reprint, Stuttgart-Bad Cannstatt: Frommann, 1969), 116.

50. "Aussi voyons nous que sainct Thomas qui a peut être eu le meilleur esprit, le plus solide, le plus constant, & le plus iudicieus de tous ceus qui l'ont suiui." Mersenne, *La vérité des sciences*, 125–26.

51. "n'a pas voulu établir d'autres principes que ceus d'Aristote, tant il les a iugez veritables." Mersenne, *La vérité des sciences*, 126.

52. Mersenne, *Quaestiones in Genesim*, col. 399; "Veritablement ce passage m'empesche quasi de croire qu'Aristote ait nié que la providence de Dieu s'étende sur les moindres choses du monde." Mersenne, *La vérité des sciences*, 122.

53. Pierre Charron, *Le trois veritez* (Paris, 1595), 54.

54. Mersenne, *La vérité des sciences*, 111–13.

55. John M. Headley, "Tommaso Campanella and Jean de Launoy: The Controversy over Aristotle and His Reception in the West," *Renaissance Quarterly* 43 (1990): 531–43.

56. Germana Ernst, *Tommaso Campanella: The Book and the Body of Nature*, trans. David L. Marshall (Dordrecht: Springer, 2010), 243–66.

57. Daniel 12:4.

58. "quanvis in aliquo erraverint, indicant tamen quod tota Philosophia debet renovari." Tommaso Campanella, *De gentilismo non retineando* (Paris, 1636), 6.

59. "Igitur si philosophia Graecorum habet radices infectas, & falsa Dogmata non solum in Theologicis, sed in Moralibus, & naturalibus, fomentanda non est, sed alia verax construenda, & undique seligenda; & qui aliter sapit, amat mendacium, aut haeresim, aut periculum mentiendi, & haereticandi, & illatos errores & damna nutrit." Campanella, *De gentilismo*, 3.

60. "Sed contra est Concilium Lateranense sub Leone X. sessione VIII. affirmans Philosophiam & Poësim Gentilium habere radices infectas, id est, principia falsa." Campanella, *De gentilismo*, 2.

61. "Ergo nova Philosophia optatur a sacro Concilio, quae conformis sit sacrae Theologiae Canonibusque sacris, & consuetudo perniciosa cedat rationi & utilitati." Campanella, *De gentilismo*, 2.

62. "Et quanvis S. Thomas, & Albertus, & alii, in bonum sensum Aristotelem ver-

tere conati sint, & ubi manifeste adversatur fidei, ipsum impugnaverint, tamen non modo Graeci, & Arabes, sed Christiani dicunt S. Thomam Latinum non intellexisse Aristotelem Graecum, vel torsisse." Campanella, *De gentilismo*, 8.

63. "Et ex suspicione contra D. Thomam natus est ex Aristotele, & Averroë, & aliis, Machiavellismus radix malorum faciens religionem *di ragion di stato*, & occupavit thronos Regum & capita populorum." Campanella, *De gentilismo*, 8.

64. Germana Ernst, "Campanella e il 'De tribus impostoribus,'" *Nouvelles de la république des lettres* 2 (1986): 144–70. For *On the Three Impostors*, see George Minois, *The Atheist's Bible: The Most Dangerous Book That Never Existed*, trans. Lys Ann Weiss (Chicago: University of Chicago Press, 2012).

65. "dicunt se credere, ut Christianos, immortalitatem animae; sed tamen aliter dicendum esse secundum philosophiam." Campanella, *De gentilismo*, 21.

66. "Religio est ars regnandi, & retinendi populos in officio, & obedientia." Campanella, *De gentilismo*, 17. Aristotle, *Politics*, 6.2.1319b24–25, discusses religious rites in a democracy; Aristotle, *Politics*, 6.4.1321a35–40, treats public sacrifices and the power of an oligarchy; Aristotle, *Politics*, 6.5.1322b17–20, treats how public religion should be controlled by officials; Aristotle, *Politics*, 7.8.1329a27–35, maintains, "It is fitting that the people worship the Gods."

67. "Adde Calvinistas, Lutheristas & Puritanos: proceres enim eorum Machiavellistae sunt, & Atheistae, sicut Calvinus & Averroes." Campanella, *De gentilismo*, 21.

68. "Ergo casu dico Aristoteles introductus est in scholas Christianorum." Campanella, *De gentilismo*, 33.

69. Campanella, *De gentilismo*, 32.

70. Campanella, *De gentilismo*, 33. Campanella based this view on: 1 Maccabees 1:14 and 2 Maccabees 4:12.

71. "Tanta erat huius saeculi ruditas, quando Aristoteles intrusus est; cum non reperirentur aliorum Philosophorum Codices, excepto Aristotele, & ab Arabibus advecto." Campanella, *De gentilismo*, 33.

72. Campanella, *De gentilismo*, 37.

73. "Nec testatur iuramento Aristotelem esse veracem; sed tantum exposuit, veluit si quis Virgilium, aut Mahometum exponerit." Campanella, *De gentilismo*, 39.

74. "Iurare in verba Philosophorum Gentilium, videlicet, Aristotelis, & Platonis, vel Senecae, vel Parmenidis, vel aliorum, est haeresis & periurium, & impietas maxima." Campanella, *De gentilismo*, 52.

75. Pintard, *Libertinage*, 156–73.

76. "ut videar mihi perpetuo hisce oculis intueri, & audire de Cathedra docentem redivivum quendam in Terris Pomponatium, adeo secure & intrepide Aristotelis mentem aperit, docet, inculcat, tuetur & defendit." Gabriel Naudé, *Epistolae* (Geneva, 1667), 28; Françoise Charles-Daubert, "La fortune de Cremonini chez libertins érudits du XVIIe siècle," in vol. 1 of *Cesare Cremonini: Aspetti del pensiero e scritti*, ed. Ezio Riondato and Antonino Poppi (Padua: La Garangola, 2000), 178.

77. Naudé, *Epistolae*, 105–6.

78. Maryanne Cline Horowitz, "Doubts about 'Witches' and 'Magicians' in Reginald Scot and Gabriel Naudé," in *History Has Many Voices*, ed. Lee Palmer Wandel (Kirksville, MO: Truman State University Press, 2003), 7–22.

79. For Naudé's historical method, see Lorenzo Bianchi, *Rinascimento e liberti-*

nismo: Studi su Gabriel Naudé (Naples: Bibliopolis, 1996), 35–62; Paul Nelles, "The Library as an Instrument of Discovery: Gabriel Naudé and the Uses of History," in *History and the Disciplines: The Reclassification of Knowledge in Early Modern Europe,* ed. Donald R. Kelley (Rochester, NY: University of Rochester Press, 1997), 41–57.

80. Gabriel Naudé, *Apologie pour tous les grands personnages qui on esté faussement soupçonnez de magie* (Paris, 1669), 11.

81. Plato, *Symposium*, 219b3–c2.

82. Naudé, *Apologie*, 227.

83. Ibid., 232–33.

84. Ibid., 243. Aristotle, *De caelo*, 1.1.268a13–15.

85. Naudé, *Apologie*, 243–44.

86. Ibid., 244, 246.

87. "Eodem quoque modo Pomponatius, cum multorum adversus suum de animae immortalitate ex Aristotelis mente libellum, censuras dicteriis quibusdam aut excipiat, aut prorsus eludat." Garbriel Naudé, *Iudicium de Augustino Nipho*, in Agostino Nifo, *Opuscula moralia et politica* (Paris, 1645), ee1v.

88. "Hinc igitur Philosophi, libere de rebus cunctis & loqui, & scribere consueverant . . . quibus nunc assentiri nemo posset, qui non statim temeritatis, & violatae religionis rebus sisteretur." Naudé, *Iudicium*, ii4r.

89. For La Mothe Le Vayer's life, see Pintard, *Libertinage*, 131–47.

90. François La Mothe Le Vayer, *Petit discours chrestien de l'immortalité de l'ame*, vol. 4 of *Oeuvres* (Paris, 1669), 129.

91. Ruth Whelan, "The Wisdom of Simonides: Bayle and La Mothe Le Vayer," in *Scepticism and Irreligion in the Seventeenth and Eighteenth Centuries*, ed. Richard H. Popkin and Arjo Vanderjagt (Leiden: Brill, 1993), 230–53; Popkin, *History of Scepticism*, 82–87.

92. La Mothe Le Vayer, *Petit discours chrestien de l'immortalité de l'ame*, 4:180.

93. Ibid., 4:180–81.

94. Ibid., 4:181–82.

95. Ibid., 4:182.

96. Ibid., 4:191–93.

97. Ibid., 4:182.

98. Stephen Gaukroger, *Descartes' System of Natural Philosophy* (Cambridge: Cambridge University Press, 2002), 21.

99. Kahn, "Entre atomisme, alchimie et théologie," 244.

100. "Cette proposition est fausse, téméraire, scandaleuse, et attaque d'une certaine façon le sacrement sacro-saint de l'Eucharistie." Kahn, "Entre atomisme, alchimie et théologie," 250.

101. Kahn, "Entre atomisme, alchimie et théologie," 243–52.

102. Lynn Sumida Joy, *Gassendi the Atomist: Advocate of History in an Age of Science* (Cambridge: Cambridge University Press, 1987), 25–29.

103. For Gassendi's Christianization of Epicureanism, see Margaret J. Osler, "Providence and Divine Will in Gassendi's Views on Scientific Knowledge," *Journal of the History of Ideas* 44 (1983): 549–60.

104. Pierre Gassendi, *Exercitationes paradoxicae adversus Aristoteleos* (Amsterdam, 1649), *6r.

105. Joy, *Gassendi the Atomist*, 33.

106. "Quod immerito Aristotelei libertatem sibi philosophandi ademerint. Novi certe celebrem Philosophiae simul, ac Theologiae professorem, qui asseuerit se existimare magnum iri Deo praestitum obsequium, si proprio sanguine obsignasset, confirmassetque verissima esse quaecumque Aristotelis operibus continentur. Legerat forte Commentatorem, qui tam aperte pronunciavit nullum per mille, & quinquentos annos animadversum fuisse errorem in textu Aristotelis." Gassendi, *Exercitationes paradoxicae*, 29–30. For the history and meaning of *libertas philosophandi*, see Robert B. Sutton, "The Phrase *Libertas Philosophandi*," *Journal of the History of Ideas* 14 (1953): 310–16; "The 'Sceptical Crisis' Reconsidered: Galen, Rational Medicine and the Libertas Philosophandi," *Early Science and Medicine* 11 (2006): 247–74.

107. Gassendi, *Exercitationes paradoxicae*, 31–32.

108. "Principio ille non Iudaeus, non Christianus fuit." Gassendi, *Exercitationes paradoxicae*, 53.

109. Gassendi, *Exercitationes paradoxicae*, 54.

110. "voluptatibusque obscoenis adeo impense deditu." Gassendi, *Exercitationes paradoxicae*, 55–56.

111. For Frey's career, see Ann Blair, "The Teaching of Natural Philosophy in Early Seventeenth-Century Paris: The Case of Jean Cécile Frey," *History of Universities* 12 (1993): 91–158.

112. Jean-Cécile Frey, *Cribrum philosophorum*, in *Opuscula varia* (Paris, 1646), 69–70.

113. Ibid., 71–73.

114. Monnette Martinet, "Chronique des relations orageuses de Gassendi et de ses satellites avec Jean-Baptiste Morin," *Corpus, revue de philosophie* 20/21 (1992): 47–64.

115. Headley, "Tommaso Campanella and Jean de Launoy," 539–41.

116. Jean de Launoy, *De varia Aristotelis in academia parisiensi fortuna* (Paris, 1653), 5–35.

117. Ibid., 33.

118. "extremam Aristotelis fortunam." Launoy, *De varia Aristotelis fortuna*, 70.

119. "Historice loquor, nec meum est tantas inter se componere lites." Launoy, *De varia Aristotelis fortuna*, 70.

120. "Theologiae ancillam recentes novique Doctores, nominant Gentilium Philosophiam; tametsi non inveniam Christianos veteres ita locutos." Launoy, *De varia Aristotelis fortuna*, 76.

121. "Scriptura & traditio, sacrae doctrinae principia duo cum sufficiant sibi." Launoy, *De varia Aristotelis fortuna*, 76.

122. Ibid., 78–79.

CHAPTER 8: The New Sciences, Religion, and the Struggle over Aristotle

1. "Atheismus est culpa, qua nulla alia aeque lacessit iram Numinis. Atheismus est malum, quo nullum est perniciosius generis humano." Valeriano Magni, *De atheismo Aristotelis*, in *Principiorum philosophiae. Editio duo* (Cologne, 1661), 122.

2. "Tyrannus est, qui premit genus humanum perniciosius ullo Heresiarcha, ullove hominum, quos tulerit aetas ulla. . . . Laudatur ab Ethnicis, a Christianis, a Catholicis ab Hereticis." Magni, *De atheismo Aristotelis*, 122.

3. "Aristoteles inducit atheismum iis argumentis, quibus nulla efficaciora sint excogitabilia." Magni, *De atheismo Aristotelis*, 122.

4. "tamen docet, Mundum esse aeternum, & increatum, qui regatur fatali necessitate motus Coelorum, Deo nil penitus efficiente. His adde ex ejus sententia mortalitatem Animae rationalis." Magni, *De atheismo Aristotelis*, 122.

5. "Est porro infidelis, qui detrahit Deo creationem, & regimen mundi. Eum vero, qui non legit in textu Stagiritae mundum aeternum, & increatum. ignoro, an vivat: laudo tamen, zelum, ac pietatem nonullorum, qui virum, aemulum auctoritatis divinae apud homines torquent ad dogmata Fidei." Magni, *De atheismo Aristotelis*, 123.

6. "Iurare in verba Philosophorum Gentilium . . . est haeresis & periurium, & impietas maxima, ut probatum est in quaestione, utrum liceat novam cudere Philosophiam." Campanella, Tommaso. *De gentilismo non retineando* (Paris, 1636), 52.

7. Valeriano Magni, *Experimenta de incorruptibilitate aquae*, in *Principiorum philosophiae. Editio duo* (Cologne, 1661), 116–21.

8. Valeriano Magni, *Demonstratio ocularis loci sine locato*, in *Principiorum philosophiae. Editio duo* (Cologne, 1661), 73–79; Massimo Bucciantini, "La discussione sul vuoto in Italia: Il caso di Valeriano Magni," in *Discussioni sul nulla tra medioevo ed età moderna*, ed. Massimiliano Lenzi and Alfonso Maierù (Florence: Olschki, 2009), 285–301.

9. Charles B. Schmitt, *John Case and Aristotelianism in Renaissance England* (Kingston, Ontario: McGill–Queen's University Press, 1983), 13–17.

10. Francis A. Yates, "Giordano Bruno's Conflict with Oxford," *Journal of the Warburg and Courtauld Institutes* 2 (1939): 227–42; Mordechai Feingold, "Giordano Bruno in England, Revisited," *Huntington Library Quarterly* 67 (2004): 329–46; Giordano Bruno, *La cena de le ceneri*, in *Dialoghi italiani*, ed. Giovanni Aquilecchia, 3rd ed. (Florence: Sansoni, 1958), 1:20–21, 1:137.

11. *Statuta antiqua universitatis oxoniensis*, ed. Strickland Gibson (Oxford: Clarendon Press), 1931, 437; Yates, "Bruno's Conflict with Oxford," 230.

12. P. M. Rattansi, "Intellectual Origins of the Royal Society," *Notes and Records of the Royal Society of London* 23 (1968): 129–43; Mordechai Feingold, "Gresham College and London Practitioners: The Nature of the English Mathematical Community," in *Gresham College: Studies in the Intellectual History of London in the Sixteenth and Seventeenth Centuries*, ed. Francis Ames-Lewis (Aldershot, UK: Ashgate, 1999), 174–88. For criticism of Aristotle within sixteenth-century English universities see Mordechai Feingold, *The Mathematicians' Apprenticeship: Science, Universities and Society in England, 1560–1640* (Cambridge: Cambridge University Press, 1984), 102–3.

13. Barbara J. Shapiro, "Latitudinarianism and Science in Seventeenth-Century England," *Past and Present* 40 (1968): 16–41.

14. Douglas Hedley, "Persons of Substance and the Cambridge Connection: Some Roots and Ramifications of the Trinitarian Controversy in Seventeenth-Century England," in *Socinianism and Arminianism: Antitrinitarians, Calvinists and Cultural Exchange in Seventeenth-Century Europe*, ed. Martin Mulsow and Jan Rohls (Leiden: Brill, 2005), 225–40.

15. Paola Zambelli, "Magic and Radical Reformation in Agrippa of Nettesheim," *Journal of the Warburg and Courtauld Institutes* 39 (1976): 92.

16. Sarah Mortimer, *Reason and Religion in the English Revolution* (Cambridge:

Cambridge University Press, 2010), 61; Nicholas S. Davidson, "Unbelief and Atheism in Italy, 1500–1700," in *Atheism from the Reformation to the Enlightenment*, ed. Michael Hunter and David Wootton (Oxford: Oxford University Press, 1992), 66; George Hunston Williams, *The Radical Reformation* (Philadelphia: Westminster, 1962), 20–24.

17. Carlos M. N. Eire, *War against the Idols: The Reformation of Worship from Erasmus to Calvin* (Cambridge: Cambridge University Press, 1986), 195–233.

18. Jonathan Sheehan, "Sacred and Profane: Idolatry, Antiquarianism and the Polemics of Distinction in the Seventeenth Century," *Past and Present* 192 (2006): 35–66; Martin Muslow, "Antiquarianism and Idolatry: The *Historia* of Religions in the Seventeenth Century," in *Historia: Empiricism and Erudition in Early Modern Europe*, ed. Gianna Pomata and Nancy G. Siraisi (Cambridge, MA: MIT Press, 2005).

19. Erik Jorink, *Reading the Book of Nature in the Dutch Golden Age, 1575–1715* (Leiden: Brill, 2010), 68–70; Nicholas Wickenden, *G. J. Vossius and the Humanist Concept of History* (Assen: Van Gorcu, 1993), 27–30.

20. Philippe de Mornay [Du Plessis], *De veritate religionis Christianae, adversus Atheos, Epicureos, Ethnicos, Iudaeos, Mahumedistas, & caeteros infideles* (Leiden, 1605), 39–40, 378–79.

21. Ibid., 104.

22. Peter Harrison, *The Fall of Man and the Foundation of Science* (Cambridge: Cambridge University Press, 2007), 91.

23. R. B. [Richard Bostocke], *Auncient Phisicke* (London, 1585), *****r.

24. Ibid., *****v.

25. Ibid., 6r.

26. [Bostocke], *Auncient Phisicke*, 6r; Aristotle, *Physics*, 2.2.194b13. For the sun and humans generating, see Aristotle, *De generatione et corruptione*, 2.10.336a15–337a34.

27. [Bostocke], *Auncient Phisicke*, 7r–7v.

28. "quandoquidem hoc videretur nihil aliud quam Theologiam exponere ludibrio hominum atheorum." Thomas Lydiat, *Praelectio astronomica de natura coeli & conditionibus elementorum* (London, 1605), A5r; Feingold, *The Mathematicians' Apprenticeship*, 72; Ann Blair, "Mosaic Physics and the Search for a Pious Natural Philosophy in the Late Renaissance," *Isis* 91 (2000): 45.

29. "mentes Ethnicorum Philosophorum praesertim Aristotelicorum." Lydiat, *Praelectio*, A5r–A5v.

30. "id praecipue operam dans ut demonstrarem idem esse verum Physice ac Theologice." Lydiat, *Praelectio astronomica*, A5v.

31. "At non major quam impostorum maximus. Imposturae enim, atque adeo Principis Imposturae Antichristi, haec praerogativa singularis est." Francis Bacon, *Redargutio philosophiarum*, vol. 3 of *The Works of Francis Bacon*, ed. James Spedding et al. (London: Longman, 1857), 567. Paolo Rossi, *Francis Bacon: From Magic to Science*, trans. Sacha Rabinovitch (London: Routledge, 1968), 47. For Bacon's revival of Democritus, see Reid Barbour, "Bacon, Atomism, and Imposture: The True and the Useful in History, Myth, and Theory," in *Francis Bacon and the Refiguring of Early Modern Thought*, ed. Julie Robin Solomon and Catherine Gimelli Martin (Aldershot, UK: Ashgate, 2005), 17–43. For his atomism and Democritean thought, see Benedino Gemelli, *Aspetti dell'atomismo classico nella filosofia di Francis Bacon e nel Seicento* (Florence: Olschki, 1996),

141–95; Antonio Clericuzio, *Elements, Principles and Corpuscles: A Study of Atomism and Chemistry in the Seventeenth Century* (Dordrecht: Kluwer, 2000), 78–79.

32. Francis Bacon, *De augmentis scientiarum*, vol. 4 of *The Works of Francis Bacon*, ed. James Spedding et al. (London: Longman, 1860), 363–65.

33. Steven Matthews, *Science and Theology in the Thought of Francis Bacon* (Aldershot, UK: Ashgate, 2008), 27–50.

34. Lisa Jardine, *Francis Bacon: Discovery and the Art of Discourse* (Cambridge: Cambridge University Press, 1974), 59–75.

35. Howard Hotson, *Commonplace Learning: Ramism and its German Ramifications, 1543–1630* (Oxford: Oxford University Press, 2007), 283.

36. Stephen Gaukroger, *Francis Bacon and the Transformation of Early-Modern Philosophy* (Cambridge: Cambridge University Press, 2001), 74–83; Matthews, *Science and Theology in the Thought of Francis Bacon*, 109; Virgil K. Whitaker, "Francesco Patrizi and Francis Bacon," *Studies in the Literary Imagination* 4 (1971): 107–20.

37. John Wilkins, *The Discovery of a World in the Moone* (London, 1638), 25.

38. Ibid., 31–32.

39. Barbara J. Shapiro, *John Wilkins, 1614–1672: An Intellectual Biography* (Berkeley: University of California Press, 1969), 31.

40. Richard W. Serjeantson, "Hobbes, the Universities and the History of Philosophy," in *The Philosopher in Early Modern Europe: The Nature of a Contested Identity*, ed. Conal Condren et al. (Cambridge: Cambridge University Press, 2006), 126–27.

41. Thomas Hobbes, *Leviathan* (Latin version), vol. 3 of *Opera philosophica quae latine scripsit omnia*, ed. William Molesworth (London: Bohn, 1841), 496.

42. Thomas Hobbes, *Leviathan* (English version), vol. 3 of *Collected Works*, ed. William Molesworth (London: Routledge, 1994), 670; Serjeantson, "Hobbes, the Universities and the History of Philosophy," 130–35.

43. Hobbes, *Leviathan* (English version), 3:670.

44. Hobbes, *Leviathan* (English version), 3:675; Gianni Paganini, "Hobbes's Critique of the Doctrine of Essences and Its Sources," in *The Cambridge Companion to Hobbes's Leviathan*, ed. Patricia Springborg (Cambridge: Cambridge University Press, 2007), 337–57.

45. John Webster, *Academiarum examen, or The Examination of Academies* (London, 1654), 53–54.

46. Seth Ward and John Wilkins, *Vindiciae academiarum* (Oxford, 1654), 38.

47. Mordechai Feingold, "Aristotle and the English Universities in the Seventeenth Century: A Re-evaluation," in *European Universities in the Age of Reformation and Counter Reformation*, ed. Helga Robinson-Hammerstein (Dublin: Four Courts, 1998), 136–48.

48. Ward and Wilkins, *Vindiciae academiarum*, 38.

49. John Amos Comenius, *Naturall Philosophie Reformed by Divine Light, or, A Synopsis of Physicks* (London, 1651), a8r.

50. Ibid., a8v.

51. Blair, "Mosaic Physics," 32–58.

52. Robert Fludd, *Mosaicall Philosophy Grounded upon the Essential Truth or Eternal Sapience* (London, 1659), 28, 33, 123.

53. Henry More, *Divine Dialogues* (Glasgow, 1743), 109.

54. Ibid.

55. Ralph Cudworth, *The True Intellectual System of the Universe: Wherein All the Reason and Philosophy of Atheism Is Confuted* (Andover, MA: Gould & Newman, 1837), 1:114–15.

56. Ibid., 1:116.

57. Walter Charleton, *The Darknes of Atheism Dispelled by the Light of Nature. A Physico-Theological Treatise* (London, 1652), 1.

58. Ibid., 2.

59. Walter Charleton, *Physiologia Epicuro-Gassendo-Charltoniana: or, A Fabrick of Science Natural, upon the Hypothesis of Atoms* (London, 1654), 2, 3.

60. Joseph Glanvill, *The Vanity of Dogmatizing* (London, 1661), 151.

61. Ibid., 152–53.

62. Glanvill, *Vanity of Dogmatizing*, 183–84; Henry G. Van Leeuwen, *The Problem of Certainty in English Thought, 1630–1690* (The Hague: Nijhoff, 1963), 75–76.

63. Nicholas H. Steneck, "'The Ballad of Robert Crosse and Joseph Glanvill' and the Background to 'Plus Ultra,'" *British Journal for the History of Science* 14 (1981): 59–74.

64. Joseph Glanvill, *Plus ultra: or, the Progress and Advancement of Knowledge* (London, 1668), 124–25.

65. "ipsis Impietatis fundamentis Theologiam suam extruxerunt." Samuel Parker, *Disputationes de deo, et providentia divina* (London, 1678), v.

66. "quem Atheorum omnium principem atque ipso Epicuro apertiorem religionis hostem singulari disputatione demonstrabimus." Parker, *Disputationes de deo*, 65.

67. Parker, *Disputationes de deo*, 64–66.

68. Robert Boyle, *The Early Essays and Ethics of Robert Boyle*, ed. John T. Harwood (Carbondale: Southern Illinois University Press, 1991), 260.

69. Michael Hunter, "How Boyle Became a Scientist," *History of Science* 33 (1995): 74–77.

70. Robert Boyle, *Essay of the Holy Scriptures*, vol. 13 of *Works of Robert Boyle*, ed. Michael Hunter and Edward B. Davis (London: Pickering & Chatto, 2000), 187.

71. Ibid., 13:189.

72. Timothy Shanahan, "God and Nature in the Thought of Robert Boyle," *Journal of the History of Philosophy* 26 (1988): 547–69.

73. J. R. Jacob, "Boyle's Atomism and the Restoration Assault on Pagan Naturalism," *Social Studies of Science* 8 (1978): 213–14.

74. Robert Boyle, *Usefulness of Natural Philosophy*, vol. 3 of *Works of Robert Boyle*, ed. Michael Hunter and Edward B. Davis (London: Pickering & Chatto, 2000), 244.

75. Robert Boyle, *Reason and Religion*, vol. 8 of *Works of Robert Boyle*, ed. Michael Hunter and Edward B. Davis (London: Pickering & Chatto, 2000), 237.

76. Ibid.

77. Michael Hunter and Edward B. Davis, "The Making of Robert Boyle's *Free Enquiry into the Vulgarly Receiv'd Notion of Nature*," in *The Boyle Papers: Understanding the Manuscripts of Robert Boyle*, ed. Michael Hunter (Aldershot, UK: Ashgate, 2007), 251–71.

78. Dmitri Levitin, "The Experimentalist as Humanist: Robert Boyle on the History

of Philosophy," *Annals of Science*, forthcoming; Muslow, "Antiquarianism and Idolatry," 698–703.

79. Robert Boyle, *Free Enquiry into the Vulgarly Received Notion of Nature*, vol. 10 of *Works of Robert Boyle*, ed. Michael Hunter and Edward B. Davis (London: Pickering & Chatto, 2000), 447.

80. Robert Boyle, *The Christian Virtuoso*, vols. 11–12 of *Works of Robert Boyle*, ed. Michael Hunter and Edward B. Davis (London: Pickering & Chatto, 2000), 11:298, 12:412–13.

81. Boyle, *Christian Virtuoso*, 11:300–301.

82. John Ray, *The Wisdom of God Manifested in the Works of the Creation* (London, 1714), 51.

83. Robert Hooke, *Micrographia* (London, 1665), 198.

84. For Cartesians, Aristotelians, and atheism in France, see Alan Kors, *The Orthodox Sources of Disbelief*, vol. 1 of *Atheism in France, 1650–1729* (Princeton, NJ: Princeton University Press, 1990), 265–96. For Aristotelian physics in seventeenth-century France, see Laurence W. B. Brockliss, *French Higher Education in the Seventeenth and Eighteenth Centuries: A Cultural History* (Oxford: Oxford University Press, 1987), 337–50.

85. Theo Verbeek, *Descartes and the Dutch: Early Reactions to Cartesian Philosophy, 1637–1650* (Carbondale: Southern Illinois University Press, 1992), 6.

86. Franco Burgersdijk, *Institutionum metaphysicarum libri duo* (Leiden, 1640), 2.

87. Verbeek, *Descartes and the Dutch*, 7; J. A. van Ruler, *The Crisis of Causality: Voetius and Descartes on God, Nature and Change* (Leiden: Brill, 1995), 71–72.

88. Ruler, *Crisis of Causality*, 9–35.

89. Verbeek, *Descartes and the Dutch*, 13–38.

90. Theo Verbeek, *La querelle d'Utrecht* (Paris: Impressions nouvelles, 1988), 38.

91. "qui statuunt aliquod verum in Theologia falsum esse in Philosophia." Gisbertus Voetius, *Selectarum disputationum theologicarum, pars prima –[quinta]* (Utrecht, 1648–67), 4:840. Andreas J. Beck, *Gisbertus Voetius (1589–1676): Sein Theologieverständnis und seine Gotteslehre* (Göttingen: Vandenhoeck & Ruprecht, 2007), 153; Antonella Del Prete, "Against Descartes: Maarten Schoock's *De scepticismo*," in *The Return of Scepticism from Hobbes and Descartes to Bayle*, ed. Gianni Paganini (Dordrecht: Kluwer, 2003), 138–39.

92. Voetius, *Selectarum disputationum*, 1:1–12.

93. Del Prete, "Against Descartes," 138–39.

94. Voetius, *Selectarum disputationum*, 1:206.

95. Voetius, *Selectarum disputationum*, 1:198; Beck, *Gisbertus Voetius*, 153.

96. Antonio Rocco, *De immortalitate animae rationalis via peripatetica* (Venice, 1645), [122–23]; Claude Bérigard, *Circulus pisanus, de veteri & peripatetica philosophia* (Udine, 1643), 5–6.

97. "Et philosophos illos, qui in Scholis Italiae cum Atheismos contra Religionem Christianam, immo & contra Theologiam naturalem effutirent, effugium quaerebant in distinctione hac, quod in via naturae falsum aut incertum esset, e. gr. animam esse immortalem, dari daemonas, dari miracula a solo Deo immediate factas, esse providentiam Dei specialem circa omnes singulatim creaturarum & hominum actiones &c. sed in via Scripturae & traditionis Ecclesiasticae aut concedebant." Voetius, *Selectarum disputationum*, 4:840.

98. Josef Bohatec, *Die cartesianische Scholastik in der Philosophie und reformierten Dogmatik des 17. Jahrhunderts* (Leipzig: Deichert, 1912), 149; Edward G. Ruestow, *Physics at Seventeenth and Eighteenth-Century Leiden: Philosophy and the New Science in the University* (The Hague: Martinus Nijhoff, 1973), 38; Verbeek, *Descartes*, 34–51.

99. Ruestow, *Physics*, 44–45.

100. "At Christiana haec est libertas, Christianis omnibus concessa, eorumque animis innata, in iis, quae rerum spectant naturam & Philosophicae sunt contemplationis, nec *Aristotelis* nec ullius hominis authoritati mentem suam subjicere, judicium suum submittere." Adriaan Heereboord, *Meletemata philosophica* (Nijmegen, 1665), 333.

101. Heereboord, *Meletemata philosophica*, 330.

102. Ladislaus Lukács, ed., *Monumenta paedagogica Societitatis Iesu*, in *Monumenta historica Societiatis Iesu* (Rome: Institutum historicum Societatis Iesu, 1965–92), 5:129, 5:283.

103. Bruno Nardi, "La fine dell'averroismo," in *Saggi sull'aristotelismo padovano dal secolo XIV al XVI* (Florence: Sansoni, 1958), 443–55.

104. Massimo Campanini, "Edizioni e traduzioni di Averroè tra XIV e XVI secolo," in *Lexiques et glossaires philosophiques à la Renaissance*, ed. J. Hamesse and M. Fattori (Louvain-la-Neuve: FIDEM, 2003), 21–42.

105. Dag Nikolaus Hasse, "Aufstieg und Niedergang des Averroismus in der Renaissance: Niccolò Tignosi, Agostino Nifo, Francesco Vimercato," in *Herbst des Mittelalters? Fragen zur Berwertung des 14. und 15. Jahrhunderts*, ed. Jan A. Aertsen and Martin Pickavé (Berlin: De Gruyter, 2004), 447–73, and "The Attraction of Averroism: Vernia, Achillini, Prassicio," in vol. 2 of *Philosophy, Science and Exegesis in Greek, Arabic and Latin Commentaries*, ed. Peter Adamson et al. (London: Institute of Classical Studies, 2004), 131–47.

106. René Descartes, *Oeuvres*, ed. Charles Adam and Paul Tannery (Paris: Vrin, 1983), 7:3.

107. Thomas Browne, *Pseudodoxia epidemica*, ed. Robin Robbins (Oxford: Clarendon Press, 1981), 1:431.

108. Ibid., 1:586.

109. Christoph Clavius, *In sphaeram Ioannis de sacro bosco commentarius* (Rome, 1585), 455.

110. "Nunc Averroes in scholis depontanus evasit." Théophile Raynaud, *De malis ac bonis libris* (Lyon, 1653), 200.

111. Gottfried Wilhelm Leibniz, *Philosophische Schriften*, in *Sämtliche Schriften und Briefe* (Berlin: Gmbh, 1923–), 6/4a:464.

112. "Son autorité est nulle, & personne ne perd du tems à le lire." Pierre Bayle, *Dictionnaire historique et critique*, 5th ed. (Amsterdam, 1740), article on "Averroes," com. F, 1:387.

113. Piotr Skarga, *Artes duodecim sacramentariorum seu Zvingliocalvinistarum, quibus oppugnat* (Vilnius, 1550), 317.

114. Charles Drelincourt, *Neuf dialogues contre les missionnaires sur le service des eglises reformées* (Geneva, 1655), 305.

115. Jean Daillè, *Replique aux deux livres que messieurs Ada et Cottiby ont pubile contre luy* (Geneva, 1669), 116; Ernest Renan, *Averroès et l'averroïsme: Essai historique*, 2nd ed. (Paris: Lévy, 1861), 431–32.

116. Numbers 23:10; Gerardus Johannes Vossius, *De philosophia et philosophorum*

sectis (The Hague, 1657–58), 91; Voetius, *Selectarum disputationes*, 1:206; Louis Moreri, *Le grand dictionnaire historique* (Lyon, 1674), 178; Bayle, *Dictionnaire*, article on "Averroes," 1:387; *Naudaeana et Patiniana*, 2nd ed. (Amsterdam, 1703), 24.

117. For the character of late Scholasticism, see Dennis Des Chene, *Physiologia: Natural Philosophy in Late Aristotelian and Cartesian Thought* (Ithaca, NY: Cornell University Press, 1996).

118. "Ut enim iam non dicam de incredibili hominis impietate & crassissima rerum divinarum ignorantia." Johannes De Raei, *Clavis philosophiae naturalis, seu introductio ad naturae contemplationem* (Leiden, 1654), *3r; Marjorie Grene, "Aristotelico-Cartesian Themes in Natural Philosophy: Some Seventeenth-Century Cases," *Perspectives on Science* 1 (1993): 66–87.

119. De Raei, *Clavis philosophiae naturalis*, *5r.

120. "Aujourd'huy on entend ce Philosophe d'une autre maniere. Ce n'est point proprement sa Philosophie qui regne dans les Ecoles, c'est celle des Arabes." Bernard Lamy, *Entretiens sur les sciences*, ed. François Girbal and Pierre Clair (Paris: PUF, 1966), 236.

121. "Après eux étoient rangés les Commentateurs Arabes, entre lesquels Averroës est le plus considerable." Lamy, *Entretiens sur les sciences*, 236.

122. "Pour entendre la Philosophie des Arabes, il n'est question que d'appliquer aux termes d'Aristote, les préventions de l'Enfance." Lamy, *Entretiens sur les sciences*, 255.

123. Trevor McClaughlin, "Censorship and Defenders of the Cartesian Faith in Mid-Seventeenth Century France," *Journal of the History of Ideas* 40 (1979): 571; Roger Ariew, "Damned If You Do: Cartesians and Censorship, 1663–1706," *Perspectives on Science* 2 (1994): 257.

124. Charles Du Plessis d'Argentré, *Collectio judiciorum de novis erroribus* (Paris: Cailleau, 1728–36), 3:303.

125. Ariew, "Damned If You Do," 257–58.

126. Ibid., 261.

127. Argentré, *Collectio judiciorum*, 3:339–40; McClaughlin, "Censorship and Defenders," 567.

128. McClaughlin, "Censorship and Defenders," 569.

129. Yves Marie André, *La vie du R. P. Malebranche prête de l'oratoire* (Paris: Poussielgue, 1886), 5; Steven Nadler, *The Best of All Possible Worlds: A Story of Philosophers, God, and Evil in the Age of Reason* (Princeton, NJ: Princeton University Press, 2008), 42.

130. Nicolas Malebranche, *The Search after Truth [Recherche de la vérité]*, trans. Thomas M. Lennon and Paul J. Olscamp (Cambridge: Cambridge University Press, 1997), 145.

131. Ibid., 148–49.

132. [François Bernier]. *Requeste des maistres ès arts* ([Paris], 1671), 7–8; Nicholas Dew, *Orientalism in Louis XIV's France* (Oxford: Oxford University Press, 2009), 143–44. Emile Magne, *Une amie inconnue de Molière, suive de Molière et l'université* (Paris: Emile-Paul, 1922), 101–31.

133. [Bernier], *Requeste des maistres*, 6.

134. For the authorship of this treatise, see Ariew, "Damned If You Do," 261–63.

135. *Plusiers raisons pur empêcher la censure ou la condemnation de la philosophie de Descartes*, in *Oeuvres de Boileau Despreaux*, ed. Charles-Hugues Le Febvre de Saint-Marc (Paris, 1747), 119–24, 130.

136. Ibid., 132–33.

137. Sébastien Basso, *Philosophiae naturalis adversus Aristotelem libri XII* (Geneva, 1621, 198–201); Bérigard, *Circulus pisanus*, 3.

138. Roger Ariew, *Descartes and the Last Scholastics* (Ithaca, NY: Cornell University Press, 1999), 140–54.

139. Thomas Stanley, *The History of Philosophy, in Eight Parts* (London, 1656), part 6: 1–99.

140. Muzio Pansa, *De osculo seu consensu ethnicae & Christianae philosophiae* (Marburg, 1605), 265–66.

141. Jill Kraye, "Daniel Heinsius and the Author of *De mundo*," in *The Uses of Greek and Latin*, ed. Anthony Grafton et al. (London: Warburg Institute, 1988), 171–97, and "Aristotle's God and the Authenticity of *De mundo*: An Early Modern Controversy," *Journal of the History of Philosophy* 28 (1990): 339–58.

142. Aristotle, *De mundo*, 6.397b30–398a1.

143. Fortunio Liceti, *De pietate Aristotelis erga Deum & homines* (Udine,, 1645), 5, 69–70, 88.

144. Niccolò Cabeo, *Commentaria in quatuor libros meteorologicorum Aristotelis* (Rome, 1646), 1:419.

145. "Multa dicis, & nihil probas." Cabeo, *Commentaria in libros meteorologicorum*, 1:396.

146. "In impiam mundi aeternitatem, & Aristotelis atheismum inducit, Deum fingens ex necessitate naturae agentem, nihil providentem, aut praevidentem." Cabeo, *Commentaria in libros meteorologicorum*, 1:396.

147. Rapin, *Comparaison de Platon et d'Aristote*, 41–48.

148. Ibid., 219.

149. Ibid., 150.

150. Laurence W. B. Brockliss, *French Higher Education in the Seventeenth and Eighteenth Centuries: A Cultural History* (Oxford: Oxford University Press, 1987), 271–72.

CONCLUSION

1. Silvia Berti, "At the Roots of Unbelief," *Journal of the History of Ideas* 56 (1995): 559; Ruth Whelan, "The Wisdom of Simonides: Bayle and La Mothe Le Vayer," in *Scepticism and Irreligion in the Seventeenth and Eighteenth Centuries*, ed. Richard H. Popkin and Arjo Vanderjagt (Leiden: Brill, 1993), 232; Paul Oskar Kristeller, "The Myth of Renaissance Atheism and the French Tradition of Free Thought," *Journal of the History of Philosophy* 6 (1968): 233–43.

2. Louis Moreri, *Le grand dictionnaire historique* (Lyon, 1674), 1113.

3. Pierre Bayle, *Dictionnaire historique et critique*, 5th ed. (Amsterdam, 1740), 1:323.

4. Bayle, *Dictionnaire*, article on "Aristote," 1:324–26

5. Bayle, *Dictionnaire*, article on "Aristote," com. I, 1:326.

6. Bayle, *Dictionnaire*, article on "Aristote," com. N, 1:327.

7. Bayle, *Dictionnaire*, article on "Aristote," 1:327.

8. Bayle, *Dictionnaire*, article on "Aristote," 1:328–29.

9. Bayle, *Dictionnaire*, article on "Aristote," 1:329.

10. Bayle, *Dictionnaire*, article on "Pomponace," 3:779.

11. Bayle, *Dictionnaire*, article on "Pomponace," 3:781.

12. Bayle, *Dictionnaire*, article on "Pomponace," 3:781–82.

13. Bayle, *Dictionnaire*, article on "Pomponace," com. F, 3:780.

14. Bayle, *Dictionnaire*, article on "Pomponace," com. F, 3:780; Gregorio Piaia, "Gli aristotelici padovani al vaglio del *Dictionnaire historique et critique*," in *La presenza dell'aristotelismo padovano nella filosofia della prima modernità*, ed. Gregorio Piaia (Padua, Italy: Antenore, 2002), 430–35.

15. Bayle, *Dictionnaire*, article on "Pomponace," com. H, 3:782–83.

16. Bayle, *Dictionnaire*, article on "Pomponace," com. F, 3:780.

17. Bayle, *Dictionnaire*, article on "Zabarella," com. G, 4:530; Piaia, "Aristotelici padovani," 439.

18. Bayle, *Dictionnaire*, article on "Zabarella," com. G, 4:530.

19. Mori, *Bayle philosophe*, 165.

20. Bayle, *Dictionnaire*, article on "Cesalpin," com. C, 2:118.

21. Bayle, *Dictionnaire*, article on "Averroes," com. D, 1:386; Gianluca Mori, "Sullo Spinoza di Bayle," *Giornale critico della filosofia italiana*, 2nd ser., 67 [69] (1988): 348–67.

22. Bayle, *Dictionnaire*, article on "Averroes," com. D. 1:386; Harold Stone, "Why Europeans Stopped Reading Averroës: The Case of Pierre Bayle," *Alef: Journal of Comparative Poetics* 16 (1996): 77–95.

23. Moreri, *Grand dictionnaire*, 178; Bayle, *Dictionnaire*, article on "Averroes," com. F, 1:387.

24. Johannes Henricus Hottinger, *Bibliothecarius quadripartitus* (Zurich, 1664), 273.

25. Taylor Corse, "Dryden's 'Vegetarian' Philosopher: Pythagoras," *Eighteenth-Century Life* 34 (2010): 1–28.

26. [Thomas Tryon], *Averroeana, Being a Transcript of Several Letters* (London, 1695), A2r. On linking Arabia with Phoebus, see Nabil I. Matar, *Islam in Britain, 1558–1665* (Cambridge: Cambridge University Press, 1998), 91–97.

27. [Tryon], *Averroeana*, 5, 12,

28. Ibid., 58.

29. Denis Diderot and Jean Le Rond d'Alembert, eds. *Encyclopédie, ou dictionnaire raisonnée des sciences, des artes et des métiers* (Paris, 1751–65), 1:659–60.

30. Ernest Renan, *Averroès et l'averroïsme: Essai historique*, 2nd ed. (Paris: Lévy, 1861), 432–33, 479; Francesco Fiorentino, *Pietro Pomponazzi: Studi storici su la scuola bolognese e padovana del secolo* XVI (Florence: Le Monnier, 1868), 474–505; John Addington Symonds, *Renaissance in Italy: Italian Language* (New York: Holt, 1888), part 2, 479.

31. W. D. Ross, *Aristotle* (London: Methuen, 1923), 184.

32. Richard Bodéüs, *Theology of the Living Immortals*, trans. Jan Garrett (Albany: State University of New York Press, 2000), 7–41.

33. Martha C. Nussbaum and Amélie Oksenberg Rorty, eds., *Essays on Aristotle's De anima* (Oxford: Clarendon Press, 1992).

34. Jacob Burckhardt, *Die Kultur der Renaissance in Italien: Ein Versuch*, 14th ed. (Leipzig: Kröner, 1925), 464–81.

35. Renan, *Averroès et l'averroïsme*, 432–33; Symonds, *Renaissance in Italy*, 479; Eckhard Kessler, "Metaphysics or Empirical Science? The Two Faces of Aristotelian Natural Philosophy in the Sixteenth Century," in *Readings of the Corpus Aristotelicum*, ed. Marianne Pade (Copenhagen: Museum Tusculum, 2001), 79–101.

Principal Primary Sources

This listing contains primary sources used in this book. Consult the notes for references to relevant secondary sources.

Manuscripts and Archival Sources

Balsamo, Annibale. *Dubia aliquot in posteriora circa mentem Averrois*. MS, D. 129 inf. Biblioteca Ambrosiana, Milan.

Corti, Matteo. *Recollectae in septimum colliget Averrois*. MS, Lat. VII, 50 (=3570). Biblioteca Marciana, Venice.

Cremonini, Cesare. *Quaestio de animi moribus et facultatibus*. MS, 2075. Biblioteca universitaria, Padua.

Mainardi, Pietro. *Colliget Averois cum explanationes super V, VI, VII libri*. MS, II 84. Biblioteca Ariostea, Ferrara.

Pomponazzi, Pietro. *Commentarii in Aristotelis octo physicorum libros*. MS, Lat. 6533. Bibliothèque nationale de France, Paris.

———. *In libros meteorum*. MS, R. 96 sup. Biblioteca Ambrosiana, Milan.

Porzio, Simone. *Prologus Averrois super primum phisicorum Aristotelis*. MS, A. 153 inf. Biblioteca Ambrosiana, Milan.

Statuta collegium artium. Studio b. 216. Archivio di Stato, Bologna.

Storella, Franceso. *Animadversionum in Averroem, pars prima logicales locos comprehendens*. MS, I. 166 inf., fols. 123r–156r. Biblioteca Ambrosiana, Milan.

———. *Observationum in Averroem liber secundus locos ad naturalem, medicinam, atque super naturalem philosophiam attinensque amplectens*. MS, I. 166 inf., fols. 158r–214v. Biblioteca Ambrosiana, Milan.

Vernia, Nicoletto. *Quaestio utrum anima intellectiva humano corpori coniuncta tanquam unita forma substantialis*. MS, Lat. VI, 105 (=2656), fols. 156r–160v. Biblioteca Marciana, Venice.

Printed Primary Sources

Achillini, Alessandro. *De elementis*. In *Opera omnia*. Venice, 1568.

———. *De intelligentiis*. In *Opera omnia*. Venice, 1568.

———. *De orbibus*. In *Opera omnia*. Venice, 1568.

Agrippa, Cornelius. *De incertitudine et vanitate scientiarum et artium*. Paris, 1531.

Alberic of Trois-Fontaines. *Chronicon*. Edited by P. Scheffer-Boichorst. Vol. 23 of *Monumenta Germaniae Historica*, edited by Georg Heinrich Pertz. Hanover: Hahn, 1874

Albertus Magnus. *De causis proprietatum elementis.* Edited by Paul Hossfeld. Vol. 5, part 2, of *Opera omnia,* edited by Bernhard Geyer and Wilhelm Kübel. Münster: Aschendorf, 1980.

———. *De generatione et corruptione.* Edited by Paul Hossfeld. Vol. 5, part 2, of *Opera omnia,* edited by Bernhard Geyer and Wilhelm Kübel. Münster: Aschendorf, 1980.

———. *De natura loci.* Edited by Paul Hossfeld. Vol. 5, part 2, of *Opera omnia,* edited by Bernhard Geyer and Wilhelm Kübel. Münster: Aschendorf, 1980.

———. *De somno et vigilia.* Vol. 9 of *Opera omnia,* edited by August Borgnet. Paris: Vives, 1890.

———. *Metaphysica.* Edited by Bernhard Geyer. Vol. 16, part 2, of *Opera omnia,* edited by Bernhard Geyer and Wilhelm Kübel. Münster: Aschendorff, 1964.

Alexander of Aphrodisias. *Enarratio de anima.* Translated by Girolamo Donato. Brescia, 1495.

Altomare, Donato Antonio. *Omnia opera.* Lyon, 1565.

Arias, Francesco. *Dell'immitatione di Christo.* Translated by Tiberio Putignano. 3 vols. Rome, 1609–15.

Aristotle. *Aristotle's Physics.* Edited by W. D. Ross. Oxford: Clarendon Press, 1936.

———. *Meteorologicorum libri quattuor.* Edited by F. H. Fobes. Cambridge, MA: Harvard University Press, 1919.

Averroes. *Collectanea.* Supplement 1 of *Aristotelis opera cum Averrois commentariis.* Venice, 1562–74. Reprint, Frankfurt: Minerva, 1962.

———. *Commentarium magnum in Aristotelis de anima libros.* Edited by F. Stuart Crawford. Cambridge, MA: Medieval Academy of America, 1953.

———. *Commentary on Plato's Republic.* Translated by Erwin I. J. Rosenthal. Cambridge: Cambridge University Press, 1966.

———. *De anima.* Supplement 2 of *Aristotelis opera cum Averrois commentariis.* Venice, 1562–74. Reprint, Frankfurt: Minerva, 1962.

———. *De coelo.* Vol. 5 of *Aristotelis opera cum Averrois commentariis.* Venice, 1562–74. Reprint, Frankfurt: Minerva, 1962.

———. *Ibn Rushd's Metaphysics.* Translated by Charles Genequand. Leiden: Brill, 1984.

———. *Metaphysica.* Vol. 8 of *Aristotelis opera cum Averrois commentariis.* Venice, 1562–74. Reprint, Frankfurt: Minerva, 1962.

———. *Meteorologica.* Vol. 5 of *Aristotelis opera cum Averrois commentariis.* Venice, 1562–74. Reprint, Frankfurt: Minerva, 1962.

———. *Physica.* Vol. 4 of *Aristotelis opera cum Averrois commentariis.* Venice, 1562–74. Reprint, Frankfurt: Minerva, 1962.

———. *Prooemium in libros physicorum Aristotelis.* Vol. 4 of *Aristotelis opera cum Averrois commentariis.* Venice, 1562–74. Reprint, Frankfurt: Minerva, 1962.

———. *Tahafut al-tahafut.* Translated by Simon van den Bergh. 2 vols. Oxford: Oxford University Press, 1954.

Avicenna. *Liber canonis.* Lyon, 1522.

Bacon, Francis. *De augmentis scientiarum.* Vol. 4 of *The Works of Francis Bacon,* edited by James Spedding et al. London: Longman, 1860.

———. *Redargutio philosophiarum.* Vol. 3 of *The Works of Francis Bacon,* edited by James Spedding et al. London: Longman, 1857.

Bacon, Roger. *Communia naturalium*. Vol. 4 of *Opera hactenus inedita*, edited by Robert Steele. Oxford: Clarendon Press, 1913.

———. *Opus majus*. Translated by Robert Belle Burke. Philadelphia: University of Pennsylvania Press, 1928.

Balduini, Girolamo. *Expositio aurea in libros aliquot physicorum Aristotelis, et Averrois super eiusdem commentationem; et in prologum physicorum eiusdem Averrois*. Venice, 1573.

Barbaro, Ermolao. *Epistolae, orationes et carmina*. Edited by Vittore Branca. 2 vols. Florence: Bibliopolis, 1943.

Basso, Sébastien. *Philosophiae naturalis adversus Aristotelem libri XII*. Geneva, 1621.

Bayle, Pierre. *Dictionnaire historique et critique*. 4 vols. 5th ed. Amsterdam, 1740.

Beati, Giovanni Francesco. *Quaesitum in quo Averois ostendit quomodo verificatur corpora coelestia cum finita sint, et possibilia ex se acquirant aeternitatem ab alio*. [Padua], 1542.

Bembo, Pietro. *Lettere*. 2 vols. Milan: Classici Italiani, 1809.

Benvenuto da Imola. *Commentum super Dantis Aldigherii comoediam*. Edited by James Philip Lacaita. 5 vols. Florence: Barbera, 1887.

Bérigard, Claude. *Circulus pisanus, de veteri & peripatetica philosophia*. Udine, 1643.

Bernardi, Antonio. *Eversiones singularis certaminis*. Basel, 1562.

[Bernier, François]. *Requeste des maistres ès arts*. [Paris], 1671.

Bessarion, Basilios. *In calumniatorem Platonis*. Vol. 2 of *Kardinal Bessarion als Theologe, Humanist und Staatsman: Funde und Forschungen*, edited by Ludwig Mohler. Paderborn: Schöningh, 1927.

Boccaccio, Giovanni. *Boccaccio's Expositions on Dante's Comedy*. Translated by Michael Papio. Toronto: University of Toronto Press, 2009.

Boccadiferro, Lodovico. *Explanatio libri primi physicorum Aristotelis*. Venice, 1558.

———. *Lectiones super tres libros de anima*. Venice, 1566.

Bonaventure. *In Hexaemeron*. Vol. 5 of *Opera omnia*. Quaracchi: Collegium S. Bonaventurae, 1882–1902.

Borro, Girolamo. *De motu gravium, et levium*. Florence, 1575.

[Bostocke, Richard], R. B. *Auncient Phisicke*. London, 1585.

Boyle, Robert. *Essay of the Holy Scriptures*. Vol. 13 of *Works of Robert Boyle*, edited by Michael Hunter and Edward B. Davis. London: Pickering & Chatto, 2000.

———. *Free Enquiry into the Vulgarly Received Notion of Nature*. Vol. 10 of *Works of Robert Boyle*, edited by Michael Hunter and Edward B. Davis. London: Pickering & Chatto, 2000.

———. *Reason and Religion*. Vol. 8 of *Works of Robert Boyle*, edited by Michael Hunter and Edward B. Davis. London: Pickering & Chatto, 2000.

———. *The Christian Virtuoso*. Vols. 11–12 of *Works of Robert Boyle*, edited by Michael Hunter and Edward B. Davis. London: Pickering & Chatto, 2000.

———. *Usefulness of Natural Philosophy*. Vol. 3 of *Works of Robert Boyle*, edited by Michael Hunter and Edward B. Davis. London: Pickering & Chatto, 2000.

Browne, Thomas. *Pseudodoxia epidemica*. Edited by Robin Robbins. 2 vols. Oxford: Clarendon Press, 1981.

Bruni, Leonardo. *Dialoghi ad Petrum Histrum*. In *Prosatori Latini del Quattrocento*, edited by Eugenio Garin. Milan: Ricciardi, 1952.

———. *The Humanism of Leonardo Bruni: Selected Texts.* Translated by Gordon Griffiths, James Hankins, and David Thompson. Binghamton, NY: Medieval & Renaissance Texts & Studies, 1987.

Bruno, Giordano. *La cena de le ceneri.* In *Dialoghi italiani,* edited by Giovanni Aquilecchia. 3rd ed. Florence: Sansoni, 1958.

Burgersdijk, Franco. *Institutionum metaphysicarum libri duo.* Leiden, 1640.

Cabeo, Niccolò. *Commentaria in quatuor libros meteorologicorum Aristotelis.* Rome, 1646.

Campanella, Tommaso. *De gentilismo non retineando.* Paris, 1636.

Cano, Melchior. *De locis theologicis.* Salamanca, 1563.

Cardano, Girolamo. *Contradicentium medicorum liber primus [-secundus].* Vol. 6 of *Opera omnia.* Lyon, 1663.

———. *De immortalitate animorum liber.* Vol. 2 of *Opera omnia.* Lyon, 1663.

Casmann, Otto, *Marinarum quaestionum tractatio philosophica.* Frankfurt, 1596.

Cesalpino, Andrea. *Daemonum investigatio peripatetica.* Venice, 1593.

Champier, Symphorien. *Annotamenta errata et castigationes in Avicenne opera.* Lyon, 1522.

———. *Cribratio, lima et annotamenta in Galeni, Avicennae et Consiliatoris opera.* [Paris], 1516.

———. *Epistola responsiva pro Graecorum defensione in Arabum errata.* Lyon, 1533.

———. *Hortus gallicus, pro Gallis in Gallia scriptus.* Lyon, 1533.

Charleton, Walter. *Physiologia Epicuro-Gassendo-Charltoniana: or, A Fabrick of Science Natural, upon the Hypothesis of Atoms.* London, 1654.

———. *The Darknes of Atheism Dispelled by the Light of Nature. A Physico-Theological Treatise.* London, 1652.

Charpentier, Jacques. *Platonis cum Aristotele in universa philosophia comparatio.* Paris, 1573.

Charron, Pierre. *Le trois veritez.* Paris, 1595.

Chartularium universitatis parisiensis. 4 vols. Paris: Delalain, 1889–1897.

Clavius, Christoph. *In sphaeram Ioannis de sacro bosco commentarius.* Rome, 1585.

Collegium Conimbricense. *In quatuor libros de coelo.* Lyon, 1617.

———. *In tres libros de anima.* Cologne, 1617.

Comenius, Johann Amos. *Naturall Philosophie Reformed by Divine Light, or, A Synopsis of Physicks.* London, 1651.

Conciliorum oecumenicorum decreta. Edited by Giuseppe Alberigo et al. Bologna: Istituto per le scienze religiose, 1973.

Confalonieri, Giovanni Battista. *Averrois libellus de substantia orbis expositus.* Venice, 1525.

Contarini, Gaspare. *De homocentricis ad Hieronymum Fracastorium.* In *Opera omnia.* Venice, 1589.

Cottunio, Giovanni. *De triplici statu animae rationalis.* Bologna, 1628.

Cremonini, Cesare. *Apologia dictorum Aristotelis de quinta caeli substantia adversus Xenarcum, Ioannem Grammaticum, & alios.* Venice, 1616.

———. *Le orazioni.* Edited by Antonino Poppi. Padova, 1998.

Crinito, Pietro. *De honesta disciplina.* Paris, 1508.

Cudworth, Ralph. *The True Intellectual System of the Universe: Wherein All the Reason and Philosophy of Atheism is Confuted.* 2 vols. Andover: Gould & Newman, 1837.

Daillè, Jean. *Replique aux deux livres que messieurs Ada et Cottiby ont pubile contre luy.* Geneva, 1669.

Da Monte, Giovanni Battista. *In primi libri canonis Avicennae primam fen, profundissima commentaria.* Venice, 1558.

Dandini, Girolamo. *De corpore animato.* Paris, 1611.

David of Dinant. *I testi di David di Dinant: filosofia della natura e metafisica a confronto col pensiero antico.* Edited by Elena Cadei. Spoleto: Centro italiano di studi sull'alto medioevo, 2008.

Delrio, Martin. *Disquisitionum magicarum libri sex.* Mainz, 1617.

De Raei, Johannes. *Clavis philosophiae naturalis, seu introductio ad naturae contemplationem.* Leiden, 1654.

Descartes, René. *Oeuvres.* Edited by Charles Adam and Paul Tannery. 11 vols. Paris: Vrin, 1983.

De' Vieri, Francesco. *Vere conclusioni di Platone conformi alla dottrina Christiana. Et a quella d'Aristotile.* Florence, 1590.

De Vio, Tommaso. *Commentaria in libros Aristotelis de anima.* Venice, 1514.

Diderot, Denis, and Jean Le Rond d'Alembert, eds. *Encyclopédie, ou dictionaire raisonnée des sciences, des artes et des métiers.* 17 vols. Paris, 1751–65.

Dominicus of Flanders. *Acutissimae quaestiones super tres libros de anima & Sancti Thomae commentaria.* Venice, 1560.

Drelincourt, Charles. *Neuf dialogues contre les missionnaires sur le service des eglises reformées.* Geneva, 1655.

Du Val, Guillaume. *Synopses analyticae universae doctrinae peripateticae.* In *Aristotelis operum,* edited by Isaac Casaubon and Giulio Pace. 2 vols. Paris, 1619.

Elia del Medigo. *Parafrasi della "Repubblica" nella traduzione latina di Elia del Medigo.* Edited by Annalisa Coviello and Paolo Edoardo Fornaciari. Florence: Olschki, 1992.

Erasmus, Desiderius. *Opera omnia.* Amsterdam: North-Holland, 1969.

———. *Opus epistolarum.* Edited by P. S. Allen. 12 vols. Oxford: Clarendon, 1906–58.

Eusebius. *De evangelica praeparatione.* Translated by George Tapezuntius. Venice, 1497.

Fernel, Jean. *On the Hidden Causes of Things.* Edited and translated by John M. Forrester and John Henry. Leiden: Brill, 2005.

Ficino, Marsilio. *Opera.* 2 vols. Basel, 1576.

———. *Platonic Theology.* Translated by Michael J. B. Allen and edited by James Hankins. 6 vols. Cambridge, MA: Harvard University Press, 2001–6.

Fludd, Robert. *Mosaicall Philosophy Grounded upon the Essential Truth or Eternal Sapience.* London, 1659.

Fonseca, Pedro de. *In libros metaphysicorum Aristotelis Stagiritae, cui praemissi sunt institutionum dialecticarum libri octo.* Lyon, 1597.

Foresti, Giacomo Filippo. *Supplementum cronicarum.* Venice, 1486.

Frey, Jean-Cécile. *Cribrum philosophorum.* In *Opuscula varia.* Paris, 1646.

Frytsche, Marcus. *Meteorum, hoc est impressionum aerearum et mirabilium naturae operum.* Wittenberg, 1598.

Fuligatti, Giacomo. *Vita Roberti Bellarmini politiani.* Liege, 1626.

Garasse, François. *La doctrine curieuse des beaux esprits de ce temps, ou prendus tel.* 2 vols. Paris, 1623. Reprint, Westmead: Gregg, 1971.

Gassendi, Pierre. *Exercitationes paradoxicae adversus Aristoteleos.* Amsterdam, 1649.

Gaudenzio, Paganino. *De pythagoraea animarum transmigratione*. Pisa, 1641.

Genua, Marcantonio. *Disputatio de intellectus humani immortalitate*. Mondovì, 1565.

Gesner, Konrad. *Bibliotheca universalis*. Zurich, 1545.

Giles of Rome. *Errores philosophorum*. Edited by Josef Koch. Milwaukee: Marquette University Press, 1944.

Giovio, Paolo. *Gli elogi degli uomini illustri*. Vol. 8 of *Opera*, edited by Renzo Meregazzi. Rome: Istituto poligrafico dello stato, 1972.

Glanvill, Joseph. *Plus ultra: or, the Progress and Advancement of Knowledge*. London, 1668.

——. *The Vanity of Dogmatizing*. London, 1661.

Gozze, Nicolò Vito di. *In sermonem Averrois de substantia orbis*. Bologna, 1580.

Heereboord, Adriaan. *Meletemata philosophica*. Nijmegen, 1665.

Hobbes, Thomas. *Leviathan* (English version). Vol. 3 of *Collected Works*, edited by William Molesworth. London: Routledge, 1994.

——. *Leviathan* (Latin version). Vol. 3 of *Opera philosophica quae latine scripsit omnia*, edited by William Molesworth. London: Bohn, 1841.

Hooke, Robert. *Micrographia*. London, 1665.

Hottinger, Johannes Henricus. *Bibliothecarius quadripartitus*. Zurich, 1664.

Javelli, Crisostomo. *Solutiones rationum animi mortalitatem probantium*. Venice, 1525.

John Duns Scotus. *Philosophical Writings: A Selection*. Translated by Allan Wolter. Indianapolis: Hackett, 1990.

John of Jandun. *In libros Aristotelis de coelo et mundo quaestiones subtilissimae: quibus adiecimus Averrois sermonem de substantia orbis cum commentario ac quaestionibus*. Venice, 1552.

——. *Subtilissime quaestiones in octo libros Aristotelis de physico auditu*. Venice, 1544.

——. *Quaestiones super Parvis naturalibus ad Aristotelis et Averrois intentionem*. Venice, 1589.

John of Wales. *Florilegium de vita et dictis illustrium philosophorum et breviloquium de sapientia sanctorum*. Rome, 1655.

La Galla, Giulio Cesare. *De immortalitate animorum ex Aristotelis sententia*. Rome, 1621.

La Mothe Le Vayer, François. *Petit discours chrestien de l'immortalité de l'ame*. Vol. 4 of *Oeuvres*. Paris, 1669.

Lamy, Bernard. *Entretiens sur les sciences*. Edited by François Girbal and Pierre Clair. Paris: PUF, 1966.

Lando, Ortensio. *Paradossi cioe sententie fuori del comun parere*. Venice, 1544.

Launoy, Jean de. *De varia Aristotelis in academia parisiensi fortuna*. Paris, 1653.

Leibniz, Gottfried Wilhelm. *Philosophische Schriften*. In *Sämtliche Schriften und Briefe*, 20 vols. Berlin: Gmbh, 1923.

Leone, Ambrogio. *Castigationes adversus Averroem*. Venice, 1532.

Liceti, Fortunio. *De pietate Aristotelis erga Deum & homines*. Udine, 1645.

Ligorio, Pirro. *Libro di diversi terremoti*. Edited by Emanuela Guidoboni. Rome: De Luca, 2005.

Llull, Ramon. *Opera latina*. Edited by Josep Enric Rubio Albarracin. Vol. 32 of *Corpus christianorum: continuatio mediaevalis*. Turnholt: Brepols, 1975.

Longo, Giovanni Bernardino. *Dilucida expositio in prologum Averrois in posteriora Aristotelis*. Naples, 1551.

Lydiat, Thomas. *Praelectio astronomica de natura coeli & conditionibus elementorum.* London, 1605.

Magni, Valeriano. *De atheismo Aristotelis.* In *Principiorum philosophiae editio duo.* Köln, 1661.

———. *Demonstratio ocularis loci sine locato.* In *Principiorum philosophiae editio duo.* Köln, 1661.

———. *Experimenta de incorruptibilitate aquae.* In *Principiorum philosophiae editio duo.* Köln, 1661.

Mainetti, Mainetto. *Commentarii in librum primum Aristotelis de coelo. Necnon librum Averrois de substantia orbis.* Bologna, 1570.

Maioragio, Marcantonio. *In quatuor Aristotelis libros de coelo paraphrasis.* Basel, 1554.

Malebranche, Nicolas. *The Search after Truth [Recherche de la vérité].* Translated by Thomas M. Lennon and Paul J. Olscamp. Cambridge: Cambridge University Press, 1997.

Mansi, Giovanni Domenico. *Sacrorum conciliorum nova et amplissima collectio.* 54 vols. Paris: Welter, 1901–27. Reprint, Graz: Akademische Druck- u. Verlagsanstalt, 1960–62.

Marta, Giacomo Antonio. *Pugnaculum Aristotelis adversus principia Bernardini Telesii.* Rome, 1587.

Matthew of Paris. *Chronica majora.* Edited by Henry Richard Luard. 7 vols. London: Longman, 1872–83.

Mazzoni, Jacopo. *De triplici hominum vita, activa nempe, contemplativa, & religiosa methodi tres.* Cesena, 1576.

———. *In universam Platonis et Aristotelis philosophiam praeludia, sive de comparatione Platonis et Aristotelis.* Edited by Sara Matteoli. Naples: M. D'Auria, 2010.

Melanchthon, Philip. *Commentarius de anima.* Wittenberg, 1550.

———. *De Aristotele.* Vol. 11 of *Opera quae supersunt omnia,* edited by Karl Gottlieb Bretschneider. Halle: Schwetschke, 1834–60.

Mercuriale, Girolamo. *Praelectiones patavinae, de cognoscendis, et curandis humani corporis affectibus.* Venice, 1613.

Mersenne, Marin. *L'impiété des déistes, athées et libertins de ce temps.* Paris, 1624. Reprint, Stuttgart-Bad Cannstatt: Frommann, 1975.

———. *La vérité des sciences. Contre les sceptiques ou pyrrhoniens.* Paris, 1625, Reprint, Stuttgart-Bad Cannstatt: Frommann, 1969.

———. *Quaestiones celeberrimae in Genesim.* Paris, 1623.

Milich, Jakob. *Liber II, C. Plinii de mundi historia, cum commentariis.* Frankfurt, 1543.

Morcillo, Sebastián Fox. *De naturae philosophia, seu de Platonis et Aristotelis consensione.* Louvain, 1554.

More, Henry. *Divine Dialogues.* Glasgow, 1743.

Moreri, Louis. *Le grand dictionnaire historique.* Lyon, 1674.

Mornay, Philippe de [Du Plessis]. *De veritate religionis Christianae, adversus Atheos, Epicureos, Ethnicos, Iudaeos, Mahumedistas, & caeteros infideles.* Leiden, 1605.

Musso, Cornelio. *Delle prediche quadragesimali.* 2 vols. Venice, 1603.

Nacchiante, Giacomo. *Opus doctum ac resolutum.* Venice, 1557.

Naudé, Gabriel. *Apologie pour tous les grands personnages qui on esté faussement soupçonnez de magie.* Paris, 1669.

——. *Epistolae.* Geneva, 1667.

——. *Iudicium de Augustino Nipho.* In Agostino Nifo, *Opuscula moralia et politica.* Paris, 1645.

Naudaeana et Patiniana. 2nd ed. Amsterdam, 1703.

Nifo, Agostino. *Averroys de mixtione defensio.* Venice, 1505.

——. *Commentationes in librum de substantia orbis.* Venice, 1508.

——. *De immortalitate animae libellus.* Venice, 1518.

——. *Expositio super octo Aristotelis Stagiritae libros de physico auditu, Averrois etiam Cordubensis in eosdem libros prooemium, ac commentaria.* Venice, 1552.

——. *In Averrois de animae beatitudine.* Venice, 1524.

——. *In duodecimum metaphysices Aristotelis [et] Auerrois volumen commentarii.* Venice, 1518.

——. *In librum destructio destructionum Averrois commentarii.* Lyon, 1542.

Nizolio, Mario. *De veris principiis et vera ratione philosophandi contra pseudophilosophos.* Edited by Quirinus Breen. 2 vols. Rome: Fratelli Bocca, 1956.

Nuñez, Pedro Juan. *Vita Aristotelis per Ammonium, seu Philoponum.* Leiden, 1621.

Odoni, Rinaldo. *Discorso per via peripatetica ove si dimostra se l'anima, secondo Aristotele, è mortale, o immortale.* Venice, 1557.

[Ottimo]. *L'ultima forma dell'Ottimo commento: Chiose sopra La Comedia di Dante Alleghieri fiorentino tracte da diversi ghiosatori.* Edited by Claudia Di Fonzo. Ravenna: Longo, 2008.

Panigarola, Francesco. *Prediche quadragesimali.* Venice, 1617.

Pansa, Muzio. *De osculo seu consensu ethnicae & Christianae philosophiae.* Marburg, 1605.

Parker, Samuel. *Disputationes de deo, et providentia divina.* London, 1678.

Patrizi, Francesco. *Apologia contra calumnias Theodori Angelutii.* Ferrara, 1584.

——. *Discussiones peripateticae.* Venice, 1571.

——. *Nova de universis philosophia.* Ferrara, 1591.

——. *Nova de universis philosophia: Materiali per un'edizione emendata.* Edited by Anna Laura Puliafito Bleuel. Florence: Olschki, 1993.

Patrologiae cursus completus, series graeca posterior. Edited by J. P. Migne. 221 vols. Paris: Garnier, 1844–66.

Perera, Benito. *De communibus omnium rerum naturalium principiis et affectionibus.* Paris, 1579.

Petrarca, Francesco. *Invectives.* Translated by David Marsh. Cambridge, MA: Harvard University Press, 2003.

——. *Rerum senilium libri.* Edited by Ugo Dotti. 3 vols. Torino: Nino Aragno, 2010.

Piccolomini, Francesco. *Librorum ad scientiam de natura attinentium pars prima.* Venice, 1596.

——. *Librorum ad scientiam de natura attinentium pars secunda.* Venice, 1600.

Pico della Mirandola, Gianfrancesco. *Examen vanitatis doctrinae gentium, & veritatis Christianae disciplinae.* Mirandola, 1520.

——. *Vita Hieronymi Savonarolae.* Edited by Elisabetta Schisto. Florence: Olschki, 1999.

Pico della Mirandola, Giovanni. *Syncretism in the West: Pico's 900 Theses (1486).* Edited by S. A. Farmer. Tempe, AZ: Medieval & Renaissance Texts & Studies, 1998.

Pietro d'Abano. *Conciliator.* Venice, 1565. Reprint, Padua: Antenore, 1985.

———. *Trattati di astronomia.* Edited by Graziella Federici Vescovini. Padua: Programma, 1992.

Plusiers raisons pur empêcher la censure ou la condemnation de la philosophie de Descartes. In Boileau, *Oeuvres de Boileau Despreaux.* Edited by Saint-Marc. Paris, 1747.

Pomponazzi, Pietro. *Apologiae libri tres.* Bologna, 1518.

———. *Corsi inediti dell'insegnamento padovano.* Edited by Antonino Poppi. 2 vols. Padua: Antenore, 1970.

———. *Defensorium autoris.* Venice, 1525.

———. *De naturalium effectuum admirandorum causis, seu de incantationibus.* In *Opera.* Basel, 1567.

———. *Libri quinque de fato.* Edited by Richard Lemay. Lugano: Thesaurus mundi, 1957.

———. *Tractatus acutissimi, utillimi et mere peripatetici.* Venice, 1525.

———. *Tractatus de immortalitate animae.* Bologna, 1516.

Pontano, Girolamo. *De immortalitate animae ex sententia Aristotelis.* Rome, 1597.

Porzio, Simone. *De humana mente disputatio.* Florence, 1551.

Possevino, Antonio. *Bibliotheca selecta.* 2 vols. Rome, 1593.

Postel, Guillaume. *Eversio falsorum Aristotelis dogmatum.* Paris, 1552.

———. *Liber de causis.* Paris, 1552.

Prassicio, Luca. *Questio de immortalitate anime.* Naples, 1521.

Pseudo-Burley, Walter. *Liber de vita et moribus philosophurm.* Edited by Hermann Knust. Tübingen: Litterarischer Verein, 1886.

Raguseo, Giorgio. *Peripateticae disputationes.* Venice, 1613.

Ramus, Petrus. *Aristotelicae animadversiones.* Paris, 1553.

———. *Commentariorum de religione christiana libri quatuor.* Frankfurt, 1576.

Rapin, René. *La comparaison de Platon et d'Aristote.* Paris, 1671.

Ray, John. *The Wisdom of God Manifested in the Works of the Creation.* London, 1714.

Raynaud, Theophile. *De malis ac bonis libris.* Lyon, 1653.

Robbe, L. ed. *Vita Aristotelis ex codice marciano.* Leiden: Van Leeuwen, 1861.

Rocco, Antonio. *De immortalitate animae rationalis via peripatetica.* Venice, 1645.

Rodotà, Pietro Pampilio. *Dell'origine progresso, e stato presente del rito greco in Italia.* 3 vols. Rome, 1758–63.

Rubio, Antonio. *Commentarii in libros Aristotelis de anima.* Brescia, 1626.

Salutati, Coluccio. *De fato et fortuna.* Edited by Concetta Bianca. Florence: Olschki, 1985.

———. *De laboribus Herculis.* Edited by Berthold L. Ullman. Zurich: Thesaurus mundus, 1951.

———. *Epistolario.* Edited by Francesco Novati. 4 vols. Rome: Forzani, 1893.

Sanudo, Marino. *I diarii.* Edited by Rinaldo Fulin. 58 vols. Venice: Visentini, 1879–1903.

Sarpi, Paolo. *Istoria del Concilio Tridentino.* 2 vols. Florence: Sansoni, 1966.

Savonarola, Girolamo. *Prediche sopra l'Esodo.* 2 vols. Rome: Belardetti, 1955–56.

———. *Triumphus crucis.* Edited by Mario Ferrara. Rome: Belardetti, 1961.

Siger of Brabant. *Quaestiones in tertium de anima. De anima intellectiva. De aeternitate mundi.* Edited by Bernardo Bazán. Louvain: Publications universitaires, 1972.

Scaliger, Julius Caesar. *Exotericarum exercitationum libri XV.* Paris, 1557.

Schott, Andreas. *Vitae comparatae Aristotelis ac Demosthenis.* Augsburg, 1603.

Skarga, Piotr. *Artes duodecim sacramentariorum seu Zvingliocalvinistarum, quibus op-pugnat.* Vilnius, 1550.

Simplicius. *In Aristotelis categorias commentarium.* Vol. 8 of *Commentaria in Aristo-telem graeca,* edited by Karl Kalbfleisch. Berlin: Reimer, 1907.

Soto, Domenico de. *Quaestiones in octo libros physicorum Aristotelis.* Douai, 1613.

Speroni, Sperone. *Opere.* 5 vols. Venice, 1740.

Spina, Bartolomeo. *Opuscula.* Venice, 1519.

Stanley, Thomas. *The History of Philosophy, in Eight Parts.* London, 1656.

Statuta almae universitatis dd. philosophorum, et medicorum cognomento artistarum pa-tavini gymnasii. Padua, 1607.

Statuta antiqua universitatis oxoniensis. Edited by Strickland Gibson. Oxford: Claren-don Press, 1931.

Statuti delle università e dei collegi dello studio bolognese. Edited by Carlo Malagola. Bologna: Zanichelli, 1888.

Talon, Omer. *Dialecticae praelectiones in Porphyrium.* Paris, 1550.

Tasso, Torquato. *Dialoghi.* In *Opere,* edited by Bruno Maier. Milan: Rizzoli, 1965.

Taurellus, Nicolaus. *Alpes caesae, hoc est, Andr. Caesalpini Itali monstrosa & superba dogmata discussa & excussa.* Frankfurt, 1597.

———. *De rerum aeternitate.* Marburg, 1614.

———. *Kosmologia.* Hamburg, 1603.

———. *Medicae praedictionis methodus.* Frankfurt, 1581.

———. *Philosophiae triumphus, hoc est metaphysica philosophandi methodus.* Arnheim, 1617.

Telesio, Bernardino. *De rerum natura iuxta propria principia.* Edited by Luigi de Franco. 2 vols. Cosenza: Casa del libro, 1965.

Theophrastus redivivus. Edited by Guido Canziani and Gianni Paganini. 2 vols. Flor-ence: La Nuova Italia, 1981.

Thomas Aquinas. *De aeternitate mundi contra murmurantes.* In *Opuscula philosophica,* edited by Raimondo M. Spiazzi. Rome: Marietti, 1954.

———. *Tractatus de unitate intellectus contra Averroistas.* Edited by Bruno Nardi and Paolo Mazzantini. Spoleto: Centro italiano di studi sull'Alto Medioevo, 1988.

Tignosi, Niccolò. *In libros Aristotelis de anima commentarii.* Florence, 1551.

Titelmans, Frans. *Compendium philosophiae naturalis.* Lyon, 1558.

Toletus, Francisco. *Commentaria una cum quaestionibus in octo libros Aristotelis de physica auscultatione.* Venice, 1580.

Tomasini, Giacomo Filippo. *Bibliothecae patavinae manuscriptae publicae & privatae.* Udine, 1639.

Trapezuntius, George. *Comparationes phylosophorum Aristotelis et Platonis.* Venice, 1523.

Trincavelli, Vittore. *Quaestio de reactione iuxta Aristotelis sententiam et commentatoris.* Venice, 1520.

Trombetta, Antonio. *Tractatus singularis contra Averroystas de humanarum animarum plurificatione.* Venice, 1498.

[Tryon, Thomas]. *Averroeana, Being a Transcript of Several Letters.* London, 1695.

Valla, Lorenzo. *Repastinatio dialectice et philosophie*. Edited by Gianni Zippel. 2 vols. Padua: Antenore, 1982.

Valle de Moura, Emanuel do. *De incantationibus seu ensalmis*. Ebora, 1620.

Vanini, Giulio Cesare. *Amphiteatrum aeternae providentiae divino-magium*. Lyon, 1615.

——. *De admirandis naturae*. Paris, 1616.

Varchi, Benedetto. *Lezzioni sopra diverse materie, poetiche, e filosofiche*. Florence, 1590.

Venier, Francesco. *I discorsi sopra i tre libri dell'anima d'Aristotele*. Venice, 1555.

Vera Cruz, Alfonso de la. *Physica speculatio*. Salamaca, 1569.

Vernia, Nicoletto. *Contra perversam Averroys opinionem de unitate intellectus et de anime felicitate*. Venice, 1505.

Vimercati, Francesco. *De anima rationali peripatetica disceptatio*. Venice, 1566.

——. *In eam partem duodecimi libri Metaphys. Aristotelis, in qua de deo & caeteris mentibus divinis disseritur, commentarii*. Paris, 1551.

Vives, Juan Luis. *Über die Gründe des Verfalls der Künste/De causis corruptarum artium*. Translated and edited by Wilhelm Sendner. Munich: Fink Verlag, 1990.

Voetius, Gisbertus. *Selectarum disputationum theologicarum*. 4 vols. Utrecht, 1648–67.

Vossius, Gerhard Johannes. *De philosophia et philosophorum sectis*. The Hague, 1657–58.

Ward, Seth, and John Wilkins. *Vindiciae academiarum*. Oxford, 1654.

Webster, John. *Academiarum examen, or The Examination of Academies*. London, 1654.

Wier, Johann. *De praestigiis daemonum et incantationibus ac venificiis*. Basel, 1566.

Wilkins, John. *The Discovery of a World in the Moone*. London, 1638.

Zabarella, Giacomo. *De rebus naturalibus*. Frankfurt, 1597.

Index

Rapin, René, 167–68
Ray, John, 157
Raynaud, Théophile, 162
Regius, Henricus, 159
Remonstrants, 158–59
Renaissance, the, 8, 52–53, 71, 176–77
Renan, Ernest, 23, 176
resurrection: bodily, 20, 63, 104, 113, 125, 141, 152, 154; of Lazarus, 21–22
revelation, 36, 46, 64, 97, 126, 157; in Middle Ages, 18, 20–21
rhetoric, 107; and humanists, 28, 34, 37, 39, 59, 95, 102–3; and Ramus, 111–12, 150
Richelieu, Cardinal, 123, 127
Rinuccini, Alamanno, 43
Robert de Courçon, 23
Rocco, Antonio, 159
Rome, 3, 54, 66, 93, 100, 116, 119
Roger Bacon, 18, 23, 133
Ross, W. D., 176
Royal Society, the, 131, 147, 150, 154–55, 157
Rubio, Antonio, 95
Russiliano, Tiberio, 79

Sabians, 157
Saint-Cyran, 128
Salutati, Coluccio, 10, 32–33, 84, 161
Santa Caterina, Pisa, 22
Santa Maria Novella (Florence), 22
Santorio, Santorio, 159
Sarpi, Paolo, 86, 126
Savonarola, Girolamo, 45–46, 48
Scaliger, Julius Caesar, 83–84, 166
Schegk, Jacob, 96
Scholasticism, 4, 10, 29, 50, 147, 175–76; early modern, 123, 154, 156, 158
Schoock, Martin, 158–59
Schott, Andreas, 91
scientific revolution, the, 1, 3, 7, 144, 176–77
Scotists, 4, 80, 89, 107
scripture, 84, 97, 116–17, 131, 142, 151; and Aristotle, 3, 36, 68, 116, 167; authority of, 47, 88, 109–10, 158, 160; and Bessarion, 41–42; and medievals, 17, 20, 108; and Mosaic philosophy, 147, 149, 153, 166; and Pomponazzi, 64, 68, 172
Scultetus, Abraham, 152

Scutellius, Nicolaus, 60
secularism, 2–3, 177
Seneca, 32, 38, 48, 156
sensation, 55, 58, 64, 70, 78, 93, 116
Sextus Empiricus, 45–46, 123
Siger of Brabant, 18, 20–21, 24, 26, 30
Simon of Tournai, 32
Simplicius, 12, 106, 131; and Gianfrancesco Pico, 46–47; in Renaissance, 8, 52–54, 74, 77, 85; and soul; 56, 73
sincerity, 7, 10, 118, 123, 126, 155; and Pomponazzi, 67–69, 172
Skarga, Piotr, 162
skepticism, 3, 45, 123, 128, 130, 137; attacks against, 126, 128, 159, 169; and Bayle, 170, 172; and Gianfrancesco Pico, 46–47, 111; mitigated, 150, 154
Socinianism, 147, 156, 158–59, 166, 169
Socrates, 14, 49, 93, 121, 135
Sorbonne, the, 122–23, 139, 142, 163–64. *See also* Paris
Soto, Domingo de, 93
soul: as assisting form, 55, 63, 93; and Descartes, 153, 158, 161, 164, 170; and Jesuits, 87, 89, 91, 94; immortality, 6–7, 16, 18, 20, 40–42, 44, 47–48, 60, 69, 84; material, 43, 62–63, 67, 70, 98, 119, 149, 153, 161; mortality, 62–64, 67, 78, 81–82, 85, 94, 100, 103, 106, 116, 119, 129, 132, 141, 145, 151, 161, 177; sleeping, 147; as substantial form, 18, 25–26, 55–56, 60, 158; unicity, 21, 33, 43, 55–56, 59, 61, 64, 70, 78, 95, 137–38, 161, 173
Sozzini, Fausto, 147, 156
Speroni, Sperone, 84
Spini, Giorgio, 127
Spinoza, Benedictus, 169–70, 172–73
Spinozism, 170–71, 173
spontaneous generation, 79, 84, 98, 127, 155
statutes (university), 9; Bologna, 66, 72; Oxford, 123, 146–47; Padua, 6, 119, 123; Paris, 23, 27, 122–23
Steuco, Agostino, 50, 153
Stoics, 34, 38, 67–68, 161
Storella, Francesco, 76
Strabo, 105–6, 112
Suarez, Francisco, 158, 163